Conductors, Semiconductors, Superconductors

Conductors, Semiconductors, Superconductors

Editor

Lefteris Tyler

scitus academics

Conductors, Semiconductors, Superconductors
Edited by **Lefteris Tyler**

ISBN: 978-1-68117-246-0
Library of Congress Control Number: 2016934787

SCITUS Academics LLC,
www.scitusacademics.com
Box No. 4766, 616 Corporate Way,
Suite 2, Valley Cottage,
NY 10989

Notice

Preface

In physics and electrical engineering, a conductor is an object or type of material that allow the flow of electrical current in one or more directions. A metal wire is a common electrical conductor. In metals such as copper or aluminum, the mobile charged particles are electrons. Positive charges may also be mobile, such as the cationic electrolyte(s) of a battery, or the mobile protons of the proton conductor of a fuel cell. Conduction materials include metals, electrolytes, superconductors, semiconductors, plasmas and some nonmetallic conductors such as graphite and Conductive polymers. A semiconductor is a material that has a conductivity somewhere between that of a conductor and an insulator. In other words, they are not as conductive as metals, but more conductive than plastics and other insulators. Examples include silicon, germanium, and gallium arsenic. A superconductor is a metal that acts strangely when cooled down to a certain temperature. When these materials are at that one special temperature, which we call the critical temperature, they suddenly become perfect conductors. A current in a superconductor can keep flowing without any decay, forever. Examples of superconducting materials include aluminum, tin, mercury and lead. Superconductors can be used to create powerful electromagnets, like those used in MRI scanners in hospitals. They can also be used to separate magnetic and non-magnetic materials.This book entitled Conductors, Semiconductors, Superconductors provides an introduction to the fundamentals of solid state physics, including a description of the key people in the field and the historic context. The book concentrates on the electric and magnetic properties of materials.

Table of Contents

1

CHAPTER

ORGANIC SEMICONDUCTORS AND CONDUCTORS WITH TERT-BUTYL SUBSTITUENTS

Toshiki Higashino, Yuto Akiyama, Hirotaka Kojima, Tadashi Kawamoto and Takehiko Mori *

Department of Organic and Polymeric Materials, Tokyo Institute of Technology, O-okayama 2-12-1, Meguro-ku, Tokyo 152-8552, Japan

ABSTRACT

Tetrathiafulvalene (TTF), pentacene, and quarterthiophene with tert-butyl substituents are synthesized, and the crystal structures and the transistor properties are investigated. The tetracyanoquinodimethane (TCNQ) complex of tert-butyl TTF constructs highly one-dimensional segregated columns with tetragonal crystal symmetry.

Keywords: organic transistors; organic semiconductors; crystal structures

1. INTRODUCTION

Tight crystal packing is an important factor to achieve good transport in molecular conductors, because intermolecular interaction is easily interrupted by the introduction of sterically bulky groups. However, for example, the ethylene parts of bis(ethylenedithio)tetrathiafulvalene (BEDT-TTF) are indispensable groups to realize the two-dimensional transport, by balancing the transport in the stacking and the inter-stacking directions. Recently, we have found that organic field-effect transistor (OFET) properties of TTF derivatives are much stabilized and improved by the introduction of tert-butyl groups [1]; examples are HMTTF (hexamethylene-TTF) [2], OMTTF (octamethylene-TTF) [3], and DBTTF (dibenzo-TTF) (Figure 1) [4]. Since these TTF derivatives are very strong organic donors, it is comparatively difficult to attain a small threshold voltage V_{th} and long-term stability, though HMTTF and DBTTF show high mobility exceeding 1 $cm^2 V^{-1} s^{-1}$ [5,6,7,8]. In these materials, the tert-butyl groups work like a passivation layer, where V_{th} drastically goes down to 0 V, and the transistors maintain the stable operation even after several months, without changing the energy levels. Interestingly, other s- and n-alkyl groups do not show similar effects. Prompted by these findings, we have synthesized tert-butyl substituted TTF (**1**); pentacene (**2**); and quarterthiophene (4T, **3**) molecules (Figure 1), and investigated the structural and transistor properties. The former two are new compounds, though several alkyl-substituted TTF and pentacene molecules have been reported [9,10,11,12]. Although alkyl-substituted oligothiophenes have been extensively studied [13], the tert-butyl substituted 4T has been studied to some extent only in the S-dioxide form [14]. We have also investigated the structural and conducting properties of the TCNQ complex of (**1**).

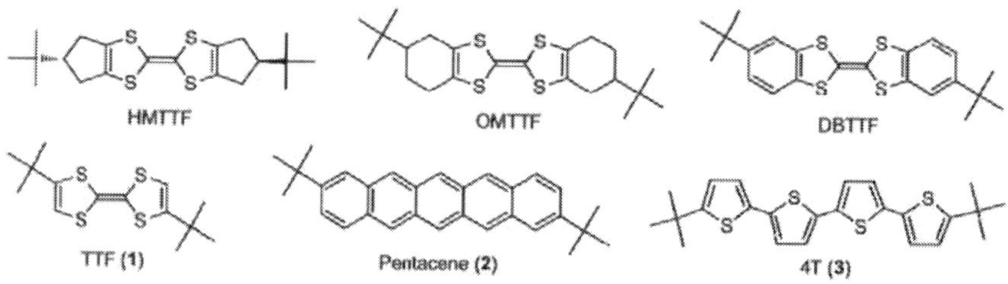

Figure 1. Molecular structures of organic semiconductors with tert-butyl substituents.

2. RESULTS AND DISCUSSION

2.1. Synthesis

Synthesis of tert-butyl substituted TTF (**1**) was achieved as shown in Scheme 1 [2,3]. The reaction of 1-chloro-3,3-dimethylbutan-2-one with potassium isopropylxanthate, followed by cyclization with sulfuric acid, afforded 4-tert-butyl-1,3-dithiol-2-one (**5**), which was coupled by triethylphosphite to give 4,4′-di-*tert*-butyl-tetrathiafulvalene (**1**) [15]. This phosphite-mediated coupling reaction potentially produced *E*- and *Z*-isomers with respect to the tert-butyl groups, but we can isolate a single isomer after repeated recrystallization from ethanol, which was verified by the X-ray crystal structure analysis. This compound is highly soluble in organic solvents; it dissolves even in hot hexane and in hot ethanol.

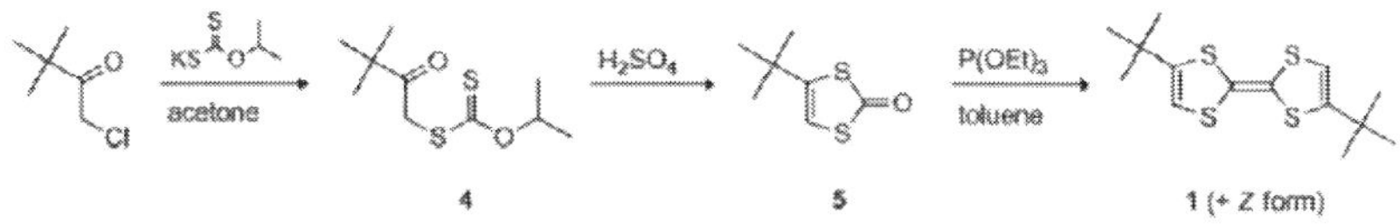

Scheme 1. Synthetic scheme for tert-butyl substituted Tetrathiafulvalene (TTF) (**1**).

Synthesis of tert-butyl substituted pentacene (**2**) was attempted according to Scheme 2. Acylation of tert-butylbenzene with pyromellitic dianhydride afforded di(p-benzoyl)benzene (**9**) [16,17], which was cyclized by sulfuric acid to give tert-butyl substituted pentacene diquinone (**10**). Compound (**2**) was, however, not obtained by the reduction of (**10**) with sodium borohydride or lithium aluminium hydride, though there were reports on the reduction of (**10**) with aluminum amalgam [18] and the reduction of (**9**) with palladium carbon under H_2 atmosphere [11].

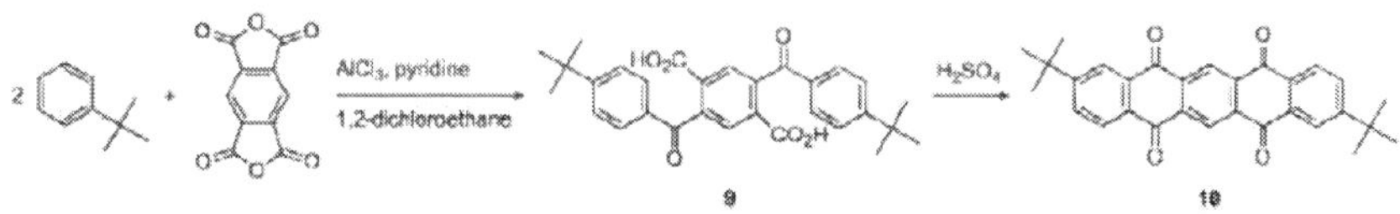

Scheme 2. Attempted synthetic scheme for (**2**).

Then synthesis of *tert*-butyl substituted pentacene (**2**) was achieved via the route shown in Scheme 3 [19]. tert-Butyl substituted pentacene quinone (**8**) was obtained from 4-*tert*-butyl-1, 2-dimethyl-benzene by bromination with N-bromosuccinimide (NBS), followed by hydrolysis with a dimethylamine aqueous solution and the successive Aldol reaction with 1,4-cyclohexanedione under basic conditions. This reaction potentially produced *E*- and *Z*-isomers, so we isolated a single isomer after repeated recrystallization from ethyl acetate. The resulting (**8**) was reduced with lithium aluminium hydride, and the subsequent hydrochloric acid treatment afforded 2,9-di-tert-butyl-pentacene (**2**) [20].

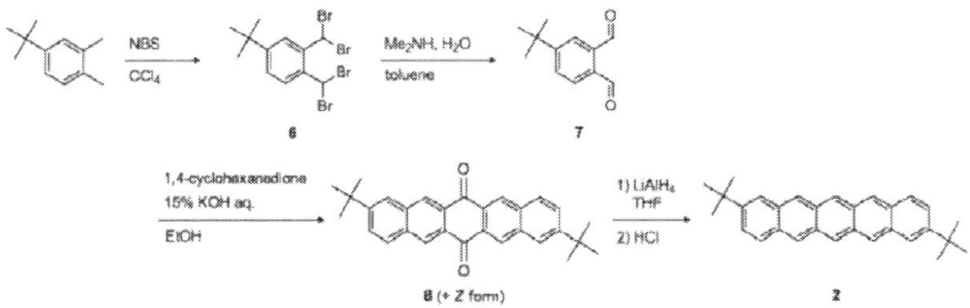

Scheme 3. Synthetic scheme for tert-butyl substituted pentacene (**2**).

Synthesis of tert-butyl substituted quarterthiophene (**3**) was achieved as shown in Scheme 4 [21,22,23]. 2-Bromo-5-*tert*-butylthiophene (**12**) was obtained by the low-temperature Friedel-Crafts alkylation of thiophene with *tert*-butylchloride [3,21], followed by the bromination with NBS at 5-position. After the Stille coupling of (**12**) with 2-(tributylstannyl)thiophene, the homo coupling reaction with lithium diisopropylamine and copper(II) chloride finally gave α,ω-di-*tert*-butyl-quaterthiophene (**3**). This compound shows much improved solubility compared with the parent 4T.

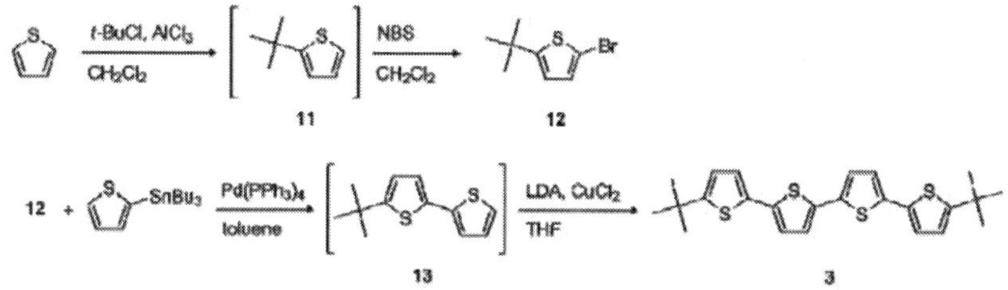

Scheme 4. Synthetic scheme for tert-butyl substituted quarterthiophene (3).

2.2. Electrochemical and Optical Properties

The electrochemical and optical properties have been investigated by cyclic voltammetry (CV) and ultraviolet/visible (UV/vis) absorption measurements as shown in Table 1. The energy gap is estimated from the absorption edge. The highest occupied molecular orbital (HOMO) energy is obtained from half-wave oxidation potential $E_{1/2}^{1}$ [24,25], and the lowest unoccupied molecular orbital (LUMO) energy is estimated by adding the energy gap to the HOMO energy [26]. Table 1 lists the results of density functional theory (DFT) calculations by Gaussian 09 program at the level of B3LYP using the 6-31G(d,p) basis set as well [27]. Since the solubility in organic solvents is improved for all compounds by the *tert*-butyl substitution, CV of these compounds including (**2**) and (**3**) are easily measured at room temperature. All compounds show two reversible oxidation peaks, indicating two one-electron transfers. The theoretically calculated HOMOs approximately coincide with the experimentally obtained HOMOs. The substituted compounds have slightly higher HOMOs than the unsubstituted materials owing to the electron donating alkyl groups.

Table 1. Redox potentials (vs. SCE [a]: saturated calomel electrode), optical gaps, and the experimentally estimated and calculated highest occupied molecular orbital (HOMO) and lowest unoccupied molecular orbital (LUMO) energies.

Compounds	Experiments					Calculation	
	$E_{1/2}^{1}$ (V)	$E_{1/2}^{2}$ (V)	Gap (eV)	HOMO (eV)	LUMO (eV)	HOMO (eV)	LUMO (eV)
1	0.16	0.70	3.20	−4.56	−1.36	−4.40	−0.70
2	0.63	1.04	2.09	−5.03	−2.94	−4.49	−2.29
3	0.75	0.99	2.68	−5.15	−2.47	−4.83	−1.76
TTF	0.22	0.73	3.06	−4.62	−1.56	−4.57	−0.92
Pentacene	-	-	-	−5.0 [b]	−3.2 [b]	−4.61	−2.40
4T	0.78	-	2.74	−5.18	−2.44	−4.97	−1.94

[a] Fc/Fc^{+} 0.40 V *vs.* SCE and −4.80 eV to vacuum; [b] From [28].

2.3. Single-Crystal X-ray Structure Analysis

Single-crystal X-ray structure analyses were carried out for **1** and **2**. The crystals were obtained by crystallization from hexane for **1** and by sublimation for **2**. Table 2 shows the crystallographic data together with those of the unsubstituted TTF and pentacene [29,30]. The tert-butyl derivatives have entirely different structures from the parent compounds. A unit cell contains two molecules for all crystals. The lattice volumes are, however, approximately doubled by the tert-butyl substitution, indicating that the volume of each molecule is doubled, though the molecular weights are not increased so much. As a consequence, the density is considerably reduced due to the loose packing of the tert-butyl substituted molecules.

The crystal structures are depicted in Figure 2 and Figure 3. The molecules are located on inversion centers, and a half of the molecule is crystallographically independent. There is no disorder on the tert-butyl position, so the crystals of both compounds contain only the E-isomers. Temperature factors of the tert-butyl groups are, however, considerably large due to the thermal motion and rotation. The TTF and pentacene skeletons are almost planar.

Table 2. Crystallographic data for TTF, **1**, pentacene, and **2**.

	TTF [29]	1	Pentacene [30]	2
Empirical formula	$C_6H_4S_4$	$C_{14}H_{20}S_4$	$C_{22}H_{14}$	$C_{30}H_{30}$
Formula weight	204.33	316.55	278.35	390.57
Temperature (K)	290	173	293	173
Crystal system	monoclinic	monoclinic	triclinic	monoclinic
Space group	$P2_1/c$	$P2_1/a$	P-1	$P2_1/a$
a (Å)	7.352(2)	9.2639(2)	6.266(1)	10.8100(2)
b (Å)	4.0181(11)	9.7388(3)	7.775(1)	6.3668(2)
c (Å)	13.901(4)	9.3381(2)	14.530(1)	16.1783(4)
α (°)	90	90	76.475(4)	90
β (°)	101.426(10)	103.263(2)	87.682(4)	97.483(1)
γ (°)	90	90	84.684(4)	90
V (Å^3)	402.5(2)	820.01(4)	685.15(15)	1111.24(4)
Z	2	2	2	2
$D_{calc.}$ (g/cm^3)	1.686	1.282	1.349	1.167
Reflections (unique)	3571 (926)	8389 (1501)	2856 (2684)	12341 (2033)
R_1	0.028	0.0768	0.069	0.0561
R_w	0.065	0.2215	0.179	0.1925
GOF	1.12	1.05	0.94	1.08

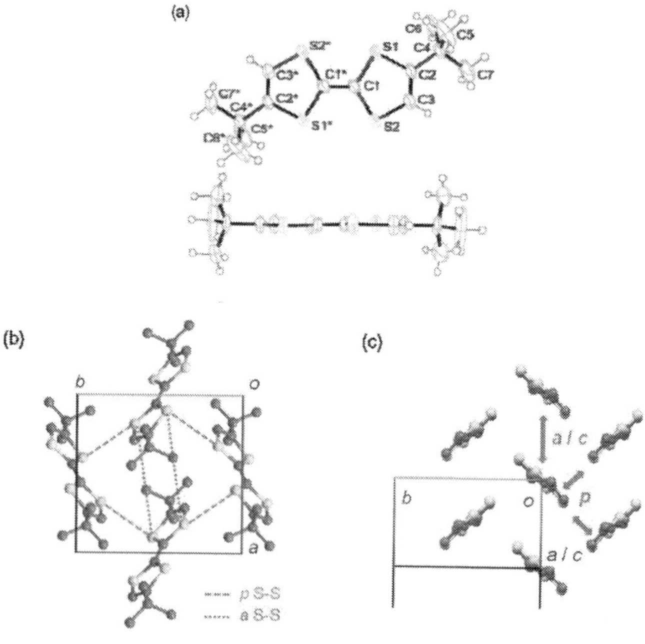

Figure 2. Crystal structure of **1**. (**a**) Top and side views of the molecular structure; (**b**) view perpendicular to the molecular plane (hydrogen atoms are not shown); and (**c**) projection along the molecular long axis (tert-butyl groups are not shown). Anisotropic displacement ellipsoids are drawn at the 50% probability level.

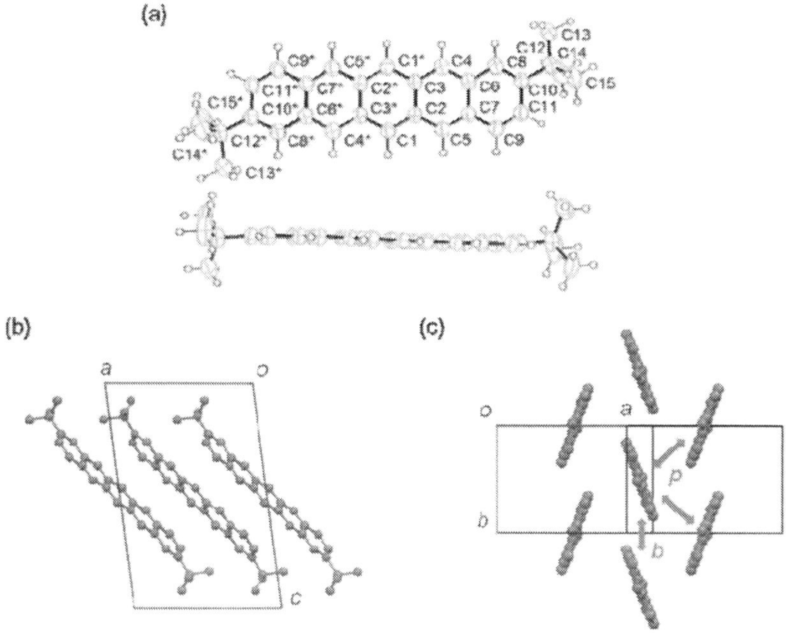

Figure 3. Crystal structure of **2**. (**a**) Top and side views of the molecular structure; (**b**) view perpendicular to the molecular plane (hydrogen atoms are not shown); and (**c**) projection along the molecular long axis (tert-butyl groups are not shown). Anisotropic displacement ellipsoids are drawn at the 50% probability level.

Compounds **1** and **2** have herringbone-type molecular packings, where the stacking directions are a for **1** and b for **2**, and the dihedral angles are 89° and 42°, respectively. In order to discuss the intermolecular interactions, transfer integrals (*t*) are shown in Table 3 [31] together with the slip distances along the molecular long axis (*D*) and the short sulfur-sulfur distances (S-S). The transfer integrals are as a whole small in spite of the herringbone-like packing. In particular, the intracolumnar (*a*) and intercolumnar (*c*) interactions in **1** are very small owing to the large slip distances along the molecular long axes (*D* = 8.1 and 5.0 Å, respectively). The two-dimensional interaction is, however, maintained by the diagonal interaction (*p*), though this transfer integral is relatively small. This is due to the large dihedral angle of 89°, which is usually disadvantageous in the herringbone structures [32]. Short S-S distances exist in this direction.

Table 3. Intermolecular transfer integrals and geometric parameters.

1	t (meV)	D (Å)	S-S (Å)	2	t (meV)	D (Å)
a	1	8.1	6.59	b	−23	1.2
c	−5	5.0	5.73	p	−1	-
p	−11	-	3.70			

The dihedral angle of **2** is small (42°), and closes to 53° of the unsubstituted pentacene [33]. The predominant interaction is, however, the intracolumnar interaction (b), which results in relatively one-dimensional transport. This is because D is small for b (1.2 Å), but as large as 4 Å for p. This reminds us of the previously reported alkyl-substituted BEDT-TTF molecules [34], where large D destroys the intermolecular networks.

In the herringbone structure, the bandwidths are estimated to be $4|t_a| + 4|t_c| + 8|t_p| = 112$ meV for **1** and $4|t_b| + 8|t_p| = 100$ meV for **2** [32]. These values are smaller than the bandwidth of TTF stacks in (TTF)(TCNQ) (0.48 eV) [35] and the unsubstituted pentacene (200 meV), on account of the steric hindrance of the tert-butyl groups, and the resulting relatively large intermolecular distances.

2.4. Thin-Film Properties

Atomic force microscopy (AFM) images of thin films deposited on hexamethylenedisilazane (HMDS)-treated SiO_2/Si substrates are shown in Figure 4. A thin film of **1** consists of block-like grains with large voids and roughness. Plate-like grains as large as more than 2 μm are formed for **2**. The grain size increases with increasing the substrate temperatures. In contrast, **3** forms many dense small grains, and the substrate is well covered.

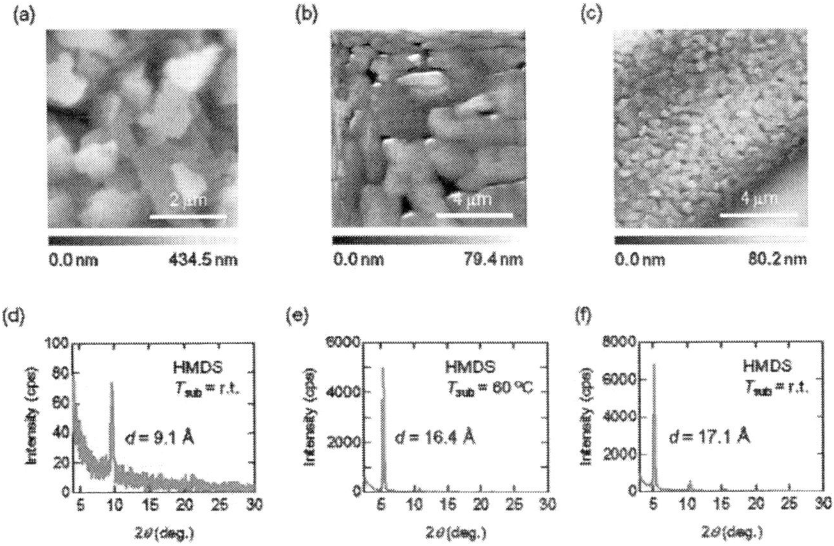

Figure 4. Atomic force microscopy (AFM) images and X-ray diffraction (XRD) patterns of the thin films deposited on hexamethylenedisilazane (HMDS)-treated SiO_2: (**a**) and (**d**) for **1**; (**b**) and (**e**) for **2** at T_{sub} = 60 °C and (**c**) and (**f**) for **3**.

The thin films of **1**, **2** and **3** show clear X-ray diffraction (XRD) peaks (Figure 4). In the thin films of **2** and **3**, the large peaks are observed, indicating formation of lamellar ordering and good crystallinity. The d values are 9.1 Å and 16.4 Å for **1** and **2**, respectively, in good agreement with the crystallographic c axes (Table 1). The thin film of **3** shows the d value of 17.1 Å, which is shorter than the molecular length (21.9 Å), indicating that the molecules are tilted by 52° from the surface normal.

2.5. Transistor Propzerties

OFETs with top-contact Au electrodes are fabricated by vacuum deposition of **1**, **2**, and **3** on the bare and HMDS-treated SiO_2/Si substrates. The OFET characteristics are measured in air (Table 4). Compound **1** does not show transistor properties. Compound **2** shows increasing performance with increasing the deposition temperature, and exhibits a mobility of 2.8×10^{-3} $cm^2\ V^{-1}\ s^{-1}$ for a thin film deposited at 60 °C (Figure 5a,b). This is due to the increasing grain size and the decreasing number of grain boundaries [36,37], as observed in AFM and XRD. After 30 days, the device shows reduced carrier mobility by two orders of magnitude, though the unsubstituted pentacene does not show transistor properties at all after this period.

Table 4. Transistor characteristics of **2** and **3**.

Compounds	SAMs		μ ($cm^2 V^{-1} s^{-1}$)	V_{th} (V)	I_{on}/I_{off}
2	HMDS	r.t.	2.5×10^{-4}	−36	1×10^3
		40 °C	2.9×10^{-4}	−22	2×10^2
		60 °C	2.8×10^{-3}	−44	2×10^3
		after 30 days	1.3×10^{-5}	−20	26
3	bare	r.t.	6.3×10^{-3}	−17	2×10^4
	HMDS	r.t.	3.5×10^{-3}	−17	1×10^4
4T	bare	r.t.	3.5×10^{-3}	−20	6×10^3
	HMDS	r.t.	7.8×10^{-3}	−13	6×10^4

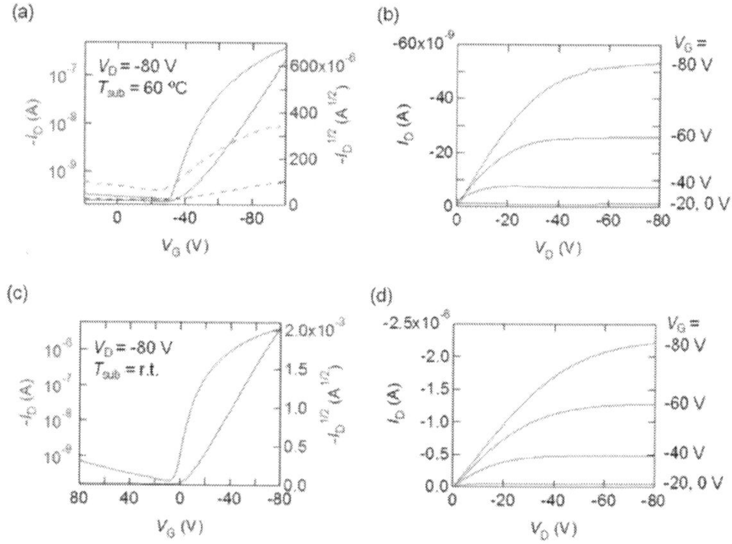

Figure 5. (**a**) Transfer and (**b**) output characteristics of **2**, for devices fabricated at T_{sub} = 60 °C with top-contact Au electrodes. The dotted curves are characteristics after 30 days. (**c**) Transfer and (**d**) output characteristics of **3**, for devices fabricated at room temperature.

Compound **3** shows comparable mobility to similarly fabricated devices of the unsubstituted 4T (Figure 5c,d). Interestingly, the device fabricated on the bare substrate has shown higher mobility than that of the HMDS-treated substrates, suggesting that the tert-butyl substitution plays a role of self-assembled monolayers (SAMs). It is a merit of **3** that this material can be used without SAMs-treatment.

2.6. Charge-Transfer Salt

Since **1** is a strong electron donor, the charge-transfer salts are prepared. Slow evaporation of an acetonitrile solution of **1** and tetracyanoquinodimethane (TCNQ) has given black needles of (**1**)(TCNQ). The crystallographic data are listed in Table 5, and the crystal structure is depicted in Figure 6. The crystal belongs to the space group $P4_2/mbc$, and each half of **1** and TCNQ is crystallographically independent. A unit cell contains respective four molecules of **1** and TCNQ, and the composition is 1:1. Both molecules are located on the 2/m sites, and form segregated stacks. This crystal contains only the E-isomer of **1**. The TTF and TCNQ molecules are located on the mirror planes ($z = 0$). There is a 4_2-screw axis parallel to the TTF stack, but is only the two-fold axis along the TCNQ stack. Among charge-transfer salts of TTF derivatives, halides of TTF and o-dimethyl-TTF have tetragonal symmetry [38,39]. In these halides, however, the halogen ions are located on the four-fold axis, and the TTF molecules form regular parallel stacks. On the contrary, in the present complex, the TTF stack is located on the 4_2-screw axis. The TTF molecules are disordered to two positions with occupancy 69% for the majority molecule (M1) and 31% for the minority molecule (M2). The M1 and M2 molecules are arranged in a staggered manner as the two molecules depicted in Figure 6(b), but share the common tert-butyl parts. The 4_2-screw axis generates perpendicular molecules M1' and M2' at the $z = 1/2$ position (Figure 6(c)). Here, the z = 1/2 position defines the next molecule in the stack, and c/2 = 3.417 Å corresponds to the interplanar distance. The sulfur atoms of the M1′ molecule are located just on the top of the sulfur atoms of the M1 molecule, so that M1-M1′ arrangement is an eclipsed conformation. M2-M2' arrangement is similarly an eclipsed stack, but M1-M2′ and M1′-M2 are staggered. Since there is no disorder on the tert-butyl groups, the M1 and M2 molecules always appear at z = 0, and M1' and M2' are only at z = 1/2. On the basis of the molecular orbital calculation [31], the eclipsed arrangements afford large transfer integrals (M1-M1′: 323 meV and M2-M2′: 315 meV), whereas the staggered arrangements afford small transfers (M1-M2′: 103 meV and M1′-M2: 96 meV). The TCNQ molecules form an ordered regular stack of slightly rotated molecules, and the transfer integral is only one (121 meV). There are no meaningful intercolumnar transfer integrals, and the electronic structure is highly one-dimensional. This is due to the steric effect of the tert-butyl groups, which reminds us conducting nanowires surrounded by supramolecular insulating networks [40]. It is noteworthy that the intercolumnar directions are completely isotropic owing to the tetragonal symmetry.

Table 5. Crystallographic data for (**1**) tetracyanoquinodimethane (TCNQ).

	(1)(TCNQ)
Empirical formula	$C_{26}H_{24}S_4N_4$
Formula weight	520.74
Temperature (K)	273
Crystal system	tetragonal
Space group	$P4_2/mbc$
a (Å)	19.7657(7)
c (Å)	6.8345(3)
V (Å^3)	2670.1(2)
Z	4
D_{calc} (g/cm^3)	1.295
Reflections (unique)	29583 (1335)
R_1	0.0938
R_w	0.2868
GOF	1.08

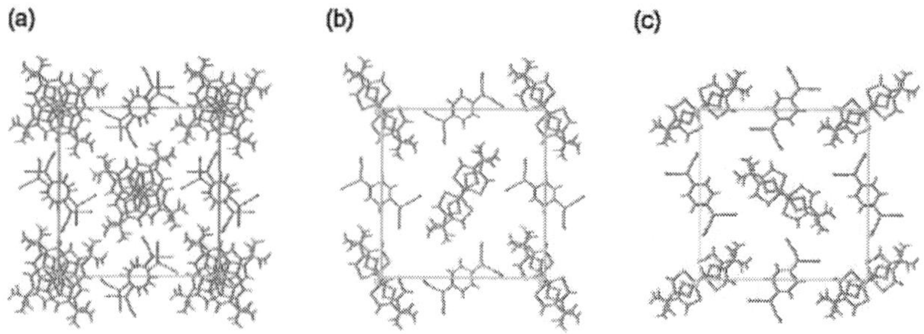

Figure 6. Crystal structure of (**1**)(TCNQ). (**a**) Projection along the c axis; and the molecules at (**b**) z = 0; and (**c**) z = 1/2.

The infrared (IR) spectrum shows the CN stretching at 2189 cm^{-1}, from which the degree of charge transfer is estimated to be 0.86 [41]. This salt shows room-temperature conductivity of 7 S cm^{-1}, and activated temperature dependence with the activation energy of 0.1 eV.

The X-ray photograph (Figure 7) shows broad satellites at the (1/3)c* positions. The three-fold periodicity is agreeable to the 2:1 disorder, and, we suppose, a stack is composed of such molecular arrangement as -M1-M1′-M2-M1″-M1-M2′-. Note that the dashed molecules have to appear alternately, and the ratio of M1 and M2 is 2:1. The eclipsed arrangements (M1-M1' and M2-M2') appear twice, and other four are staggered arrangements. This six-fold periodicity is the reason that all our attempts are unsuccessful to solve the crystal structure without disorder in a lower symmetry. We also suspect that there is no order of the M2 positions between the adjacent columns. If we can approximate the IR result to be 2/3 charge transfer, namely ($TTF^{2/3+}$)($TCNQ^{2/3-}$), we can also imagine charge order such as $M1^{+}$-$M1^{+}$-$M2^{0}$, though charge disproportionation is not obvious from the bond lengths. This charge transfer degree is slightly larger than 0.59 of (TTF)(TCNQ) [42], and approximately the same as (TTF)$Cl_{0.77}$ [38].

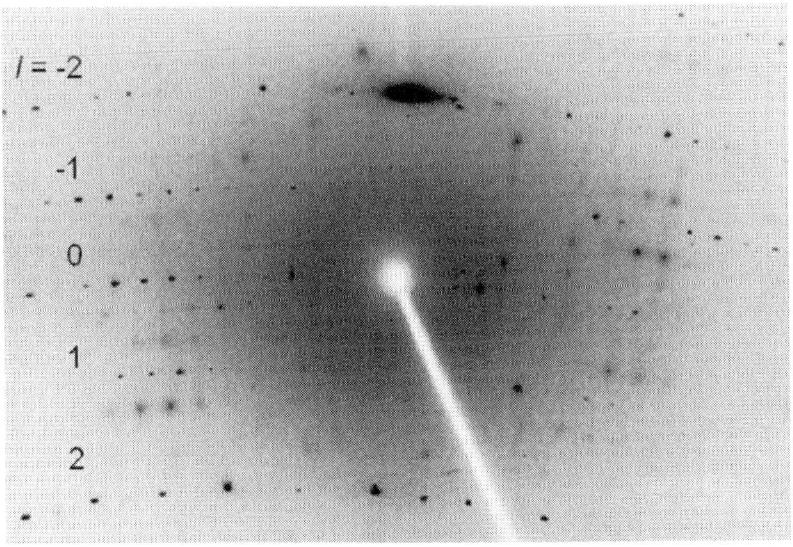

Figure 7. X-ray photograph of (**1**)(TCNQ) approximately along the *c**-axis at 273 K.

3. EXPERIMENTAL SECTION

3.1. Synthesis

Column chromatography was carried out on silica gel (Wako Wakogel C-200). For thin-layer chromatography (TLC) analysis, Merck pre-coated plates (silica gel 60 F254, 0.25 mm) were used. The ^{1}H-NMR (300 MHz) spectra were recorded using a JEOL JNM-AL300 spectrometer. Chemical shifts were given in δ values (ppm) using TMS as the internal standard. Mass spectra were taken on a Shimadzu GC–MS-QP-5000 mass spectrometer. The IR spectra were obtained on a Shimadzu FTIR-8400S spectrometer.

S-(3,3-Dimethyl-2-oxobutyl)-O-isopropylcarbonodithioate (**4**). 1-chloro-3,3-dimethylbutan-2-one (1.35 g, 10 mmol) was added dropwise to a solution of potassium isopropylxanthate (1.92 g, 11 mmol) in acetone (20 mL), and the reaction mixture was stirred for 1 h. The obtained white suspension was filtered, and the filtrate was evaporated to give a pale yellow oil **4** (2.30 g, 9.81 mmol, yield 98%). MS (EI) *m/z* 234 [M^+]; ^{1}H-NMR (300 MHz, $CDCl_3$) δ 5.73 (m, 1H), 4.26 (s, 2H), 1.40 (d, 6H), 1.25 (s, 9H).

4-tert-Butyl-1,3-dithiol-2-one (**5**). Compound **4** (4.37 g, 18.6 mmol) was added dropwise to ice-cooled sulfuric acid (20 mL), and after 2 h stirring, the resulting mixture was poured onto ice (40 g). The product was extracted with diethyl ether, washed with water, a saturated aqueous solution of sodium hydrogen carbonate and brine, and dried over magnesium sulfate. After evaporation, the product was purified by column chromatography (ethyl acetate:hexane = 1:2, R_f = 0.8), to give a yellow oil **5** (2.65 g, 15.2 mmol, yield 82%). MS (EI) *m/z* 174 [M^+]; ^{1}H-NMR (300 MHz, $CDCl_3$) *δ* 6.30 (s, 1H), 1.30 (s, 9H).

4,4'-Di(tert-butyl)tetrathiafulvalene (**1**). Distilled triethylphosphite (3.5 mL, 20.8 mmol) was added to a solution of **5** (1.49 g, 8.5 mmol) in dry toluene (5 mL) under N_2 atmosphere, and refluxed for 17 h. Distilled triethylphosphite (3.0 mL, 17.9 mmol) was further added, and refluxed for 5 h. After the triethylphosphite was removed by distillation under vacuum, the product was purified by column chromatography (dichloromethane:hexane = 1:1, R_f = 0.7), to give an orange solid **1** (0.86 g, 2.72 mmol, yield 65%). The E-isomer of **1** was isolated by repeated recrystallization from ethanol. m.p.172 °C; MS (EI) *m/z* 316 [M^+]; ^{1}H-NMR (300 MHz, $CDCl_3$) δ 5.85 (s, 2H), 1.24 (s, 18H).

4-tert-Butyl-1,2-bis(dibromomethyl)benzene (**6**). N-bromosuccinimide (37.4 g, 210 mmol) was added to a solution of 4-tert-butyl-1,2-dimethylbenzene (8.16 g, 50.3 mmol) in tetrachloromethane (100 mL) under N_2 atmosphere. The mixture was stirred for 30 min, and refluxed for 24 h. The orange suspension was filtered, and the residue was washed with diethyl ether. The resulting filtrate was concentrated under reduced pressure. The product was extracted with dichloromethane and washed with water, a saturated aqueous solution of sodium sulfite and brine. The organic layer was dried over magnesium sulfate and concentrated under reduced pressure. The product was recrystallized from hexane, to give a white solid **6** (19.34 g, 40.5 mmol, yield 81%). MS (EI) *m/z* 478 [M^+].

4-tert-Butylphthalaldehyde (**7**). 50% aqueous dimethylamine (30 mL) was added to a solution of **6** (2.73 g, 10.7 mmol) in toluene (20 mL) under N_2 atmosphere, stirred for 24 h, and refluxed for 48 h. The reaction mixture was extracted with ethyl acetate, and washed with 10% aqueous hydrochloric acid, water and brine. The organic layer was dried over magnesium sulfate and concentrated under reduced pressure. The product was purified by column chromatography (ethyl acetate:hexane = 2:1, R_f = 0.7), to give a red oil 7 (1.30 g, 6.83 mmol, yield 64%). MS (EI) *m/z* 190 [M^+]; ^{1}H-NMR (300 MHz, $CDCl_3$) δ 10.59 (s, 1H), 10.49 (s, 1H), 7.99 (d, 1H), 7.92 (d, 1H), 7.78 (dd, 1H).

2,9-Di(tert-butyl)pentacene-6,13-dione (**8**). 15% potassium hydroxide aqueous solution (50 mL) was added dropwise to a mixture of **7** (4.74 g, 25.0 mmol) and 1,4-cyclohexanedione (1.34 g, 12.5 mmol). After stirring for 2 h, the reaction mixture was poured into water (120 mL) for quenching. The resulting solid was filtered and washed with water, followed by ethanol, to give a yellow solid **8** (3.71 g, 8.82 mmol, 71%). The E-isomer of **8** was isolated by repeated recrystallization from ethyl acetate. m.p. 290 °C; MS (EI) m/z 420 [M^+].

2,9-Di(tert-butyl)pentacene (**2**). Lithium aluminium hydride (0.2 g, 5.3 mmol) was added to a solution of **8** (0.2 g, 0.48 mmol) in dry THF (15 mL) under N_2 atmosphere and refluxed for 12 h. 6 M aqueous hydrochloric acid (6 mL) was added to the reaction mixture under cooling with ice and refluxed for 1 h. The mixture was cooled to room temperature and water was added. The residue was filtered and washed with water, followed by ethanol. The product was purified by sublimation (250 °C), to give a violet solid (0.14 g, 0.35 mmol, 73%). MS (EI) *m/z* 390 [M^+].

2-Bromo-5-tert-butylthiophene (**12**). A solution of thiophene (4.60 g, 54.8 mmol) and tert-butyl chloride (5.13 g, 55.5 mmol) in dry dichloromethane (10 mL) was added dropwise to a mixture of aluminium chloride (7.30 g, 54.8 mmol) in dry dichloromethane (20 mL) at −70 °C under N_2 atmosphere and stirred for 1 h at this temperature before stirred for 40 h at room temperature. The mixture was poured onto ice, and a saturated aqueous solution of sodium hydrogen carbonate was added for quenching. The organic layer was washed with water and brine, dried over magnesium sulfate and concentrated under reduced pressure. The obtained brown oil **11** was dissolved in dichloromethane (20 mL). N-bromosuccinimide (9.75 g, 54.8 mmol) was slowly added, and stirred for 1 h. The mixture was washed with water and brine, dried over magnesium sulfate, and concentrated under reduced pressure. The product was distilled under vacuum, to give a yellow oil **12** (5.41 g, 24.7 mmol, yield 45%). MS (EI) *m/z* 219 [M^+], ^{1}H-NMR (300 MHz, $CDCl_3$) δ 1.35 (s, 9H), 6.57 (d, J = 3.6 Hz, 1H), 6.84 (d, J = 3.6 Hz, 1H).

α,ω-Di(tert-butyl)quaterthiophene (**3**). A solution of n-butyl lithium (6.90 mL, 11.0 mmol) in 1.6 M hexane was dropwise added to a solution of thiophene (0.84 g, 10.0 mmol) in dry diethyl ether (20 mL) at −40 °C under N_2 atmosphere, and stirred for 2 h. After tributyltin chloride (3.26 g, 10 mmol) was dropwise added, the mixture was refluxed overnight. The reaction mixture was extracted with dichloromethane, washed with water, and dried over magnesium sulfate. Concentration under reduced pressure gives a yellow oil (3.70 g, 9.9 mmol). The obtained oil (1.86 g, 5.0 mmol) was dissolved in dry toluene (20 mL), to which **12** (1.10 g, 5.0 mmol) and tetrakis(triphenylphosphine)paradium(0) (0.29 g, 0.25 mmol) was added, and refluxed overnight. The reaction mixture was quenched by adding a saturated aqueous solution of ammonium chloride, extracted with diethyl ether, washed with water, dried over magnesium sulfate, and concentrated under reduced pressure. The obtained brown oil was distilled under vacuum, to give a colorless oil **13** (0.89 g, 4.0 mmol, yield 80%).

A solution of n-butyl lithium (1.50 mL, 2.40 mmol) in 1.6 M hexane was dropwise added to a solution of diisopropylamine (0.22 g, 2.16 mmol) in dry THF (5 mL) at −78 °C under N_2 atmosphere. A solution of **13** (0.48 g, 2.16 mmol) in dry THF (3 mL) was dropwise added, and stirred for 30 min. A mixture of anhydrous copper(II) chloride (0.29 g, 2.16 mmol) in DMF (3 mL) was added and stirred for another 1 h. The reaction was quenched by 3% aqueous hydrochloric acid, extracted with dichloromethane, washed with water, dried over magnesium sulfate, and concentrated under reduced pressure. The product was purified by column chromatography (dichloromethane:hexane = 1:1), and recrystallization from ethanol gives a yellow solid **3** (0.23 g, 0.52 mmol, yield 49%). m.p. 222 °C; MS (EI) *m/z* 442 [M^+]; ^{1}H-NMR (300 MHz, $CDCl_3$) δ 1.23 (s, 18H), 6.74 (d, J = 3.6 Hz, 2H), 6.99 (d, J = 3.6 Hz, 2H), 7.00 (d, J = 3.9 Hz, 2H), 7.03 (d, J = 3.6 Hz, 2H).

3.2. Redox Properties

Cyclic voltammograms were measured on an ALS CHI-700C electrochemical analyzer in 0.1 M tetrabutylammonium hexafluorophosphate (n-$Bu_4N{\cdot}PF_6$) solutions of benzonitrile at a scan rate of 100 mV/s. The reference electrode was Ag/$AgNO_3$ with glassy carbon working electrode and a Pt supporting electrode. All the measured potentials were normalized with the ferrocene/ferrocenium redox potential of $E_{1/2}$ = 0.40 V vs. SCE and −4.8 eV to vacuum.

3.3. Photophysical Properties

UV-vis spectra in chloroform were measured on a Hitachi U-2800 UV-vis double beam spectrophotometer at room temperature using a quartz cuvette with a path length of 1 cm. Optical bandgaps (HOMO-LUMO gaps) were calculated from the edge of the visible-absorption bands.

3.4. Single Crystal Structures and Intermolecular Overlap Integral Calculations

The reflection data were collected on a Rigaku R-AXIS RAPID-II diffractometer using CuKα radiation (λ = 1.54187 Å) monochromated by confocal mirror (Rigaku VariMax) at 173 K. The temperature was controlled by a Rigaku cryostat system. Empirical absorption correction was applied with the ABSCOR program. The structures were solved by the direct method using the SHELX-97 [43] and SIR-92 [44], and refined by full-matrix least squares by applying anisotropic temperature factors for all non-hydrogen atoms. The hydrogen atoms were generated geometrically and were refined with the riding model for **1**, and were determined from the experimental data for **2**. The intermolecular transfer integrals t_i between the HOMOs were calculated on the basis of the extended Hückel molecular orbital calculation by using the same atomic parameters [31].

3.5. Device Fabrication

Transistors were fabricated onto n-doped Si substrates with a thermally grown SiO_2 dielectric layer (300 nm) treated with HMDS. The organic semiconductors were, after purification by sublimation, vacuum evaporated on the substrates with a thickness of 50 nm. AFM images were obtained with a Seiko Instruments SPA-300 and SPI3800 probe system by using an Si_3N_4 cantilever. XRD measurements of the thin films were carried out by Phillips X'Pert-MPD-OEC PW3050 with a monochromated CuKα radiation (λ = 1.542512 Å).

The top-contact electrodes were patterned by thermal deposition of Au (50 nm) using a shadow mask; the channel length and width were 100 μm and 1000 μm, respectively. The transistor characteristics were measured under ambient conditions by using a Keithley 4200 semiconductor parameter analyzer, and the mobility was evaluated from the transconductance in the saturated region.

3.6. Conductivity Measurement

Conductivity of (**1**)(TCNQ) was measured by the four-probe method applying a low-frequency ac current along the crystallographic c-axis.

4. CONCLUSIONS

TTF, pentacene, and 4T with bulky tert-butyl groups are prepared. Although the crystal density is remarkably reduced, the crystals have relatively simple herringbone structures. The tert-butyl groups largely destroy the transistor properties of TTF and pentacene, but the 4T derivative shows transistor performance comparable to the parent 4T, and interestingly, the performance on bare SiO_2 is improved. In the TCNQ complex of the tert-butyl TTF, each column is separated by the tert-butyl groups to construct a highly one-dimensional system.

ACKNOWLEDGMENTS

This work was partly supported by a Grant-in Aid for Scientific Research (B) (No. 23350061) from the Ministry of Education, Culture, Sports, Science, and Technology of Japan. We would like to express sincere gratitude to H. Iida, Center for Advanced Materials Analysis, Tokyo Institute of Technology, for X-ray diffraction measurements.

REFERENCES

1. Mori, T. Organic charge-transfer salts and the component molecules in organic transistors. Chem. Lett. **2011**, 40, 428–434.
2. Kanno, M.; Bando, Y.; Shirahata, T.; Inoue, J.; Wada, H.; Mori, T. Stabilization of organic field-effect transistors in hexamethylenetetrathiafulvalene derivatives substituted by bulky alkyl groups. J. Mater. Chem. **2009**, 19, 6548–6555.
3. Inoue, J.; Kanno, M.; Ashizawa, M.; Seo, C.; Tanioka, A.; Mori, T. Organic transistors based on octamethylenetetrathiafulvalenes. Chem. Lett. **2010**, 39, 538–540.
4. Nagakubo, J.; Ashizawa, M.; Kawamoto, T.; Tanioka, A.; Mori, T. Stabilization of organic field-effect transistors by tert-butyl groups in dibenzotetrathiafulvalene derivatives. Phys. Chem. Chem. Phys. **2011**, 13, 14370–14377.
5. Takahashi, Y.; Hasegawa, T.; Horiuchi, S.; Kumai, R.; Tokura, Y.; Saito, G. High mobility organic field-effect transistor based on hexamethylenetetrathiafulvalene with organic metal electrodes. Chem. Mater. **2007**, 19, 6382–6384.
6. Mas-Torrent, M.; Hadley, P.; Bromley, S.T.; Crivillers, N.; Rovira, C. Single-crystal organic field-effect transistors based on dibenzo-tetrathiafulvalene. Appl. Phys. Lett. **2005**, 86, 012110:1–012110:3.
7. Shibata, K.; Ishikawa, K.; Takezoe, H.; Wada, H.; Mori, T. Contact resistance of dibenzotetrathiafulvalene-based organic transistors with metal and organic electrodes. Appl. Phys. Lett. **2008**, 92, 023305:1–023305:3.
8. Yamada, T.; Hasegawa, T.; Hiraoka, M.; Matsui, H.; Tokura, Y.; Saito, G. Control of film morphology and its effects on subthreshold characteristics in dibenzotetrathiafulvalene organic thin-film transistors. Appl. Phys. Lett. **2008**, 92, 233306:1–233306:3.
9. Dhindsa, A.S.; Bryce, M.R.; Lloyd, J.P.; Petty, M.C.; Kobayashi, K.; Tukada, H. Synthesis of tetrathiafulvalene (TTF) derivatives bearing long alkyl chains. J. Chem. Soc. Chem. Commun. **1988**, 1391–1392.

10. Meng, H.; Bendikov, M.; Mitchell, G.; Helgeson, R.; Wudl, F.; Bao, Z.; Siegrist, T.; Kloc, C.; Chen, C.H. Tetramethylpentacene: Remarkable absence of steric effect on field effect mobility. Adv. Mater. **2003**, 15, 1090–1093.
11. Okamoto, K.; Kawamura, T.; Sone, M.; Ogino, K. Study on liquid crystallinity in 2,9-dialkylpentacenes. Liq. Cryst. **2007**, 34, 1001–1007.
12. Kunugi, Y.; Busujima, Y.; Ikari, M.; Okamoto, K.; Ogino, K. Organic field-effect transistors based on 2,9-disubstituted pentacene. ECS Trans. **2008**, 16, 273–282.
13. Perepichka, I.F.; Perepichka, D.F. Handbook of Thiophene-based Materials; Wiley: Chichester, UK, 2009.
14. Oliva, M.M.; Casado, J.; Navarrete, J.T.L.; Patchkovskii, S.; Goodson, T.; Harpham, M.R.; de Melo, J.S.S.; Amir, E.; Rozen, S. Do [all]-S, S′-Dioxide Oligothiophenes show electronic and optical properties of oligoenes and/or of oligothiophenes? J. Am. Chem. Soc. **2010**, 132, 6231–6242.
15. Alberola, A.; Bosch-Navarro, C.; Gavina, P.; Tatay, S. Synthesis of brominatedtetrathiafulvalenes via phosphite-mediated cross-coupling. Synth. Met. **2010**, 160, 1797–1799.
16. Fieser, L.F. Organic Synthesses Collected Volume I; Gilman, H., Ed.; Wiley: New York, NY, USA, 1941; pp. 517–519. [Google Scholar]
17. Fieser, L.F. Organic Experiments; D.C. Heath: Lexington, MA, USA, 1965; pp. 195–200.
18. Bénard, C.P.; Geng, Z.; Heuft, M.A.; Van Crey, K.; Fallis, A.G. Double diels-alder strategies to Soluble 2,9- and 2,9,6,13-Tetraethynylpentacenes, Photolytic [4+4] Cycloadditions, and pentacene crystal packing. J. Org. Chem. **2007**, 72, 7229–7236.
19. Okamoto, T.; Reese, C.; Senatore, M.L.; Tang, M.L.; Jiang, Y.; Parkin, S.R.; Bao, Z. 2,9-Dibromopentacene: Synthesis and the role of substituent and symmetry on solid-state order. Synth. Met. **2010**, 160, 2447–2451.
20. Vets, N.; Smet, M.; Dehaen, W. Synthesis and thermolysis of a Diels Alder adduct of pentacene and thiophosgene. Tetrahedron Lett. **2004**, 45, 7287–7289.
21. Belen'kii, L.I.; Yakubov, A.P. Stable heteroareniumions–VIII some transformations of alkylthiophenium ions and new synthesis of 2-t-butylthiophene. Tetrahedron **1984**, 40, 2471–2477.

22. Pinault, T.; Chérioux, F.; Therrien, B.; Süss-Fink, G. An iterative strategy for the synthesis of oligothiophenes by catalytic cross-coupling reactions. Heteroatom Chem. **2004**, 15, 121–126.

23. Kagan, J.; Arora, S.K.; Prakash, I.; Üstünol, A. The synthesis of 2,2′:5′,3″-terthiophene. Heterocycles **1983**, 20, 1341–1345.

24. Tang, M.L.; Reichardt, A.D.; Miyaki, N.; Stoltenberg, R.M.; Bao, Z. Ambipolar, high performance, acene-based organic thin film transistors. J. Am. Chem. Soc. **2008**, 130, 6064–6065.

25. Tang, M.L.; Reichardt, A.D.; Wei, P.; Bao, Z. Correlating carrier type with frontier molecular orbital energy levels in organic thin film transistors of functionalized acene derivatives. J. Am. Chem. Soc. **2009**, 131, 5264–5273.

26. Meng, H.; Zheng, L.; Lovinger, A.J.; Wand, B.-C.; Patten, P.G.V.; Bao, Z. Oligofluorene–thiophene derivatives as high-performance semiconductors for organic thin film transistors. Chem. Mater. **2003**, 15, 1778–1787.

27. Frisch, M.J.; Trucks, G.W.; Schlegel, H.B.; Scuseria, G.E.; Robb, M.A.; Cheeseman, J.R.; Scalmani, G.; Barone, V.; Mennucci, B.; Petersson, G.A.; et al. Gaussian 09, Revision B.01; Gaussian, Inc.: Wallingford, CT, USA, 2009.

28. Yasuda, T.; Goto, T.; Fujita, K.; Tsutsui, T. Ambipolar carrier transport in polycrystalline pentacene thin-film transistors. Mol. Cryst. Liq. Cryst. **2006**, 444, 219–224.

29. Batsanov, A.S. Tetrathiafulvalene revisited. Acta Crystallogr. **2006**, C62, o501–o504.

30. Mattheus, C.C.; Dros, A.B.; Baas, J.; Meetsma, A.; de Boer, J.L.; Palstra, T.M. Polymorphism in pentacene. Acta Crystallogr. **2001**, C57, 939–941.

31. Mori, T.; Kobayashi, A.; Sasaki, Y.; Kobayashi, H.; Saito, G.; Inokuchi, H. The intermolecular interaction of tetrathiafulvalene and bis(ethylenedithio)tetrathiafulvalene in organic metals. calculation of orbital overlaps and models of energy-band structures. Bull. Chem. Soc. Jpn. **1984**, 57, 627–633.

32. Kojima, H.; Mori, T. Dihedral angle dependence of transfer integrals in organic semiconductors with herringbone structures. Bull. Chem. Soc. Jpn. **2011**, 84, 1049–1056.

33. Campbell, R.B.; Robertson, J.M.; Trotter, J. The crystal and molecular structure of pentacene. Acta Crystallogr. **1961**, 14, 705–711.
34. Aoyagi, I.; Katsuhara, M.; Mori, T. Syntheses and structures of highly solublebis(ethylenedithio)tetrathiafulvalene molecules with alkyl chains. Sci. Technol. Adv. Mater. **2004**, 5, 443–447.
35. Shitzkovsky, S.; Weger, M.; Gutfreund, H. Band structure of TTF-TCNQ. J. Phys. (Paris) **1978**, 39, 711–717.
36. Kunugi, Y.; Takimiya, K.; Yamane, K.; Yamashita, K.; Aso, Y.; Otsubo, T. Organic field-effect transistor using oligoselenophene as an active layer. Chem. Mater. **2003**, 15, 6–7.
37. Takimiya, K.; Kunugi, Y.; Toyoshima, Y.; Otsubo, T. 2,6-Diarylnaphtho[1,8-bc:5,4-b′c′]dithiophenes as new high-performance semiconductors for organic field-effect transistors. J. Am. Chem. Soc. **2005**, 127, 3605–3612.
38. Scott, B.A.; La Placa, S.J.; Torrance, J.B.; Silverman, B.D.; Welber, B.J. The crystal chemistry of organic metals. Composition, structure, and stability in the tetrathiafulvalenium-halide systems. J. Am. Chem. Soc. **1977**, 99, 6631–6639.
39. Fourmigué, M.; Reinheimer, E.W.; Dunbar, K.R.; Auban-Senzier, P.; Pasquier, C.; Coulon, C. A series of strongly one-dimensional organic metals with strictly uniform stacks: (o-DMTTF)$_2$X (X = Cl, Br, I). Dalton Trans. **2008**, 4652–4658.
40. Yamamoto, H.M.; Kosaka, Y.; Maeda, R.; Yamaura, J.; Nakano, A.; Nakamura, T.; Kato, R. Supramolecular insulating networks sheathing conducting nanowires based on organic radical cations. ACS Nano **2008**, 2, 143–155.
41. Chappell, J.S.; Bloch, A.N.; Bryden, W.A.; Maxfield, M.; Poehler, T.O.; Cowan, D.O. Degree of charge transfer in organic conductors by infrared absorption spectroscopy. J. Am. Chem. Soc. **1981**, 103, 2442–2443.
42. Kagoshima, S.; Ishiguro, T.; Anzai, H. X-ray scattering study of phonon anomalies and superstructures in TTF-TCNQ. J. Phys. Soc. Jpn. **1976**, 41, 2061–2071.
43. Sheldrick, G.M. A short history of SHELX. ActaCrystallogr. **2008**, A64, 112–122.

44. Altomare, A.; Cascarano, G.; Giacovazzo, C.; Guagliardi, A.; Burla, M.C.; Polidori, G.; Camalli, M. SIR92—A program for automatic solution of crystal structures by direct methods. J. Appl. Crystallogr. **1994**, 27, 435.

2

CHAPTER

FLEXIBILITY OF HYDROGEN BOND AND LOWERING OF SYMMETRY IN PROTON CONDUCTOR

Yukihiko Yoshida [1], Junko Hatori [2], Hinako Kawakami [2], Yasumitsu Matsuo [2,*] and Seiichiro Ikehata [1]

[1] *Faculty of Science, Tokyo University of Science, 1-3 Kagurazaka, Shinjuku-ku, Tokyo 162-8601, Japan*

[2] *Faculty of Science & Engineering, Setsunan University, Ikeda-Nakamachi, Neyagawa, Osaka 572-8508, Japan*

ABSTRACT

In order to investigate why crystal symmetry lowers with increasing temperature by phase transition of T_{II-III} (=369 K) in $Cs_3H(SeO_4)_2$, in spite of the fact that crystal symmetry in the high-temperature phase of many ionic conductors becomes higher by the phase transition, we have studied the relation between the change in crystal symmetry and the appearance of proton motion. It was found from the analysis of domains based on crystal structure that the number of possible geometrical arrangement of hydrogen bond in phase II becomes two times larger than that in phase III, derived from the lowering of crystal symmetry with increasing temperature. These results indicate that the lowering of crystal symmetry in phase II appears by the increase of the number of geometrical arrangements and by the enhancement of the flexibility of hydrogen bond. Considering that the enhancement of the flexibility of hydrogen bond yields mobile proton in phase II, it is deduced that mobile proton in phase II appears in exchange for the lowering of crystal symmetry at II–III phase transition.

Keywords: phase transition; ionic conductor; hydrogen-bonded compound; ferroelasticity

1. INTRODUCTION

It is known that $M_3H(XO_4)_2$ compounds (M = K, Rb, Cs; X = S, Se) exhibit a superprotonic conductivity and becomes the electrolyte of the fuel cell [1,2,3,4,5,6,7,8,9,10,11]. The $Cs_3H(SeO_4)_2$ crystal, which is one of $M_3H(XO_4)_2$ superprotonic conductors, exhibits a superprotonic conductivity above 456 K ($=T_{I\text{-}II}$). At room temperature, the $Cs_3H(SeO_4)_2$ crystal shows a ferroelasticity and becomes insulator [12,13]. Moreover, $Cs_3H(SeO_4)_2$ shows the interesting feature that the phase transition from the low-temperature monoclinic-C2/m phase (phase III) to high-temperature monoclinic-A2/a phase (phase II) exists at 369 K ($=T_{II\text{-}III}$) [14,15,16,17,18]. In conjunction with the phase transition at $T_{II\text{-}III}$, some interesting phenomena are observed [18]. For example: (1) Mobile proton appears above $T_{II\text{-}III}$, as shown in the motional narrowing of NMR line width in Figure 1a; and (2) the increase of the electrical conductivity is observed above $T_{II\text{-}III}$ as a precursor effect of superprotonic conductivity (Figure 1b) [18]. Furthermore, the spatial symmetry C2/m in phase III is higher than the symmetry of A2/a in phase II. That is, in $Cs_3H(SeO_4)_2$, the crystal symmetry decreases by $T_{II\text{-}III}$ with the increase of temperature, in spite of the fact that crystal symmetry in a lot of materials becomes higher with increasing temperature. In this way, the phase transition at $T_{II\text{-}III}$ includes useful information needed to understand the relation between proton conductivity and the change in crystal symmetry. Therefore the investigation of the phase transition of $T_{II\text{-}III}$ will lead to the key factor needed for the appearance of flexibility of proton migration which determines proton conductivity. In the present study, we have examined possible proton sites before and after II–III phase transition from the geometrical arrangement of hydrogen bond, and have investigated why the lowering of crystal symmetry at $T_{II\text{-}III}$ is realized.

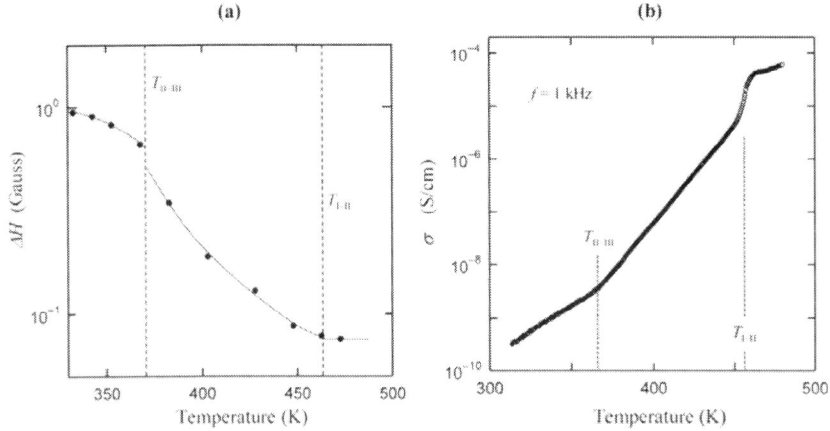

Figure 1. (**a**) Temperature dependence of the line width ΔH in the ^{1}H-NMR absorption line; (**b**) Temperature dependence of electrical conductivity along the b axis.

2. EXPERIMENTAL PROCEDURE AND CRYSTAL STRUCTURE

2.1. Experimental Procedure

$Cs_3H(SeO_4)_2$ single crystals were grown by the slow evaporation method from an aqueous solution of Cs_2SeO_4 and H_2SeO_4 with a molar ratio of Cs_2SeO_4/H_2SeO_4 = 3:1 at 35 °C. Optical observations were carried out using the polarizing microscope. The specimens used in the optical observations were approximately 0.1 to 0.2 mm in thickness.

2.2. Crystal Structure

In order to obtain the possible geometrical arrangement of hydrogen bond from the change in ferroelastic domain structure, it is necessary to use the crystal structure. Therefore, in this subsection, we show the crystal structure in phases III, II and I. Figure 2a–c show the crystal structures in phases III, II and I, respectively, based on the results reported by Merinov et al. [13,14,15]. In phase III, the $Cs_3H(SeO_4)_2$ crystal belongs to a monoclinic system with the space group C2/m. The lattice parameters are a_{III} = 10.903(3) Å, b_{III} = 6.3904(8) Å, c_{III} = 8.452(2) Å, β_{III} = 112.46(1)° and Z = 2 at room temperature [16]. As shown in Figure 2a, it is evident that in phase III the hydrogen bonds are composed between SeO_4 tetrahedrons along the direction parallel to the a-axis. In phase II, the crystal shows the monoclinic-A2/a symmetry. The lattice parameters are a_{II} = 11.037(1) Å, b_{II} = 6.415(1) Å, c_{II} = 16.042(4) Å, β_{II} = 102.69(1)° and Z = 4 at 400 K [14]. Moreover, in phase II, the hydrogen bonds are formed between SeO_4 tetrahedrons along the approximately [310] or $[3\bar{1}0]$direction, as shown in Figure 2b. In phase I, the $Cs_3H(SeO_4)_2$ crystal belongs to a trigonal system with the space group $R\bar{3}m$. The lattice constants become a_I = 6.4260(6) Å, c_I = 23.447(2) Å, Z = 3 at 470 K [15]. As shown in Figure 2c, the distance between neighboring O(2) atoms is same in time-averaged structure. Considering that hydrogen bond is composed between O(2) atoms, three equivalent positions for a proton appear in phase I, and proton conductivity is realized in phase I. In this way, from the crystal structure, we can clearly see that the phase transitions of $T_{II\text{-}III}$ and $T_{I\text{-}II}$ are structural phase transition. Especially, it is interesting property that the orientation of the hydrogen bond by the phase transition of $T_{II\text{-}III}$ mainly changes. This result implies that the difference in the geometrical arrangement of hydrogen bond observed in phases III and II includes the information for the flexibility of proton motion and the lowering of crystal symmetry at $T_{II\text{-}III}$ phase transition.

3. RESULTS AND DISCUSSION

Figure 3a–c show domain structures viewed along the direction perpendicular to the (001) plane in phases III, II and I in $Cs_3H(SeO_4)_2$, respectively. It is evident that domain structure changes at the phase transition of $T_{II\text{-}III}$ and $T_{I\text{-}II}$. As shown in Figure 3a, optical observation by means of polarizing microscope indicate that the $Cs_3H(SeO_4)_2$ crystal in phase III is optically biaxial with an interrelationship between the crystallographic axes (a_{III}, b_{III} in phase III) and the indicatrix axes (X, Y): $a_{III}//X$, $b_{III}//Y$ in phase III. In phase III, domains consist of polydomains with the two types of domain boundaries. In the present paper, the observed three kinds of domains in phase III (or phase II) are called as domains D_1, D_2 and D_3, and two types of domain boundaries are named as W- and W'-domain boundaries. The *W*- and *W'*-domain boundaries were classified as the planes of {311} and {11n} respectively, where, n is determined by the strain compatibility condition. Moreover, the adjacent domains separated by the W- or W'-domain boundary are related to the mirror symmetry on the W-domain boundary or the twofold rotational symmetry on the W'-domain boundary, respectively. We can also see that the orientation of any domain D_i is different by almost 120° from that of its adjacent domain D_j (i, j = 1, 2 and 3).

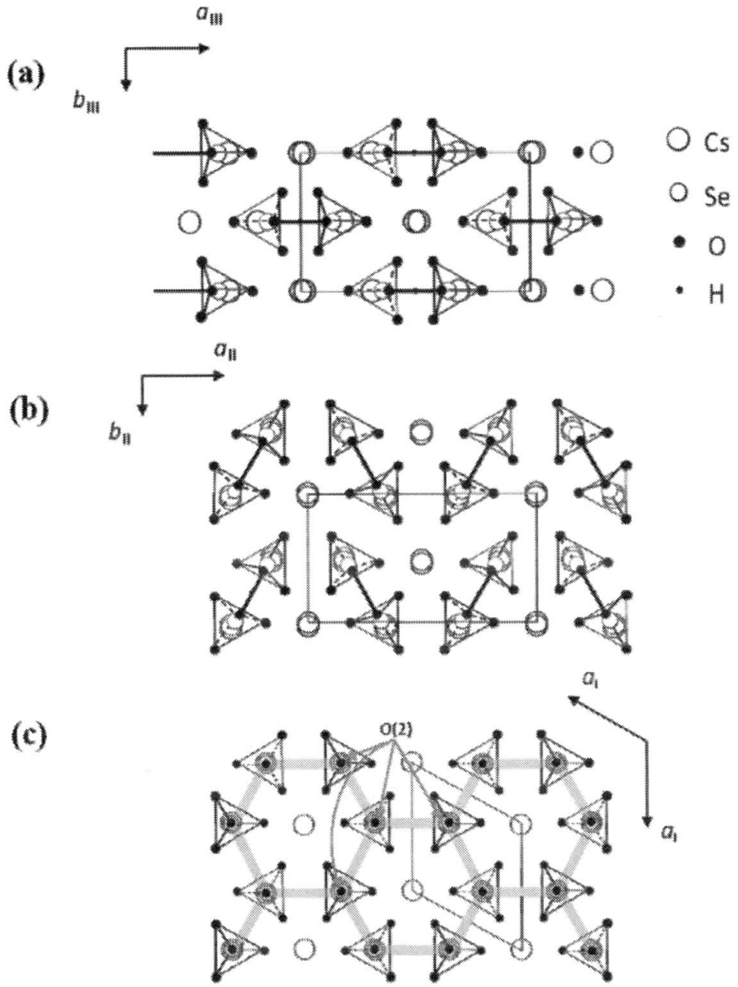

Figure 2. Crystal structure in the *a*–*b* plane (**a**) in phase III; (**b**) in phase II and (**c**) in phase I. Tetrahedrons shown by the dashed and solid lines are shown in downward and upward SeO_4 tetrahedrons, respectively. Bold gray lines in (**c**) show the possible hydrogen bonds.

The angles θ between the a_{III} (or b_{III}) axes of neighboring two domains are observed as $\theta = 119.0°$ for the W-domain boundary and $\theta = 121.0°$ for the W'-domain boundary. The angle between the a_{III} (or b_{III}) axes of neighboring two domains are calculated from the lattice constants using the equations, $\theta = 2 \tan^{-1} (a_{III}/b_{III})$ in the W-domain boundary and $\theta = \pi - 2 \tan^{-1} (a_{III}/3b_{III})$

in the W'-domain boundary. The observed angles agree satisfactorily with the calculated values θ = 119.25° in the W-domain boundary and θ = 120.74° in the W'-domain boundary. In phase II, domain structure changes from that observed in phase III by the phase transition of $T_{II\text{-III}}$, as shown in Figure 3b. From this result, we can confirm that crystal symmetry changes at $T_{II\text{-III}}$. The crystal in phase II is also optically biaxial with a relation between the crystallographic axes (a_{II}, b_{II} in phase II) and the indicatrix axes (X, Y): a_{II}//X, b_{II}//Y in phase II, respectively. Moreover, it is noted that the kinds of domain and domain boundary in phase II are the same as those in phase III, although the domain pattern changes at II–III phase transition. The angles θ between the a_{II} (or b_{II}) axes of neighboring two domains are observed as θ = 119.5° for the W-domain boundary and θ = 120.5° for the W'-domain boundary. These values are in good agreement with the values θ = 119.67° and θ = 120.33° calculated from the values of a_{II} and b_{II} in phase II. In this way, these angles in phase II are almost the same as those observed in phase III. These facts imply that the change in crystal structure is slight at II–III phase transition.

According to Sapriel [19], the orientation of domain boundaries generated by the ferroelastic phase transition from $R\bar{3}m$ to $A2/a$ is the same as those from to *C2/m*, because the orientation of the domain boundaries is determined by the change in the point group. This is consistent with our results. On the other hand, as shown in Figure 3c, in phase I, we can see clearly that the domain boundaries disappear just above $T_{I\text{-II}}$, and the crystal becomes optically uniaxial. The result in the present study is also consistent with the fact that the phase transition at $T_{I\text{-II}}$ is the ferroelastic-paraelastic phase transition and that the space group in the phase I becomes $R\bar{3}m$. Considering that in the phase transition at $T_{I\text{-II}}$ domain boundaries vanish, the averaged structure of the crystal structure in all domains in phase II becomes crystal structure in phase I. That is, the crystal structure in phase I is obtained by averaging and superposing the crystal structure in all domains with the symmetry of domain boundary.

Figure 4 shows the superposed structure obtained by the superposition of the crystal structure in all domains with the symmetries of W- and W'-domain boundaries. The averaged structure for atomic position in crystal structure of Figure 4 satisfactorily agrees with that obtained from X-ray study in phase I (Figure 2c). In this way, the averaged structure of the crystal structure in all ferroelastic domains gives the structure in high-temperature phase of ferroelastic materials. Moreover, we note that the superposed structure in Figure 4 displays the possible situation of the disordering of atoms. That is, we can obtain the information of the disordering of atom, especially the possible geometrical arrangement of hydrogen bond, from the superposed crystal structure by the ferroelastic domains.

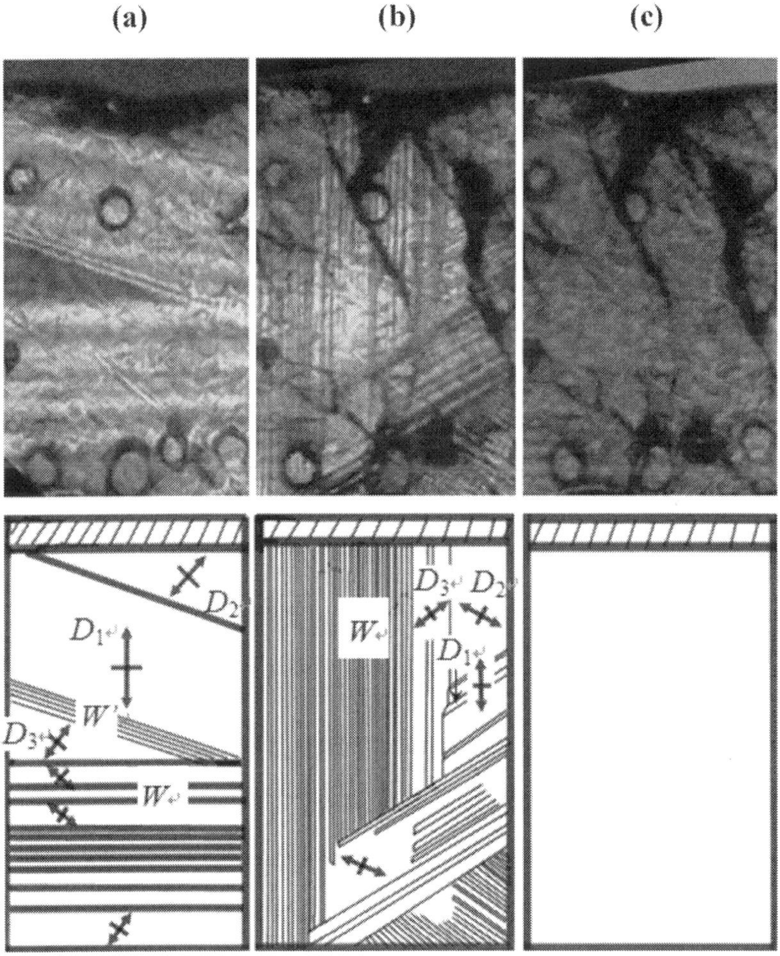

Figure 3. Domain structure (**a**) in phase III; (**b**) in phase II and (**c**) in phase I. Symbol of cross denotes the directions of crystal axes.

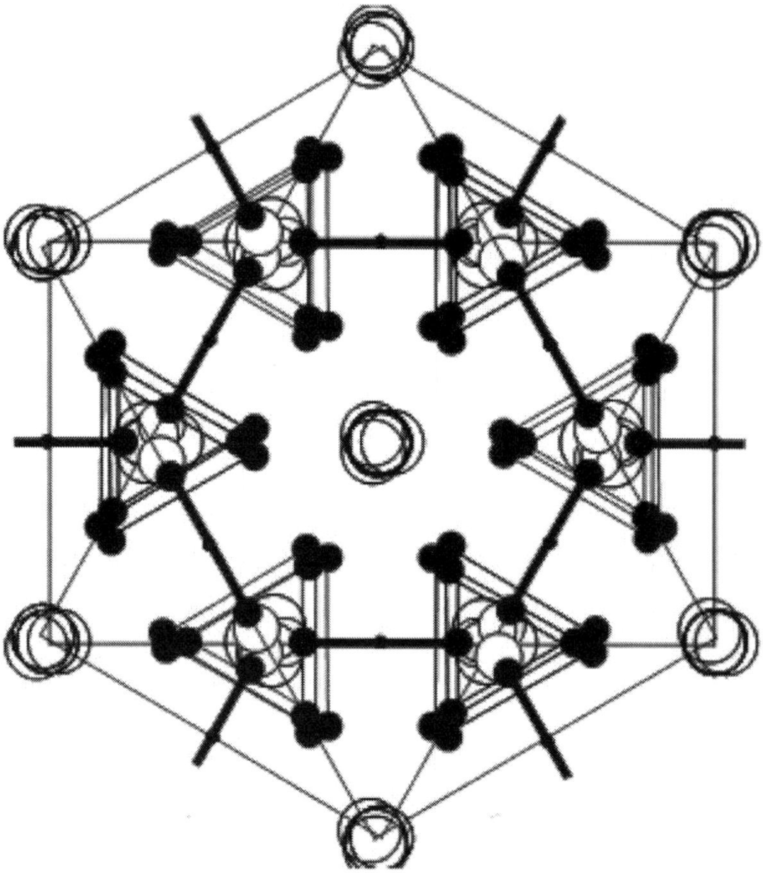

Figure 4. Superposition of all domains in phase II. Symbol of cross denotes the directions of crystal axes.

In order to investigate the possible geometrical arrangement of hydrogen bond in phases III and II, we have carried out the superposition of crystal structure in domains in phases III and II. Figure 5a,b show the crystal structures in phases III and II including three domains D_1, D_2 and D_3, respectively. In Figure 5a,b, three domains D_1, D_2 and D_3separated by the W-boundary in phases III and II are shown as an example. It is evident that the main difference in phases III and II is the hydrogen-bond arrangement. As described above, the possible hydrogen-bond arrangement is observed from the superposition of all domains. Figure 5c shows the possible hydrogen-bond arrangement obtained by thc superposition of crystal structure of $D_1$and D_3in phase III, as an example. The hydrogen bonds in

Figure 5c are realized by the break and recombination of hydrogen bond in the case of the existence of mobile proton with the position between two hydrogen bonds of phase III kept. Therefore, as shown in Figure 5c, if proton at i-site moves to *k*-site, *j*-site proton must be needed to move to *l*-site, in order to realize the crystal symmetry in phase III. In this way, in phase III, the recombination of two hydrogen bonds is simultaneously needed for one proton transport. That is, when proton is fixed in hydrogen bond, the hydrogen-bond arrangement in phase III is stable. Actually, in phase III, proton migration cannot be observed.

On the other hand, it is known that proton in phase II begins to move as a precursor motion of superprotonic motion, as shown in Figure 1. As described above, when proton moves from one hydrogen bond to another hydrogen bond, the geometrical arrangement of hydrogen bond in phase III is not appropriate. That is, when proton moves between the hydrogen bonds, the recombination of hydrogen bond is difficult in the crystal structure of phase III, because the simultaneous transport of two protons is needed in the crystal structure in phase III. This fact means that the hydrogen-bond pattern should be rearranged by II–III phase transition, as the recombination of hydrogen bond by proton transport is easily realized in phase II. Figure 5d shows the equivalent position obtained from the superposition of domain 1 and domain 3 in phase II, as an example. Considering that protons in phase II move from one hydrogen bond to another hydrogen bond as the precursor motion of superprotonic motion, proton at *j'*-site can easily move to *k'*-site without depending on the transport of proton at i'-site by the crystal structure in phase II. This result indicates that proton can easily move with the crystal structure in phase II. That is, the crystal structure in phase II is flexible for proton motion, because proton migration is not suppressed by another hydrogen bond. From these results, it is deduced that the flexibility of hydrogen bond increases in phase II in exchange for the lowering of crystal symmetry. In addition, these results imply that, in order to obtain higher proton conductivity in solid acid electrolyte, it is necessary to design a crystal structure with the flexibility of hydrogen bond. We are now planning to prepare new proton conductors, in which Cs ions are replaced by Tl and Pb ions. These results will appear in future issues.

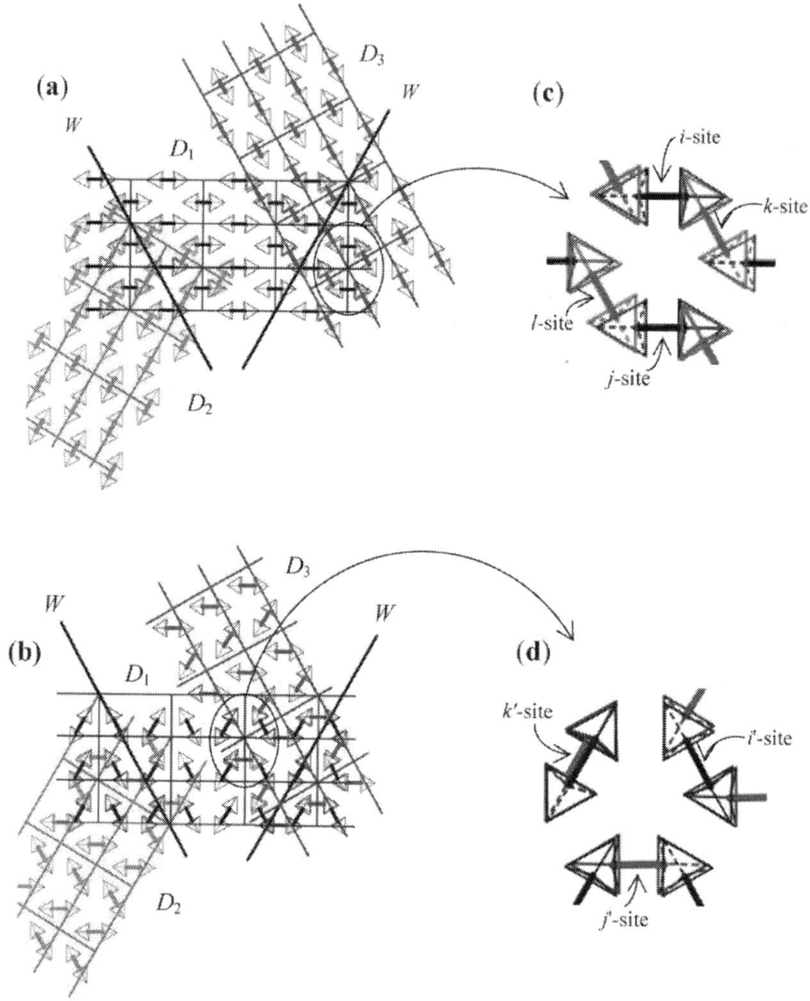

Figure 5. Schematic diagram of crystal structure (**a**) in phase III and (**b**) in phase II, including three kinds of domains. Domains D_1, D_2 and D_3 are shown by the black, red and blue colors, respectively. Averaged structure composed by the superposition of domains D_1 and D_3 (**c**) in phase III and (**d**) in phase II.

4. CONCLUSIONS

In order to investigate the origin of the lowering of crystal symmetry, we investigated the change in possible hydrogen bond patterns by phase transition of T_{II-III} on the basis of the crystal and domain structures. From the analyses of the hydrogen-bond network around T_{II-III}, we found that the number of the network patterns of hydrogen bonds in phase II is two times larger than that in phase III. These results indicate that the increase of the proton activity in phase II is caused by the increase in the number of possible geometrical arrangements of hydrogen bonds. From the analysis of the possible geometrical arrangement of hydrogen bonds, when the proton is fixed in the hydrogen bond, the hydrogen-bond arrangement in phase III is stable. On the other hand, in phase II, the proton can easily move, because the crystal structure in phase II has the flexibility for proton motion and proton migration is not suppressed. From these results, it is deduced that mobile protons in phase II appear in exchange for the lowering of crystal symmetry at II–III phase transition.

REFERENCES

1. Gesi, K. Dielectric properties and phase transitions in $X_3H(SO_4)_2$ and $X_3D(SO_4)_2$ crystals (X: K, Rb). J. Phys. Soc. Jpn. **1980**, 48, 886–889.
2. Baranov, A.I. Crystals with disordered hydrogen-bond networks and superprotonic conductivity: Review. Crystallogr. Rep. **2003**, 48, 1012–1037.
3. Pawlowski, A.; Pawlaczyk, C.; Hilzcer, B. Electric conductivity in crystal group $Me_3H(SeO_4)_2$ (Me: NH^{+4}, Rb^+, Cs^+). Solid State Ionics **1990**, 44, 17–19.
4. Kamimura, H.; Watanabe, S. A novel approach to the mechanism of ionic conductivity below and at the ferroelastic phase transition in the zero-dimensional hydrogen-bonded crystal $M_3H(XO_4)_2$ with M = Rb or Cs; X = S or Se. Phil. Mag. **2001**, B81, 1011–1019.
5. Matsuo, Y.; Takahashi, K.; Ikehata, S. Evidence for proton conduction below superionic transition on $Rb_3H(SeO_4)_2$. Solid State Commun. **2001**, 119, 79–81.

6. Matsuo, Y.; Hatori, J.; Nakajima, Y.; Ikehata, S. Superprotonic and ferroelastic phase transition in $K_3H(SO_4)_2$. Solid State Commun. **2004**, 130, 269–274.
7. Kamimura, H.; Matsuo, Y.; Ikehata, S.; Ito, T.; Komukae, M.; Osaka, T. On the mechanism of superionic conduction in the zero-dimensional hydrogen-bonded crystals $M_3H(XO_4)_2$ with M = K, Rb, Cs and X = S, Se. Phys. Status Solidi B **2004**, 241, 61–68.
8. Matsuo, Y.; Hatori, J.; Yoshida, Y.; Saito, K.; Ikehata, S. Proton conductivity and spontaneous strain below superprotonic phase transition in $Rb_3H(SeO_4)_2$. Solid State Ionics **2005**, 176, 2461–2465.
9. Hatori, J.; Matsuo, Y.; Ikehata, S. The relation between elasticity and the superprotonic phase transition temperature for $M_3H(XO_4)_2$. Solid State Commun. **2006**, 140, 452–454.
10. Ishii, T. Superprotonic phase transition in $M_3H(XO_4)_2$. Solid State Ionics **2007**, 178, 667–670.
11. Matsuo, Y.; Tanaka, Y.; Hatori, J.; Ikehata, S. Effect of uniaxial stress in superprotonic phase transition in $Cs_3H(SeO_4)_2$. Solid State Ionics **2008**, 179, 1125–1127.
12. Komukae, M.; Osaka, T.; Kaneko, T.; Makita, Y. Dielectric study of phase transitions in $Cs_3H(SeO_4)_2$ and its isotope effect. J. Phys. Soc. Jpn. **1985**, 54, 3401–3405.
13. Baranov, A.I.; Tregubchenko, A.V.; Shuvalov, L.A.; Shchagina, N.M. Structural phase transitions and proton conductivity of $Cs_3H(SeO_4)_2$ and $(NH4)3H(SeO_4)_2$ crystals. Sov. Solid State Phys. **1987**, 29, 1448–1449.
14. Merinov, B.V.; Baranov, A.I.; Shuvalov, L.A. Crystal structure of ferroelastic phase II of $Cs_3H(SeO_4)_2$. Sov. Phys. Crystallogr. **1991**, 36, 639–642.
15. Merinov, B.V.; Baranov, A.I.; Shuvalov, L.A. Crystal structure and mechanism of protonic conductivity of the superionic phase of $Cs_3H(SeO_4)_2$. Sov. Phys. Crystallogr. **1990**, 35, 200.
16. Merinov, B.V.; Bolotina, N.B.; Baranov, A.I.; Shuvalov, L.A. Crystal structure of $Cs_3H(SeO_4)_2$ (T = 295 K) and its changes at the phase transitions. Sov. Phys. Crystallogr. **1988**, 33, 824–827.
17. Komukae, M.; Sakata, K.; Osaka, T.; Makita, Y. Optical and x-ray studies in ferroelastic $Cs_3H(SeO_4)_2$. J. Phys. Soc. Jpn. **1994**, 63, 1009–1017.

18. Matsuo, Y.; Tanaka, Y.; Hatori, J.; Ikehata, S. Proton activity and spontaneous strain of $Cs_3H(SeO_4)_2$ in the phase transition at 369 K. Solid State Commun. **2005**, 134, 361–365. [Google Scholar]

19. Sapriel, J. Domain-wall orientation in ferroelastic. Phys. Rev. B **1975**, 12, 5128–5140.

3
CHAPTER

PRESSURE EFFECT ON ORGANIC CONDUCTORS

Keizo Murata [1,*], Keiichi Yokogawa [1], Sonachalam Arumugam [2] and Harukazu Yoshino [1]

[1] *Graduate School of Science, Osaka City University; 3-3-138, Sugimoto, Sumiyoshi-ku, Osaka 558-8585, Japan*

[2] *Centre for High Pressure Research, Department of Physics, Bharathidasan University, Tiruchirappalli-620-024, India*

ABSTRACT

Pressure is a powerful tool to unveil the profound nature of electronic properties in a variety of organic conductors. Starting from technology of high pressure, we plan to review what kind of physics or phenomena have previously been discussed.

Keywords: pressure; organic conductors; TTF-TCNQ; TSeF-TCNQ; HMTSF-TCNQ; $(TMTSF)_2X$; $(TMTTF)_2X$; α-$(BEDT\text{-}TTF)_2X$; β-$(BEDT\text{-}TTF)_2X$; κ-$(MeDH\text{-}TTP)_2AsF_6$; τ-type conductor

1. INTRODUCTION

It is true that the history of crystalline organic conductors is guided by the elaborate ideas and works generated by pressure experiments. The nature of a charge density wave (CDW) in tetrathiafulvalene-tetracyanoquinodimethane (TTF-TCNQ) is demonstrated by the smart analysis of conversion from conductivity, as a function of temperature at various pressures, to conductivity as a function of pressure at various temperatures [1]. By that analysis, the conductivity of CDW together with the fluctuation, as a collective motion, was well understood as phenomena at hand. Lots of trials of CDW suppression by pressure have been done from TTF-TCNQ through its family with the anticipation of superconductivity. Eventually, superconductivity was realized in $(TMTSF)_2PF_6$ by suppressing by the help of pressure the insulating phase, which was a spin density wave (SDW) instead of a CDW, where TMTSF stands for tetramethyltetraselenafulvalene [2]. Field-Induced spin density waves are observed by tuning the pressure, where superconductivity is observed [3]. The lesson of the usefulness of pressure was further succeeded when high-T_c cuprate superconductors were discovered; pressure was the guiding tool in finding further materials [4]. Thus, the study of the pressure effect of the materials is already established as an indispensable experiment to find more. In this paper, we plan to review how the pressure effect contributed to various topics of the physics of organic conductors along descriptions of the technique of pressure.

2. METHOD OF PRESSURE IN ORGANIC CONDUCTORS

The important role of the pressure as concerns solid state physics is already known for a long time. However, pressure as a tool has been developed mainly for inorganic materials at an early stage. Organic conductors can be classified into crystals and polymers. What we deal with in this paper is the organic crystals. Organic crystals are very fragile and weak mechanically. When a piston cylinder pressure cell of the clamp type is used and when the initial room temperature pressure is low, pressure becomes zero at low temperatures due to thermal contraction. In such cases, organic crystals are very likely to be broken (first problem). Voids might be produced in the pressure cell. The second problem is that organic crystals are seriously damaged at the instance of solidification of the pressure medium on cooling; for instance, if the medium is isopentane. Although the first one is a general problem for all pressure mediums and is inevitable except in the case of water, the second one due to solidification of the medium and can be solved by choosing a suitable pressure medium, which has a small volume difference at the instance of liquid-solid transition on cooling. To overcome this problem of damage to the sample on solidification by cooling, Daphne 7373 is developed [5,6]. The discontinuous drop in pressure through the solidification of Daphne 7373 is negligible, but is just observable by viewing the kink in the temperature dependence of resistance of pressure-sensitive crystals such as organic conductors. Actually, Daphne 7373 as a pressure medium was found to be a very suitable pressure medium for pressure works. Further, pressure drops or a change from room temperature to low (helium) temperature, is much smaller (typically ~0.15 GPa) than with Fluorinert 77/70 (typically ~0.5 GPa), for example. With these advantages, it was later found that Daphne 7373 is favorable even for inorganic materials. An example is the large amplitude of the de Haas van Alphen effect, whose temperature dependence determines the cyclotron mass [7,8].

In late 1990s, the pressure of the researcher's interest extended from 1 GPa to around 2–4 GPa region or up to 10 GPa. For instance, the topics of the quantum critical phenomena were discussed in 2–4 GPa region, the highest T_c-organic superconductivity [9], and 1D phase diagram of $(TMTTF)_2X$ were discussed in the 8–9 GPa region, where TMTTF is sulfur analog of TMTSF and is an abbreviation of tetramethyltetrathiafulvalene [10]. With these interests in the physics of the pressure region of the 3–4 GPa, a new piston cylinder cell other than the single walled BeCu cell was developed, which was the double-walled NiCrAl/BeCu cell. This new type of pressure cell achieved 4 GPa, in contrast to the single-walled BeCu cell with the maximum pressure of 1.5 GPa. By virtue of the stronger pressure cell, it was found that Daphne 7373 solidifies at 2.2 GPa [6,11], beyond which uniaxial compression is mixed with hydrostatic pressure. Another benefit of stronger pressure cell is that pressure values below 2.55 or 2.70 GPa became accurate for piston cylinder cell. There used to be no easy fixed points of pressure above 0.363 GPa (or 1.08 GPa but dull) of NH_4F. Therefore, the stronger pressure allowed interpolation of pressure against the load of the press. To realize hydrostatic pressure beyond the solidification pressure (2.2 GPa) of Daphne 7373 with a stronger pressure cell, one (K/M) of the authors developed a new pressure medium, which solidified at 3.7 GPa. The new medium is named Daphne 7474 [11]. The value of 3.7 GPa can be raised almost linearly up to 6.7 GPa at 100 °C.

The high quality hydrostatic pressure is achieved by choosing either a good medium with a high solidification pressure or a good apparatus. Superconducting transition temperature (T_c) and its related pressure become apparently different either with a different pressure medium or with different apparatus. By using a pressure medium with low solidification pressure achieves superconductivity with apparently low (sometimes half of the intrinsic value of) pressure in the pnictides [12]. In other words, the higher the solidification of pressure medium, the better the quality of superconductivity is. But in the case of pnictides, high quality superconductivity requires higher pressure. On the other hand, as for apparatus choosing, cubic anvil gives different but more intrinsic values T_c and P_c from other apparatus for pnictide as well [13]. Cubic anvil fulfills this purpose by realizing high quality hydrostatic pressure. The high quality pressure plays an important role for anisotropic materials whose material property is sensitive to the compression along the specific direction.

Besides the efforts of seeking the high quality hydrostatic pressure, it is worth describing the technique of uniaxial compression. Since the organic conductors are low dimensional and the overlapping integrals between the adjacent molecules are very anisotropic, the uniaxial compression is a powerful tool to find out the dominant microscopic interaction, which is associated directly with the direction. Although the idea of uniaxial compression is attractive, the extremely fragile organic conductors have prevented its realization of uniaxial compression. The breakthrough was given by Campos, Brooks et al., who buried the organic sample in an epoxy with already attached electrodes. Their method is to compress uniaxially the epoxy with a sample inside leaving two other sides of the epoxy block open, which allows expansion according to the Poisson's ratio in these two directions. This is called the uniaxial stress method [14]. Maesato and Kagoshima adopted the idea of immersing the samples into epoxy and extended the technique to piston cylinder. The uniaxial compression using a piston cylinder does not allow epoxy to expand two other directions. This effect is called the uniaxial strain method [15]. Maesato also proposed other method using the frozen oil by lowered temperature. The strain method is technically more advantageous and is free from the inevitable break of the epoxy block with the stress method. It is a matter of course that physics analysis becomes more simple when using the strain method, which deals only with the parameters of associated with one direction, whereas in stress method, with all directions. It should be noted that challenging work with direct X-ray observation under uniaxial pressure is coming out [16].

One should notice that the there is no ideal solid medium that is the same as the compressibility of the target material for the uniaxial strain method. At least, so long as the target material has an anisotropic compressibility, if the compressibility of the medium matches perfectly with that of the material along one direction, the compressibility of the medium inevitably mismatches along the other directions. Therefore, there is no perfect solid material for the ideal strain. However, although the method is not perfect, the purpose of uniaxial compression is realized at least by viewing the contrasted result between the different directions of compression. As for treatment, before it becomes hardened, the epoxy reacts chemically against samples. It reduces the chemical bond of organic salts. For example, $(TMTSF)_2X$ salt decomposes into TMTSF and others. To protect the sample from this chemical reaction by the epoxy (Stycast 1266 and its hardener before reacted), samples are protected by coating with varnish (for example X-22 (Tamiya Inc.)) and by using the somewhat reaction-processed epoxy, which is not yet hardened.

In α-$(BEDT\text{-}TTF)_2KHg(SCN)_4$ [17] and α-$(BEDT\text{-}TTF)_2I_3$[18], superconductivity is only realized by uniaxial compression, where BEDT-TTF is an abbreviation of bis(ethylenedithio)tetrathiafulvalene. Overlapping integrals are controlled not only by making the distance of the adjacent molecules along the stacking, but also the by making the relative angle of the molecules of adjacent columns. This angle is called a dihedral angle. In this case, transfer integral is suppressed by specific direction of compression, which gives the opposite result in contrast to the previous common examples (to the direction of conducting state). This is apparently observed as the tendency to insulate by increasing uniaxial pressure. A typical example is the θ-type conductors, where the insulating phase is charge ordered one, the θ-type conductors were developed by H. Mori [19]. The third most important role of uniaxial compression is to distort the symmetry. For instance, an equivalently interacted triangular lattice of dimers, which causes the spin of liquid down to absolute zero temperature, is broken into a non-equivalent triangular lattice by uniaxial compression. As a consequence, the spin liquid is suppressed and superconductivity is obtained [20,21]. This uniaxial technique is also applied to the research of inorganic materials.

With the development of pressure technique, lots of physics on organic conductors have developed drastically. There are already a lot of existing works with modern technique of pressure. This review is obliged to confine itself to the topics in which the authors were mainly involved.

3. TTF-TCNQ AND TSEF-TCNQ

The understanding of the charge density wave (CDW) has very much progressed with the widespread and smart experiments with TTF-TCNQ. TTF donates electron to accepter TCNQ, with a charge transfer ratio of 0.59 per one pair of TTF and TCNQ. The irrational value of the charge transfer ratio of 0.59 results in an incommensurate CDW, through a relation, $\rho/a = 2/\lambda$, where ρ, a and λ are charge transfer ratio, lattice constant and the wavelength of CDW. Since incommensurate CDW is associated with a second order transition, thermodynamic fluctuation of CDW is likely, and is observed as the enhancement of excess conductivity [1,22,23] and as X-ray diffuse scattering [24,25,26,27,28]. The pressure gives rise to the charge transfer ratio and accordingly changes the commensurability of CDW. The ratio of 0.59 at ambient pressure increases with pressure through the commensurate value of 0.66 (= 2/3) at 1.8 GPa, where excess conductivity is suppressed. Thus the pressure work played an important role for clarifying the nature of CDW through modifying the charge transfer ratio and then the commensurability of CDW. This is the story up to 3 GPa.

The authors extended the work up to 8 GPa and obtained a *T-P* phase diagram [29], showing that the CDW is actually suppressed and a metallic state is stabilized from 300 K to helium temperature around 9 GPa, as shown in Figure 1. Superconductivity was not found at the pressure where CDW is suppressed. One possible reason for the absence of superconductivity is that the resistivity is too large for superconductivity. The suppression of CDW by pressure cannot be simply explained by the denesting of the Fermi surface in going to higher dimension. According to S. Ishibashi, who calculated the band structure, the system becomes more 1D-like, i.e., more flat Fermi surfaces seem to be realized under pressure [30]. Therefore, the role of pressure here might be to reduce the density of state though enlarging the bandwidth, which causes a reduction of the density of states. Besides the bandwidth effect, the electron correlation in the case of CDW is known to be present, specified as the presence of $4k_F$ modulation of CDW and the primary CDW of $2k_F$. The pressure effect on the correlation is an open question.

In case of tetraselenafulvalene-tetracyanoquinodimethane (TSeF-TCNQ), which is also called as TSF-TCNQ, where sulfur atoms in TTF-TCNQ are replaced with selenium, the charge transfer ratio is 0.63. We obtained T-P phase diagram very similar to TTF-TCNQ for TSeF-TCNQ. The metallic state is stabilized around 7 GPa [32], which is lower only by 1–2 GPa than the 8–9 GPa of TTF-TCNQ [29]. It seems that this difference is caused by the difference in the charge transfer ratio at ambient pressure, i.e., 0.59 for TTF-TCNQ and 0.63 for TSeF-TCNQ. The electron correlation is much smaller in TSeF-TCNQ than TTF-TCNQ, which is confirmed by the absence of $4k_F$–CDW formation in TSeF-TCNQ, in contrast with TTF-TCNQ. But since the T-P phase diagram looks just shifted to lower pressure by replacing

TTF- with TSeF-salt, the correlation may not play an important role in the pressure effect on CDW.

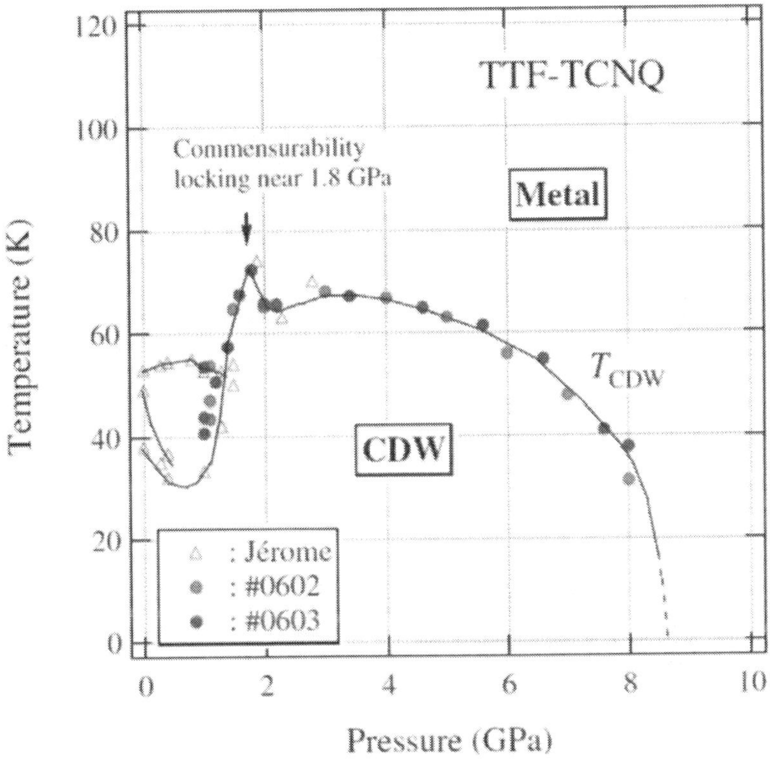

Figure 1. Temperature-Pressure phase diagram of TTF-TCNQ. Preceding work up to 3 GPa [29], including the (charge density wave) CDW commensurability peak at 1.9 GPa [31] is reproduced. The CDW is suppressed by pressure even though nesting becomes better. Reproduced with permission from JPSJ.

If the fluctuation of CDW is not taken into account, Soda and Jérome pointed out that the temperature dependence as well as pressure dependence of the resistivity perpendicular ($\rho_{\perp 1D}$) to the 1D axe obeys the same law as that along ($\rho_{//1D}$),i.e., always $\rho_{\perp 1D}$is proportional to the $\rho_{//1D}$ with a constant factor independent neither of pressure nor temperature, since both are dominated by a single parameter of a single electron scattering its lifetime along the 1D axis [33]. When fluctuation is present, conductivity along the two directions shows different temperature dependence [22,23]. The present authors examined the temperature dependence of resistivity of

TTF-TCNQ for various pressures up to 8 GPa [29]. It turned out that as it is at ambient pressure, the temperature dependence of resistivity for $\rho_{//1D}$ and $\rho_{\perp 1D}$ is not proportionally related. But the deviation between $\rho_{//1D}$ and $\rho_{\perp 1D}$ becomes smaller towards 8 GPa. It is concluded that when there is a CDW, there is a fluctuation. But the strength of fluctuation becomes smaller as T_{CDW} is reduced by increasing pressure [34].

Table 1. Comparison of four TTF-TCNQ analogues in terms of stacking pattern viewed along 1D axis, superstructures of $2k_F$ and $4k_F$, charge transfer ratio from cation to anion and the pressure, where CDW is suppressed.

	Stacking pattern (viewed along 1D)	Super-Structure	Charge transfer	P_{CDW}
TTF-TCNQ	F F F F Q Q Q Q F F F F Q Q Q Q	$2k_F$ & $4k_F$	0.59	9 GPa
TSeF-TCNQ	F F F F Q Q Q Q F F F F Q Q Q Q	$2k_F$ only	0.63	7 GPa
HMTTF-TCNQ	F Q F Q Q F Q F F Q F Q Q F Q F	$2k_F$ & $4k_F$	0.72	
HMTSF-TCNQ	F Q F Q Q F Q F F Q F Q Q F Q F	$2k_F$ only	0.74	1 GPa

4. HMTSF-TCNQ

As seen in Table 1, the stacking pattern in HMTTF-TCNQ and HMTSF-TCNQ are different from that of TTF-TCNQ and TSeF-TCNQ, where HMTTF and HMTSF stand for hexamethylenetetrathiafulvalene and hexamethylenetetraselenafulvalene. The stacking pattern is not of a layered type but of checkerboard as shown in Table 1. However, the band pictures are similar in the sense that electron and hole 1-D bands and the corresponding Fermi surface sheets are formed when the system is in the metallic state [35,36]. All of these salts undergo CDW transition. It is of interest that selenium-based TSeF-TCNQ and HMTSF-TCNQ are less electron-correlated than the sulfur-based TTF-TCNQ and HMTTF-TCNQ, which are evidenced by the absence (for non-correlated) and presence (for correlated) of $4k_F$ fluctuation of CDW viewed by X-ray.

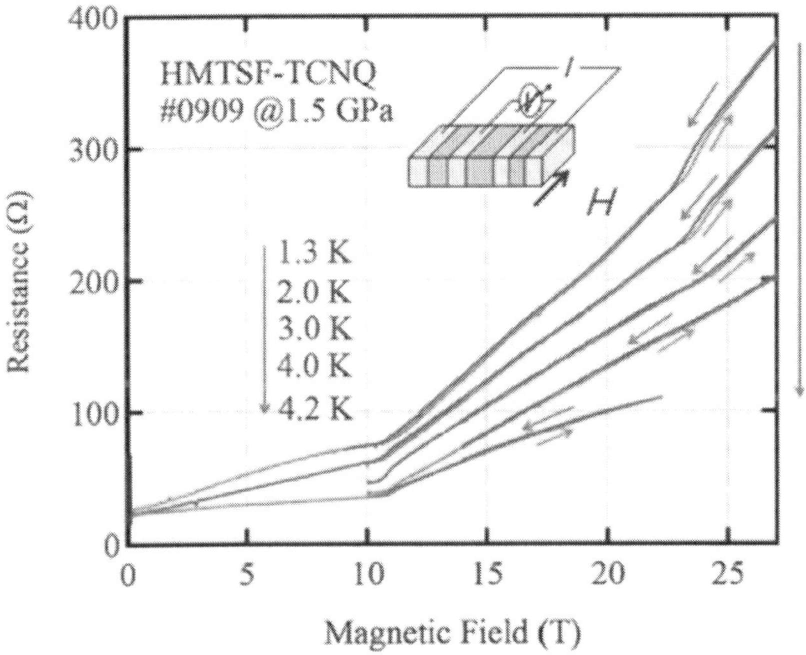

Figure 2. Magnetoresistance, which shows field-induced CDW state in HMTSF-TCNQ. Pressure at low temperature may be around 1.3 GPa at low temperature with 1.5 GPa at room temperature and with Daphne 7373 pressure medium in a piston cylinder cell [29]. Reproduced with permission from JPSJ.

The present authors revisited the HMTSF-TCNQ, and confirmed that the CDW is suppressed by a pressure of 1 GPa. In such a case where the insulating state is suppressed by pressure, the aspect of the quantum critical point in the *T-P* phase diagram is an attractive viewpoint. In cases when the insulating phase of SDW is suppressed by pressure in $(TMTSF)_2X$ salts, superconductivity at low fields, field-induced SDW and quantum Hall effect at high field were observed [3]. In this case, with CDW we expected field-induced CDW when pressure was tuned at 1 GPa. Actually, magnetoresistance at 0.4 K showed some number of kink structures in the field-sweep up to 27 Tesla (sometimes to 31 Tesla) as shown in Figure 2. These kink structures are the evidence for the field-induced phase transitions, which are of first order since they accompany hysteresis [37,38]. The latest results of the present authors are; (i) Hall voltage above around 4 Tesla shows three or four steps of plateau up to 31 Tesla suggestive of quantum Hall effect. All of them are negative, but the absolute value of the Hall voltage plateau looks to be shrinking towards a higher magnetic field. This behavior is in contradiction to the usual quantum Hall effect, wherein the Hall voltage plateau should increase with an increasing field. In the case with HMTSF-TCNQ here, it might be related to the cancellation of the effect of the electrons and holes. But we note that the Hall voltage plateau levels of our result are in the same order of magnitude of $h/e^2 = 25.8\ \mathrm{k\Omega}$.. Since we were obliged to still to the phenomena at the quantum critical point in pressure, pressure was indispensable in performing this investigation. These results will be published elsewhere.

5. $(TMTTF)_2X$ AND $(TMTSF)_2X$

The first organic superconductivity was found in $(TMTSF)_2PF_6$, under a pressure of 0.65 GPa at 1 K [2], and subsequently, in $(TMTSF)_2ClO_4$ at ambient pressure. $(TMTSF)_2X$ salts underwent a metal-insulator transition around 12 K, which was found to be the spin density wave (SDW) due to nesting and electron correlation. The transition from metallic to SDW states is very clear as shown by the divergence of $d(\log\rho)/dT$, where ρ is resistivity and T is temperature. The anisotropy of magnetic susceptibility of electron spins develops below T_{SDW}, which are viewed by static susceptibility [39] as well as by NMR [40,41]. In contrast to $(TMTSF)_2X$ salts, the resistivity of $(TMTTF)_2X$ also turns from metallic to semiconductor-like temperature dependence by lowering temperature, but this "transition" is usually very dull and broad and its origin was long considered mysterious[42]. It was much later that the nature of semiconductor-like behavior was found to be originated in charge order or charge disproportionation [43,44,45].

Jérome proposed the generic phase diagram from $(TMTTF)_2X$ to $(TMTSF)_2X$ by the temperature as a function of pressure, locating various salts for their suitable pressure axis. The salts of $(TMTTF)_2X$ are located more left than those of $(TMTSF)_2X$ [46]. Towards the left in the Figure, the more 1D-like, charge order and spin-Peierls are there, and the antiferromagnetic states are located on the commensurate (left) side. Towards the right, the antiferromagnetic state specified as a spin density wave becomes more incommensurate, superconductivity is realized and a metallic state is stabilized. Actually, superconductivity was found in $(TMTTF)_2X$ salt by using diamond anvil cell around 5 GPa [47,48]. Later on, Itoi completed a very beautiful T-P phase diagram by using a cubic anvil cell as shown in Figure 3 [10]. Remarkable result was that the superconducting T_c of this family was not around 1 K, but more than 2 K for $(TMTSF)_2AsF_6$ and the superconductivity is widespread in pressure, and seems to have two T_c peaks.

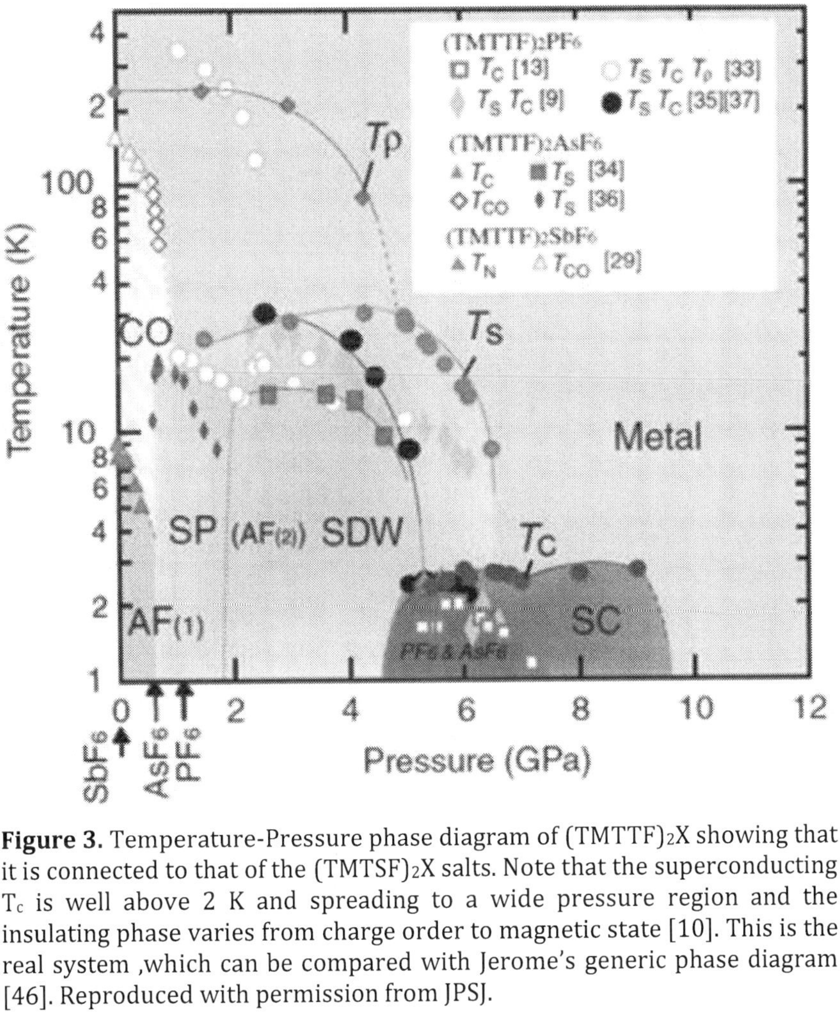

Figure 3. Temperature-Pressure phase diagram of $(TMTTF)_2X$ showing that it is connected to that of the $(TMTSF)_2X$ salts. Note that the superconducting T_c is well above 2 K and spreading to a wide pressure region and the insulating phase varies from charge order to magnetic state [10]. This is the real system ,which can be compared with Jerome's generic phase diagram [46]. Reproduced with permission from JPSJ.

It is of interest to point out that the electron correlation plays an important role all through the Jérome's phase diagram. On the left of the diagram, charge order is the consequence of the correlation. More to the right, the SDW appears. As for the SDW, the authors examine the effect of uniaxial strain along a,b and c axes on T_{SDW} in $(TMTSF)_2PF_6$. What was expected was that with *p//a*, the nesting becomes better; accordingly T_{SDW} might increase, whereas with either *p//b* or *p//c*, T_{SDW} might be suppressed in the same mechanism. The result differed from this expectation. The difference in T_{SDW} was very small as shown in Figure 4. This result suggests that T_{SDW} is determined not dominantly by the nesting, probably the bandwidth, density of state and electron correlation.

We note, in all directions of resistivity, ρ_a, ρ_b and ρ_c, the resistivity variation against the strain was remarkably different between p//a and other two ways, i.e., *p//b* and *p//c*. That is, the resistivity reduction with p//a is as moderate as that of hydrostatic pressure, whereas with *p//b* or *p//c*, the reduction is 3–4 orders of magnitude by 1 GPa. Since only the geometrical nesting fails to explain the result as mentioned above, electron correlation effect seemed to be very dominant [49,50].

The symmetry of counter anion, X of this family, frequently plays a crucial role to material properties. When X has the inversion symmetry like octahedron with PF_6^- or AsF_6^-, an anion ordering problem does not occur. However, without the inversion symmetry like tetrahedron with ClO_4^-, ReO_4^- or BF_4^-, the orientation of tetrahedron is random at high temperature and becomes either ordered or disordered depending on cooling speed, respectively. Orientational ordering causes unit cell to be doubled and the Brillouin zone to be halved, which is called zone folding [51]. The zone folding changes the material property drastically, sometimes from metal to insulator or from superconducting to insulator. Considering that this anion ordering (AO) is available in the presence of loose space at the anion site, pressure can suppress the anion ordering, resulting as described. The present author examined the T_{AO} of $(TMTSF)_2ClO_4$ by detecting the kink in resistance up to 1.1 GPa. But the strength of anion ordering becomes very weak at 0.4 GPa [52] as shown in Figure 5.

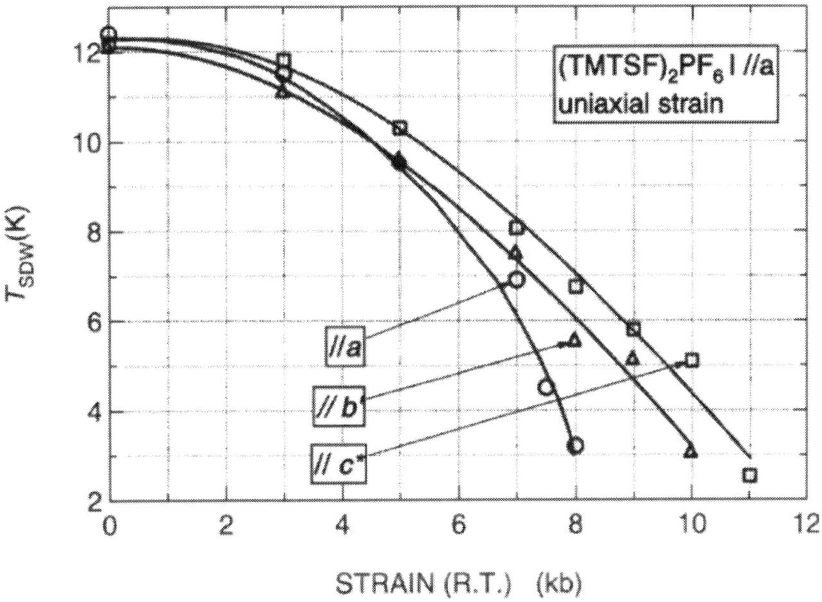

Figure 4. Temperature-Pressure phase diagram of SDW (left-bottom region of the lines) and metal in $(TMTSF)_2PF_6$ [52]. Reproduced with permission from JPSJ.

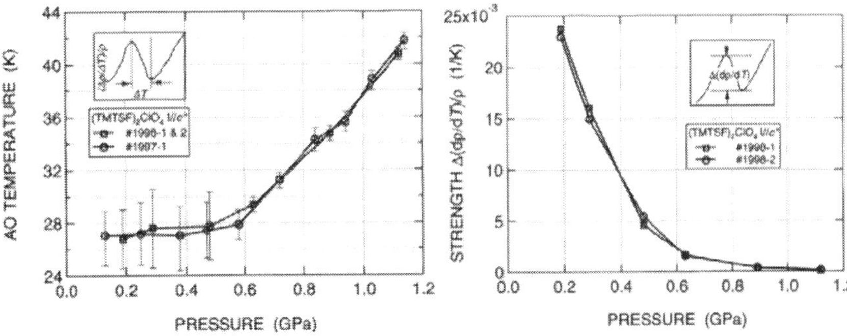

Figure 5. Temperature of anion ordering (left) and its strength (right) as a function of pressure [45]. Reproduced with permission from JPSJ.

6. B-$(BEDT\text{-}TTF)_2I_3$

This salt is a two-dimensional conductor and showed superconductivity of 1 K at ambient pressure when it was first reported [53]. Soon after, 8 K superconductivity was reported under pressure of ~0.1 GPa independently by the present author (K/M) [54,55] and Laukhin et al. [56]. Historically, the discovery of the 8 K superconductivity was an important event, because many of the researchers started to move away from this field, concluding themselves that the superconducting T_c of organic crystals should not exceed much more than 1 K. At that time, the one exception was the T_c of 3 K for $(TMTSF)_2FSO_3$[57].

The apparent existence of two T_c's as shown in Figure 6 drew attention to a new possibility. If two superconducting phases, for example, with different T_c's, coexist next to each other, it is direct evidence for non-s-wave superconductivity. In other words, the s-wave superconductivity does not have two superconducting phases. It reminded us of the two superconductivities proposed for UPt_3, where the phase boundary of two superconductivities is shown by specific measurement [58]. Later on, it was made clear that the 1 K-low T_c superconductivity is associated with an incommensurate superstructure, which appears at 195 K [59]. This superstructure is found to be suppressed by the gas pressure of 0.04 GPa [60,61], therefore, when pressure larger than 0.04 GPa is applied during cooling from room temperature through 195 K, 8 K-superconductivity is achieved even the pressure is released to zero below 195 K. As a consequence, the two phases in superconductivity supporting directly the non S-type superconductivity in terms of internal structure in superconductivity is discarded.

Application of pressure therefore simply suppresses the superconductivity. Isostructural salt like β-$(BEDT\text{-}TTF)_2IBr_2$ and β-$(BEDT\text{-}TTF)_2AuI_2$ show the superconducting T_c on the same curve of $T_c$vs. volume of β-$(BEDT\text{-}TTF)_2I_3$. This explains that the T_c is ruled by the density of state. Similar classification of superconducting T_c against volume was applied later for the first series of A_3C_{60}, where A and C_{60} are the alkali-element and fullerene, respectively.

As for the superconductivity, this family does not show superconductivity when the resistivity is higher than 150 μΩ·cm, which is the well-known value of the localization limit. This experiment was achieved by measuring the T_c against the residual resistivity of the various alloys, such as β-$(BEDT\text{-}TTF)_2(I_3)_{1-x}(IBr_2)_x$, β-$(BEDT\text{-}TTF)_2(IBr_2)_{1-x}(I_2Br)_x$, β-(BEDT-

$TTF)_2(I_2Br)_{1-x}(I_3)_x$ [62]. Further β-(BEDT-TTF)$_2$I$_2$Br never showed superconductivity. Among the trihalides, I_3, IBr_2 and I_2Br, it is known that heavy atom is located always at the center. Therefore, I_3 and IBr_2 are always symmetric, whereas I_2Br is always non-symmetric, which causes an orientational disorder and then becomes the source of scatterers. This is the reason for the absence of superconductivity in β-(BEDT-TTF)$_2$I$_2$Br. These series of experiments claim that the superconductivity is of a type that is weak against non-magnetic impurities.

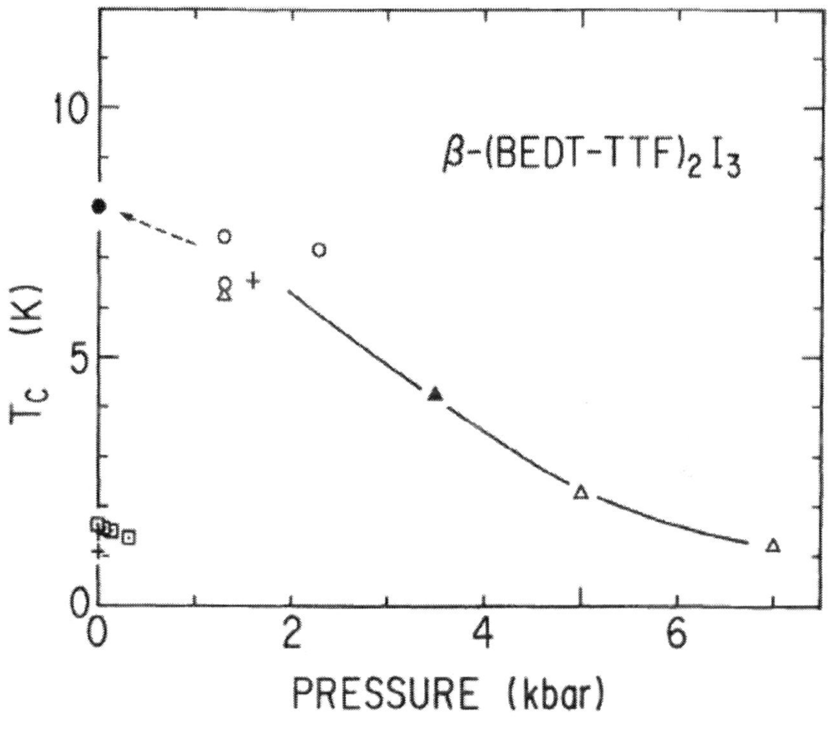

Figure 6. Temperature-Pressure phase diagram of ofβ-(BEDT-TTF)$_2$I$_3$[55]. It seemed that the two superconducting T_c's, i.e., inner structure of superconductivity was suggested. However, the different series of T_c were proved related to the structure difference. Reproduced with permission from JPSJ.

Pressure is a tool that usually reduces the electrical resistivity mainly, by the following two mechanisms: Firstly, pressure may enhance the bandwidth then mobility, in many cases. Second, pressure may enhance the dimensionality, which might help the electron pass, avoiding the scattering site. The authors tried to realized superconductivity of β-(BEDT-TTF)$_2$I$_2$Br by reducing the resistivity by means of pressure. Actually, the resistivity at room temperature was suppressed. However, the resistivity at low temperature was unchanged, which meant that the resistivity is perfectly ruled by the intrinsic and robust scatterers of I_2Br-anions at low temperature and by the band effect at high temperature [63]. It might be interesting to work in the future with an uniaxial strain from this view point.

Another way to approach superconductivity is whether or not the upper critical field, H_{c2} of superconductivity, is limited by the Clogston-Pauli value. This experiment was carried out under pressure with precise adjustment of the orientation of the magnetic field parallel to the 2D layer of the crystal by rotating the pressure cell in high magnetic field. The result was that the H_{c2} did not increase after lowering the temperature and looked saturated by the value of Pauli value, which supported the S-type superconductivity [64]. However, it was reported that the ambient pressure superconductivity of 7.1 K generated by the Ginodman's method [61] showed H_{c2}, which exceeded Pauli limit [65]. Therefore 8 K superconductivity achieved either by under pressure or by a special temperature-pressure path during cooling showed different results. It might be related to the amount of scatterers due to the perfectness of the crystal, since this superconductivity is very sensitive to impurities as mentioned above.

The Fermiology by means of the Shubnikov de Haas effect of organic conductors started with this material [66,67,68] alongside κ-(BEDT-TTF)$_2$Cu(NCS)$_2$[69]. Once the Fermi-surface of the 2D organic conductors is found to be the real objects, Fermiology by means of angular dependence of magnetoresistance developed Kajita-Yamaji oscillation [70,71,72].

7. Α-(BEDT-TTF)$_2$I$_3$

This salt has a different crystal structure from β-(BEDT-TTF)$_2$I$_3$, and both α- and β-salts were found even in the same batch grown by the electrochemical method. Without using X-ray, the structural type of crystal is distinguished by the ESR bandwidth, 70–110 Gauss (20–25 Gauss) for α (for β) crystals [73,74,75,76]. Historically and technically, α and β appeared at the same time even in the same batch in the crystal growth. And β has been focused on for study because of the existence of superconductivity and its drastically high T_c of 8 K. What was found in the α-salt was that the temperature dependence of resistivity looked metallic but very weak, the sharp metal-insulator (M-I) transition occurs at 135 K, and this M-I transition is suppressed by pressure. Kajita, Tajima et al. studied the pressure effect up to 1.9 GPa and found that this very flat temperature dependence continuing to liquid helium temperature [77]. Their Hall effect study suggested that the carrier numbers decreased from room to helium temperature by 5–6 orders of magnitude. Much later, A. Kobayashi, Katayama et al. explained the temperature dependence of resistivity and Hall effect in terms of Dirac cone structure in the electronic band structure [78].

At the same time, this material shows charge disproportionation in the metallic state and charge ordering in the insulating state, which was proved by NMR chemical shift study [79], which is also true under pressure even though the T_{MI} is lowered by pressure [80]. In the Dirac cone state, the system is in the insulating state where charge ordering is established. There are two Dirac cones in the first Brillouin zone. Montambaux predicted that the two Dirac cones approaches each other by increasing pressure and merge into one where gap opens beyond 4 GPa [81]. However, Kino pointed out that pressure just make the Dirac cones more tilt and one side of the cones will make a negative slope and gap will disappear keeping the cone structure [82]. The present authors performed the high pressure measurement up to 8 GPa using cubic anvil pressure cell, which ensures constant pressure down to 2 K. Against the theoretical prediction of semiconducting behavior above 4 GPa, our result is that the almost temperature independent resistivity behavior is unchanged up to 8 GPa. Furthermore, the sheet resistance value is very close to the universal resistance of 25.8 kΩ in the temperature-independent resistivity region in Figure 7. It can be concluded that the Dirac cone electronic structure is stable under pressure at least up to 8 GPa.

It is remarkable that the superconductivity is found under uniaxial strain along the a-axis, but not along the b-axis as shown in Figure 8. Around the superconductivity in the temperature-strain phase diagram, the charge ordered state is located. This suggested a new mechanism of superconductivity related to charge order. Historically, Dirac cone was found in graphene, the one sheet of graphite that is very fragile and hard to treat. Many of the possible experiments are still limited. From the work with α-(BEDT-TTF)$_2$I$_3$ the Dirac cones, which are of course slightly different from those of graphene, were found in a bulk system. Therefore, the research on the Dirac cone flourished.

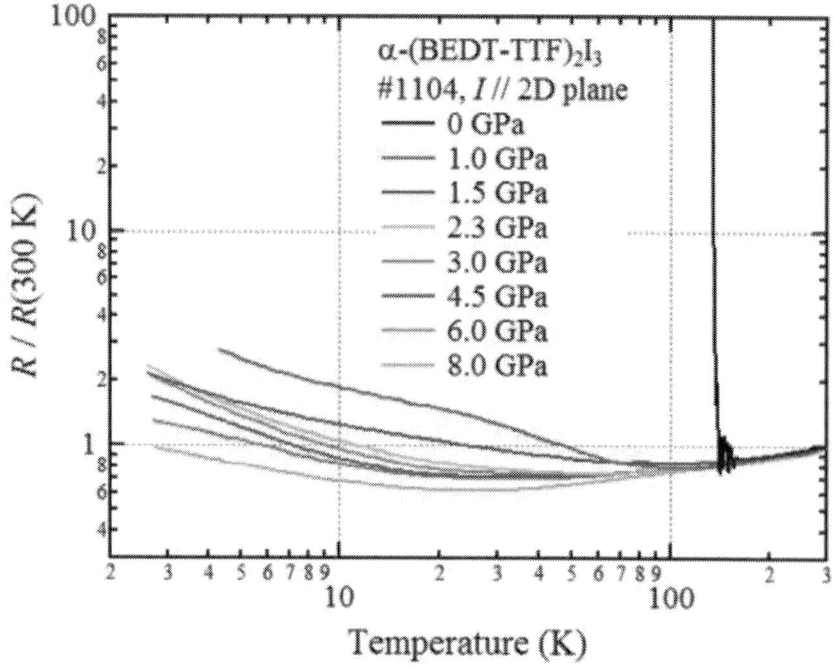

Figure 7. Temperature dependence of resistivity of α-(BEDT-TTF)$_2$I$_3$ as examined by the present authors. It is obvious that the Dirac cone state is stabilized even at 8 GPa.

8. K-(MEDH-TTP)$_2$ASF$_6$

The crystal of κ-type itself seems to form an inevitably dimerized structure. However, the dimerization energy in κ-(MeDH-TTP)$_2$AsF$_6$ is one order of magnitude smaller than the usual κ-type crystals. The dimerized structure is related to half-filled and then to Mott insulator. Then an interesting question arises whether the very weakly dimerized structure can induce a quarter-filled band and then charge ordered state [83]. With this interest, the temperature dependence of resistivity of κ-(MeDH-TTP)$_2$AsF$_6$ for various pressures up to 6 GPa was studied, where MeDH-TTP stands for 2-methyl-5-(1,3-dithiolan-2-yliden)-1,3,4,6-tetrathiapentalene [84]. At ambient pressure the system was completely insulating. But by increasing pressure, a metallic state appeared below room temperature, and then the metal-insulator transition became sharper. With further increase of pressure, insulating state was suppressed, and we anticipated superconductivity around 2.2 GPa. However, the system suddenly became Fermi liquid state without showing superconductivity. As shown by the phase diagram in Figure 9, the divergence of A, which is the prefactor of the T^2-term in resistivity from higher pressure to 2.2 GPa, was observed. In this way, the system showed a typical quantum critical point.

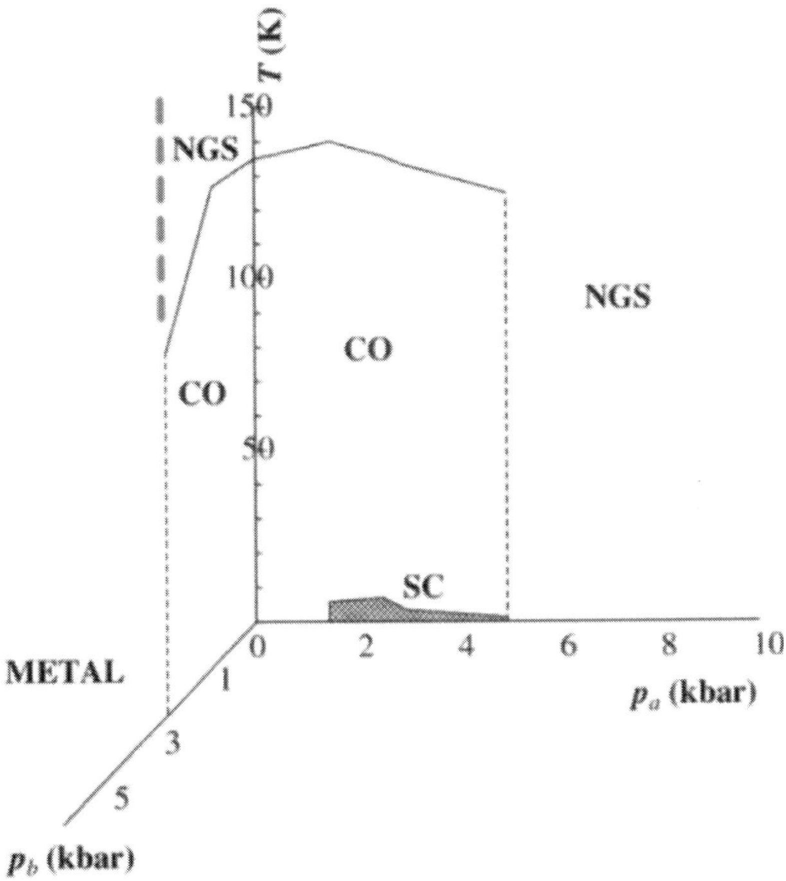

Figure 8. Temperature-Strain phase diagram of α-(BEDT-TTF)$_2$I$_3$ [18]. Superconductivity is seen only with the strain along the a-axis. Around the superconductivity, charge ordered phase is located. Reproduced with permission from JPSJ.

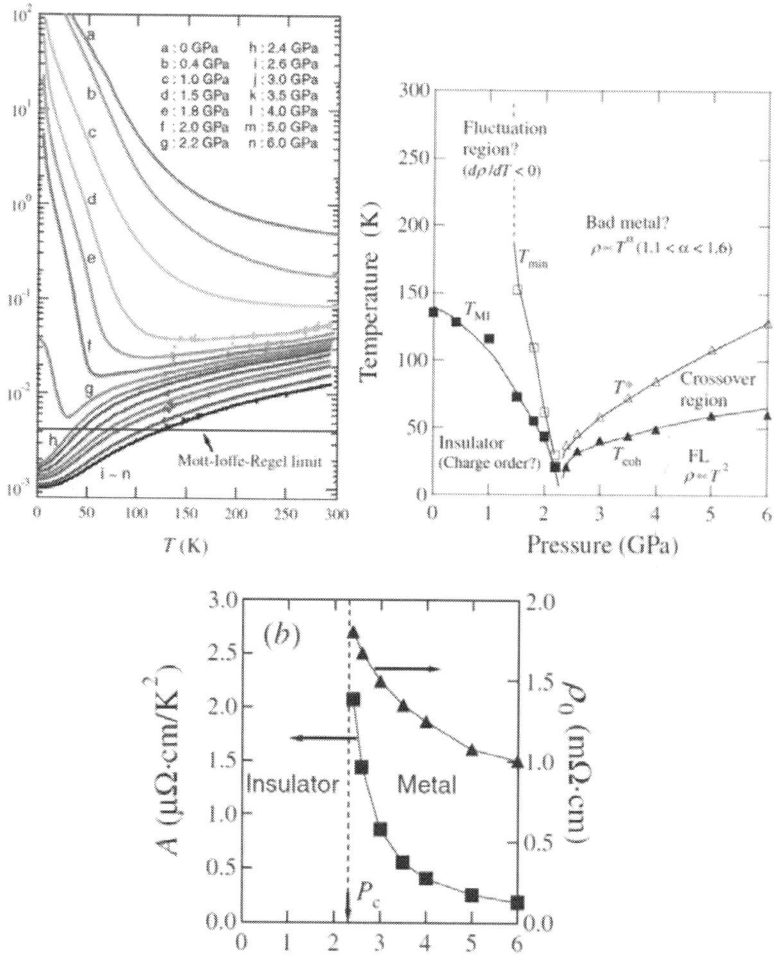

Figure 9. Temperature dependence of resistivity of κ-(MeDH-TTP)$_2$AsF$_6$ under hydrostatic pressure by cubic anvil cell with pressure medium of Daphne7373(top left). Temperature-Pressure phase diagram showing quantum critical point (top right). Divergence of the prefactor, A of T^2-term in resistivity and the residual resistivity (bottom) [84].Reproduced with permission from JPSJ.

9. T-TYPE CONDUCTORS

The chemical formula of τ-type conductor is τ-D_2A_{1+y}, where D is a chiral donor such as P-(*S,S*)-DMEDT-TTF or EDO-*S,S*-DMEDT-TTF, and A is a linear anion such as I_3^-, $AuBr_2^-$, and y is around 0.75–0.9 [85,86]. The names of P-(*S,S*)-DMEDT-TTF or EDO-S,S-DMEDT-TTF come from pyrazino-(S,S)-dimethyl-ethylenedithio-tetrathiofulvalene and ethylenedioxy-S,S-dimethylenedithio tetrathiafulvalene, respectively. Due to the low symmetry of chiral donors, one unit cell includes four layers of D_2A_{1+y} perpendicular to the conducting plane in a chiral manner. The same anion A is located in two different typical sites; one is inside the conducting layer completely as expresses by 1 in A_{1+y}, the other is in between the layers with a deficiency as described as y (< 1) in A_{1+y}. Since the unit cell structure perpendicular to the conducting layer is so complicated as four-fold, the band calculation was carried out for a single layer neglecting the four-fold structure for a unit cell in the real system. The calculated electronic structure and the Fermi surface for a single sheet give a four-fold star-shaped 2D Fermi surface in a square Brillouin zone.

The positional deficiency of the trihalide-anion at the interlayer site, with which value y is related, is one origin of disorder. Another type of disorder comes from either the two ways of bending of the non-planar ring including two oxygen atoms in the donor EDO-S,S-DMEDT-TTF or the two ways of chirality of two methyl groups in the P-S,S-DMEDT-TTF [87]. Furthermore, anions layers in between the conducting layers make different periodicity from that of the donors. Some of the conductors exhibit superstructure below ~245 K [88].

Probably due to two types of disorder, especially due to the deficiency of the I_3-anions, which are likely to dissociate from the crystals, the temperature dependence of resistivity of τ-type conductors is very sample dependent. However, roughly speaking, all of the crystals are metallic down to 30–100 K, and show resistance upturn at low temperature. Accordingly, the resistance ratio $R_{300\ K}/R_{low\ temp}$ is almost 1, which suggests the Shubnikov de Haas (SdH) effect is unlikely. However, a very clear gigantic SdH effect is observed as shown in Figure 10 [89]. From this observation, the system has at least a clean nature in spite of many disorders. The authors found two Fermi surface pockets, i.e., 0.66% and 6.1% of the FBZ (first Brillouin zone) for τ-(EDO-S,S-DMEDT-TTF)$_2$($AuBr_2$)$_{1+y}$, and 2.3% and 6.9% for τ-(P-S,S-DMEDT-TTF)$_2$($AuBr_2$)$_{1+y}$ [89,90]. Both of these values are inconsistent with the value of 12.5% calculated for y; 0.75. The small FS pocket of 0.66% is realized the Landau quantum limit, and the larger FS plays a role of reservoir of carriers at the intermediate temperature, and quantum Hall effect is realized there (~1 K) [91]. At a lower temperature, the Landau levels of the large FS pocket separate each other, and longer the large FS pocket no longer plays as a carrier reservoir. This is the reason that lowest temperature is not suitable to observe quantum Hall effect.

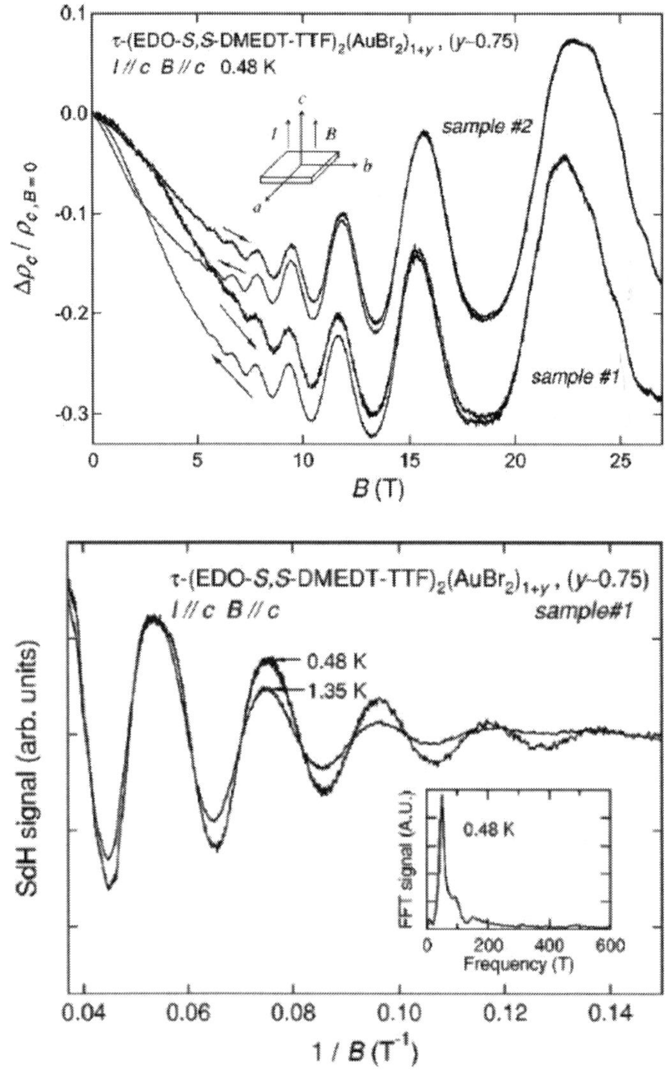

Figure 10. Giant Shubnikov-de Haas oscillation in τ-(EDO-S,S-MEEDT-TTF)$_2$(AuBr$_2$)$_{1+y}$ (top), and it 1/B-plot (bottom) [89]. Reproduced with permission from Physical Review B.

At a high field above 38 T, the system turns into a field-induced insulating state with large hysteresis [92]. The pressure effect for these materials is very striking. This field induced insulating phase is easily suppressed by extremely low pressure, namely much smaller than 0.1 GPa. The reason is unexplained. Recently, these materials drew attention by its large thermoelectric power based on the flat band in the electronic structure [93].

10. CONCLUSION

We reviewed many but limited kinds of topics, as were discussed in association with pressure works on organic conducting crystals. We started with TTF-TCNQ and finished with recent materials with which the authors were deeply involved. Since covering everything was impossible, we dared to skip even some important materials such as κ-type conductors and DCNQ-materials. We also did not touch the materials that include two competing phases of charge order and Mott insulating states found in β-$(BDA\text{-}TTP)_2I_3$, despite these being the authors' current topics [94,95]. However, although limited subjects were treated in this paper, we believe it will be realized that how much pressure has contributed to the development of this field of organic conductors.

ACKNOWLEDGEMENTS

The part of this work was supported in part by the partially supported by the Grant-in-Aid for priority area 23110722.

REFERENCES

1. Andrieux, A.; Schulz, H.J.; Jérome, D. Conductivity of the one-dimensional conductor tetrathiafulvalene-tetracyanoquinodimethane (TTF-TCNQ) near commensurability. Phys. Rev. Lett. **1979**, 43, 227–230.
2. Jérome, D.; Mazaud, A.; Ribault, M.; Bechgaard, K. Superconductivity in a synthetic organic conductor $(TMTSF)_2PF_6$. J. Phys. Lett. **1980**, 41, L95–L98.
3. Chaikin, P.M. Field Induced Spin density waves. J. Phys. I France **1996**, 6, 1875–1898.

4. Wu, M.K.; Ashburn, J.R.; Torng, C.J.; Hor, P.H.; Meng, R.L.; Gao, L.; Huang, Z.J.; Wang, Y.Q.; Chu, C.W. Superconductivity at 93 K in a new mixed-phase Y-Ba-Cu-0 compound system at ambient pressure. Phys. Rev. Lett. **1987**, 58, 908–910.
5. Murata, K.; Yoshino, H.; Yadav, H.O.; Honda, Y.; Shirakawa, N. Pt resistor thermometry and pressure calibration in a clamped pressure cell with the medium, Daphne 7373. Rev. Sci. Instrum. **1997**, 68, 2490–2493.
6. Yokogawa, K.; Murata, K.; Yoshino, H.; Aoyama, S. Solidification of high pressure medium Daphne 7373. Jpn. J. Appl. Phys. **2007**, 46, 3636–3639.
7. Shishido, H.; Ueda, T.; Hashimoto, S.; Kubo, T.; Settai, R.; Harima, H.; Onuki, Y. A de Haas–van Alphen experiment under pressure on $CeColn_5$, deviation from the quantum critical region. J. Phys. Condens. Matter **2003**, 15, L499–L504.
8. Takashima, M.; Aoki, H.; Haworth, C.J.; Matsumoto, T.; Terashima, T.; Uji, S.; Terakura, C.; Miura, T.; Uesawa, A.; Suzuki, T. De Haas-van Alphen effect of CeSb under pressure. J. Phys. Soc. Jpn. **1998**, 167, 3859–3866.
9. Taniguchi, H.; Miyashita, M.; Uchiyama, K.; Satoh, K.; Mori, N.; Okamoto, H.; Miyagawa, K.; Kanoda, K.; Hedo, M.; Uwatoko, Y. Superconductivity at 14.2 K in layered organic under extreme pressure. J. Phys. Soc. Jpn. **2003**, 72, 468–471.
10. Itoi, M.; Araki, C.; Hedo, M.; Uwatoko, Y.; Nakamura, T. Anomalously wide superconducting phase of one-dimensional organic conductor $(TMTTF)_2SbF_6$. J. Phys. Soc. Jpn. **2008**, 77, 1–4.
11. Murata, K.; Yokogawa, K.; Yoshino, H.; Klotz, S.; Munsch, P.; Irizawa, A.; Nishiyama, M.; Iizuka, K.; Nanba, T.; Okada, T.; Shiraga, Y.; Aoyama, S. Pressure transmitting medium Daphne 7474 solidifying at 3.7 GPa at room temperature. Rev. Sci. Instrum. **2008**, 79, 1–8.
12. Kotegawa, H.; Kawazoe, T.; Sugawara, H.; Murata, K.; Tou, H. Effect of uniaxial stress for pressure-induced superconductor, $SrFe_2As_2$. J. Phys. Soc. Jpn. **2009**, 78, 1–4.
13. Yamazaki, T.; Takeshita, N.; Kobayashi, R.; Fukazawa, H.; Kohori, Y.; Kihou, K.; Lee, C.H.; Kito, H.; Iyo, A.; Eisaki, H. Appearance of pressure-induced superconductivity in $BaFe_2As_2$ under hydrostatic conditions and its extremely high sensitivity to uniaxial stress. Phys. Rev. B **2010**, 81, 1–5.

14. Campos, C.E.; Brooks, J.S.; van Bentum, P.J.M.; Perenboom, J.A.A.J.; Rook, J.; Klepper, S.J.; Tokumoto, M. Uniaxial stress method for delicate crystals: Application to Shubnikov-de Haas and superconductivity studies in organic conductors. Rev. Sci. Instrum. **1995**, 66, 1061–1064.
15. Maesato, M.; Kaga, Y.; Kondo, R.; Kagoshima, S. Uniaxial strain method for soft crystals: Application to the control of the electronic properties of organic conductors. Rev. Sci. Instrum. **2000**, 71, 176–181.
16. Kondo, R.; Kagoshima, S.; Harada, J. Crystal structure analysis under uniaxial strain at low temperature using a unique design of four-axis X-ray diffractometer with a fixed sample. Rev. Sci. Instrum. **2005**, 76, 1–7.
17. Campos, C.E.; Brooks, J.S.; van Bentum, P.J.M.; Perenboom, J.A.A.J.; Klepper, S.J.; Sandhu, P.S.; Valfells, S.; Tanaka, Y.; Kinoshita, T.; Kinoshita, N.; Tokumoto, M.; Anzai, H. Uniaxial-Stress-Induced superconductivity in organic conductors. Phys. Rev. B **1995**, 52, R7014–R7017.
18. Tajima, N.; Ebina-Tajima, A.; Tamura, M.; Nishio, Y.; Kajita, K. Effects of uniaxial strain on transport properties of organic conductor alpha-$(BEDT\text{-}TTF)_2I_3$ and discovery of superconductivity. J. Phys. Soc. Jpn. **2002**, 71, 1832–1835.
19. Maesato, M.; Shimizu, Y.; Saito, G.; Miyagawa, K.; Kanoda, K. Spin-Liquid behavior and superconductivity in κ-$(BEDT\text{-}TTF)_2X$: The role of uniaxial strain. J. Phys. IV France **2004**, 114, 227–231.
20. Mori, H.; Tanaka, S.; Mori, T. Systematic study of the electronic state in θ-type BEDT-TTF organic conductors by changing the electronic correlation. Phys. Rev. B **1998**, 57, 12023–12029.
21. Shimizu, Y.; Miyagawa, K.; Kanoda, K.; Maesato, M.; Saito, G. Spin liquid state in an organic Mott insulator with a triangular lattice. Phys. Rev. Lett. **2003**, 91, 1–4.
22. Andrieux, A.; Schulz, H.J.; Jérome, D.; Bechgaard, K. Fluctuation conductivity in 1-D conductor tetrathiafulvalene-tetracyanoquinodimethane (TTF-TCNQ). J. Phys. Lett. **1979**, 40, L385–L389.
23. Jérome, D.; Schulz, H.J. Organic conductors and superconductors. Adv. Phys. **1982**, 31, 299–490.

24. Pouget, J.-P. X-Ray diffuse scattering as precursor of incommensurate Peierls transitions in one-dimensional organic charge transfer conductors. Z. Kristallogr. (Munich) **2004**, 219, 711–718.
25. Khanna, S.K.; Pouget, J.P.; Comes, R.; Garito, A.F.; Heeger, A.J. X-Ray studies of $2k_F$ and $4k_F$ anomalies in tetrathiafulvalene-tetracyanoquinodimethane (TTF-TCNQ). Phys. Rev. B **1977**, 16, 1468–1479.
26. Pouget, J.-P.; Khanna, S.K.; Denoyer, F.; Comes, R.; Garito, A.F.; Heeger, A.J. X-Ray observation of $2k_F$ and $4k_F$ scatterings in tetrathiafulvalene-tetracyanoquinodimethane (TTF-TCNQ). Phys. Rev. Lett. **1976**, 37, 437–440.
27. Kagoshima, S.; Ishiguro, T.; Anzai, H. X-Ray scattering study of phonon anomalies and superstructures in TTF-TCNQ. J. Phys. Soc. Jpn. **1976**, 41, 2061–2071.
28. Jérome, D. Organic conductors: From charge density wave TTF-TCNQ to superconducting $(TMTSF)_2PF_6$. Chem. Rev. **2004**, 104, 5565–5591.
29. Yasuzuka, S.; Murata, K.; Arimoto, T.; Kato, R. Temperature-Pressure phase diagram in TTF-TCNQ: Strong suppression of charge-density-wave state under extremely high pressure. J. Phys. Soc. Jpn. **2007**, 76, 1–4.
30. Ishibashi, S.; Kohyama, M. Ab initio pseudopotential calculation for TTF-TCNQ and TSeF-TCNQ. Phys. Rev. B **2000**, 62, 7839–7844.
31. Friend, R.H.; Miljak, M.; Jérome, D. Pressure dependence of the phase transitions in tetrathiafulvalene-tetracyanoquinodimethane (TTF-TCNQ): Evidence for a longitudinal lock-in at 20 kbar. Phys. Rev. Lett. **1978**, 40, 1048–1051.
32. Murata, K. Temperature-Pressure phase diagram of TSeF-TCNQ. Osaka City University, Osaka, Japan, Unpublished work. 2006.
33. Soda, G.; Jérome, D.; Weger, M.; Fabre, J.M.; Giral, L. Interchain coupling and nuclear magnetic relaxation in the organic conductor tetrathiafulvalene-tetracyanoquinodimethane TTF-TCNQ. Solid State Commun. **1976**, 18, 1417–1421.

34. Murata, K.; Weng, Y.; Seno, Y.; Tamilselvan, N.R.; Kobayashi, K.; Arumugam, S.; Takashima, Y.; Yoshino, H.; Kato, R. Fluctuation of the charge density wave in TTF-TCNQ under high pressure. Phys. B Phys. Condens. Matter **2009**, 404, 373–375. [Google Scholar] [CrossRef]
35. Weger, M. A model for the electronic band structure of HMTSeF-TCNQ. Solid State Commun. **1976**, 19, 1149–1155.
36. Ishibashi, S. Electronic band structure calculation under pressure of TTF-TCNQ. Private Communication, Agency of Industrial Science and Technology: Tsukuba, Japan, 2008.
37. Murata, K.; Yokogawa, K.; Kobayashi, K.; Masuda, K.; Sasaki, T.; Seno, Y.; Tamilselvan, N.R.; Yoshino, H.; Brooks, J.S.; Jérome, D.; et al. Field-Induced successive phase transitions in the CDW organic conductor HMTSF-TCNQ. J. Phys. Soc. Jpn. **2010**, 79, 1–4.
38. Murata, K.; Kang, W.; Masuda, K.; Kuse, T.; Sasaki, T.; Yokogawa, K.; Yoshino, H.; Brooks, J.S.; Choi, E.S.; Kiswandhi, A.; Kato, R. Field-Induced CDW in HMTSF-TCNQ. Phys. B **2012**, 407, 1927–1929.
39. Mortensen, K.; Tomkiewicz, Y.; Bechgaard, K. Antiferromagnetism in the organic conductor, bis-tetramethyltetraselenafulvalene hexafluoroarsenate, [$(TMTSF)_2AsF_6$]: Static magnetic susceptibility. Phys. Rev. B **1982**, 25, 3319–3325.
40. Takahashi, T.; Maniwa, Y.; Kawamura, H.; Saito, G. Determination of SDW characteristics in $(TMTSF)_2PF_6$ by ^{1}H-NMR analysis. Phys. B + C **1986**, 143, 417–421.
41. Delrieu, J.M.; Roger, M.; Toffano, Z.; Moradpour, A.; Bechgaard, K. NMR proton lineshape in $(TMTSF)_2X$: Incommensurability of nesting vector and order parameter. J. Phys. France **1986**, 47, 839–861.
42. Murata, K.; Anzai, H.; Ukachi, T.; Ishiguro, T. Metal-Insulator transition in $(TMTTF)_2BF_4$ under pressure. J. Phys. Soc. Jpn. **1984**, 53, 491–494.
43. Chow, D.S.; Zamborszky, F.; Alavi, B.; Tantillo, D.J.; Baur, A.; Merlic, C.; Brown, S.E. Charge ordering in the TMTTF Family of molecular conductors. Phys. Rev. Lett. **2000**, 85, 1698–1701.
44. Zamborszky, F.; Raas, W.Y.W.; Brown, S.E.; Alavi, B.; Merlic, C.A.; Baur, A. Competition and coexistence of bond and charge orders in $(TMTTF)_2AsF_6$. Phys. Rev. B **2002**, 66, 1–4.

45. Takahashi, T.; Nogami, Y.; Yakushi, K. Charge ordering in organic conductors. J. Phys. Soc. Jpn. **2006**, 75, 1–17.
46. Jérome, D. The physics of organic superconductors. Science **1991**, 252, 1509–1514.
47. Adachi, T.; Ojima, E.; Kato, K.; Kobayashi, H.; Miyazaki, T.; Tokumoto, M.; Kobayashi, A. Superconducting transition of $(TMTTF)_2PF_6$ above 50 kbar [TMTTF = Tetramethyltetra-thiafulvalene. J. Am. Chem. Soc. **2000**, 122, 3238–3239.
48. Jaccard, D.; Wilhelm, H.; Jérome, D.; Moser, J.; Carcel, C.; Fabre, J.M. From spin-Peierls to superconductivity: $(TMTTF)_2PF_6$ under high pressure. J. Phys. Condens. Matter **2001**, 13, L89–L95. [Google Scholar] [CrossRef]
49. Guo, F.Z.; Murata, K.; Oda, A.; Mizuno, Y.; Yoshino, H. Spin density wave transition in $(TMTSF)_2PF_6$ under uniaxial strain. J. Phys. Soc. Jpn. **2000**, 69, 2164–2169.
50. Murata, K.; Mizuno, Y.; Guo, FZ.; Shodai, S.; Iwashita, K.; Yoshino, H. Spin density wave in a reduced anisotropy in $(TMTSF)_2PF_6$. Mol. Cryst. Liq. Cryst. **2002**, 380, 85–92.
51. Grant, P.M. Electronic structure of the 2:1 charge transfer salts of TMTCF. J. Phys. Colloques **1983**, 44, 847–857.
52. Guo, F.Z.; Murata, K.; Yoshino, H.; Maki, S.; Tanaka, S.; Yamada, J.; Nakatsuji, S.; Anzai, H. Anion-Ordering transition in $(TMTSF)_2ClO_4$ under pressure. J. Phys. Soc. Jpn. **1998**, 67, 3000–3002.
53. Yagubskii, E.B.; Shchegolev, I.F.; Laukhin, V.N.; Kononovich, P.A.; Karatsovnik, M.V.; Zvarykina, A.V.; Buravov, L.I. Normal-Pressure superconductivity in an organic metal $(BEDT\text{-}TTF)_2I_3$. JETP Lett. **1984**, 39, 12–16.
54. Murata, K.; Tokumoto, M.; Anzai, H.; Bando, H.; Saito, G.; Kajimura, K.; Ishiguro, T. Superconductivity with the onset at 8 K in β-$(BEDT\text{-}TTF)_2I_3$ under pressure. J. Phys. Soc. Jpn. **1985**, 54, 1236–1239.
55. Murata, K.; Tokumoto, M.; Anzai, H.; Bando, H.; Saito, G.; Kajimura, K.; Ishiguro, T. Pressure phase diagram of the organic superconductor β-$(BEDT\text{-}TTF)_2I_3$. J. Phys. Soc. Jpn. **1985**, 54, 2084–2087.
56. Laukhin, V.N.; Kostyuchenko, E.E.; Sushko, V.Y.; Shchegolev, I.F.; Yagubskii, E.B. Effect of pressure on the superconductivity of beta-$(BEDT\text{-}TTF)_2I_3$. JETP Lett. **1985**, 41, 81–84.

57. Lacoe, R.C.; Wolf, S.A.; Chaikin, P.M.; Wudl, F.; Aharon-Shalom, E. Metal-Insulator transitions and superconductivity in ditetramethyltetraselenafulvalenium fluorosulfonate [$(TMTSF)_2FSO_3$]. Phys. Rev. B **1983**, 27, 1947–1950.
58. Fisher, R.A.; Kim, S.; Woodfield, B.F.; Phillips, N.E.; Taillefer, L.; Hasselbach, K.; Flouquet, J.; Giorgi, A.L.; Smith, J.L. Specific heat of UPt_3: Evidence for unconventional superconductivity. Phys. Rev. Lett. **1989**, 62, 1411–1414.
59. Leung, P.C.W.; Emge, T.J.; Beno, M.A.; Wang, H.H.; Williams, J.M.; Petricek, V.; Coppens, P. Novel structural modulation in the ambient-pressure sulfur-based organic superconductor β-$(BEDT\text{-}TTF)_2I_3$: Origin and effects on its electrical conductivity. J. Am. Chem. Soc. **1985**, 107, 6184–6191.
60. Ravy, S.; Moret, R.; Pouget, J.P. Explanation of the phase diagram of beta-di[bis(ethylene-dithio)tetrathiafulvalene] triiodide based on the two configurations of the organic molecule. Phys. Rev. B **1988**, 38, 4469–4480.
61. Ginodman, V.B.; Gudenko, A.V.; Kononovich, P.A.; Laukhin, V.N.; Shchegolev, I.F. Direct detection of the beta-1.5 beta-8 phase transition in beta-$(ET)_2I_3$. JETP Lett. **1986**, 44, 673–677.
62. Tokumoto, M.; Anzai, H.; Murata, K.; Kajimura, K.; Ishiguro, T. Effect of Alloying on the superconductivity in organic metals β-$(BEDT\text{-}TTF)_2$ trihalides. Synthet. Metal. **1988**, 27, A251–A256.
63. Murata, K.; Tokumoto, M.; Anzai, H.; Honda, Y.; Kinoshita, N.; Ishiguro, T.; Toyota, N.; Sasaki, T.; Muto, Y. Superconductivity of BEDT-TTF salts; (I) effect of pressure and alloying and (II) Shubnikov de Haas Effect. Synthet. Metal. **1988**, 27, A263–A270.
64. Murata, K.; Toyota, N.; Tokumoto, M.; Anzai, H.; Saito, G.; Kajimura, K.; Morita, S.; Muto, Y.; Ishiguro, T. Upper critical field of β-$(BEDT\text{-}TTF)_2I_3$ in the High T_c superconducting state. Phys. B **1986**, 143, 366–368.
65. Laukhin, V.N.; Pesotskii, S.I.; Yagubskii, E.B. Exceeding the paramagnetic limit of H_{c2} in an organic superconductor beta-$(ET)_2I_3$ with T_c = 7.1 K. JETP Lett. **1987**, 45, 501–504.
66. Murata, K.; Toyota, N.; Honda, Y.; Sasaki, T.; Tokumoto, M.; Bando, H.; Anzai, H.; Muto, Y.; Ishiguro, T. Magnetoresistance in β-$(BEDT\text{-}TTF)_2I_3$ and β-$(BEDT\text{-}TTF)_2IBr_2$, Shubnikov de Haas effect. J. Phys. Soc. Jpn. **1988**, 57, 1540–1543.

67. Toyota, N.; Sasaki, T.; Murata, K.; Honda, Y.; Tokumoto, M.; Bando, H.; Anzai, H.; Ishiguro, T.; Muto, Y. Cyclotron mass and Dingle temperature of conduction electrons moving in a layered plane of organic superconductors β-(BEDT-TTF)$_2$IBr$_2$, β-(BEDT-TTF)$_2$I$_3$ and κ-(BEDT-TTF)$_2$Cu(NCS)$_2$. J. Phys. Soc. Jpn. **1988**, 57, 2616–2619.
68. Kang, W.; Montambaux, G.; Cooper, J.R.; Jérome, D.; Batail, P.; Lenoir, Ċ. Observation of giant magnetoresistance oscillation in the high-T$_c$ phase of the two-dimensional organic conductor beta-(BEDT-TTF)$_2$I$_3$. Phys. Rev. Lett. **1989**, 62, 2559–2562.
69. Oshima, K.; Mori, T.; Inokuchi, H.; Urayama, H.; Yamochi, H.; Saito, G. Shubnikov-de Haas effect and Fermi surface in an ambient pressure organic superconductor (BEDT-TTF)$_2$Cu(NCS)$_2$. Phys. Rev. B **1988**, 38, 938–941.
70. Kajita, K.; Nishio, Y.; Takahashi, T.; Sasaki, W.; Kato, R.; Kobayashi, H.; Kobayashi, A.; Iye, Y. A new type oscillatory phenomenon in the magnetotransport of θ-(BEDT-TTF)$_2$I$_3$. Solid State Commun. **1989**, 70, 1189–1193.
71. Yamaji, K. On the angle dependence of the magnetoresistance in quasi-two-dimensional organic superconductors. J. Phys. Soc. Jpn. **1989**, 58, 1520–1523.
72. Kartsovnik, M.V.; Kononovich, P.A.; Laukhin, V.N.; Schegolev, I.F. Anisotropy of magnetoresistance and the Shubnikov-de Haas oscillations in the organic metal beta-(ET)$_2$IBr$_2$. Pis'ma Zh. Eksp. Teor. Fiz. **1988**, 48, 498–501.
73. Hennig, I.; Bender, K.; Schweitzer, D.; Dietz, K.; Endres, H.; Keller, H.J.; Gleitz, A.; Helberg, H.W. Alpha- and beta-(BEDT-TTF)$^{2+}$I^{3-} two dimensional organic metals. Mol. Cryst. Liq. Cryst. **1985**, 11, 337–441.
74. Kinoshita, N.; Tokumoto, M.; Anzai, H.; Saito, G. Anisotropy in ESR 9 Factors and Linewidths for α-and β-(BEDT-TTF)$_2$I$_3$. J. Phys. Soc. Jpn. **1985**, 54, 4498–4501.
75. Sugano, T.; Saito, G.; Kinoshita, M. Spin relaxation and diffusion in quasi-two-dimensional organic metals: The bis(ethylenedithiolo)tetrathiafulvalene compounds β-(BEDT-TTF)$_2$X (X = I$_3$ and IBr$_2$). Phys. Rev. B **1987**, 35, 6554–6559.

76. Sugano, T.; Saito, G.; Kinoshita, M. Conduction-Electron-Spin resonance in organic conductors: α and β phases of di[bis(ethylenedithiolo)tetrathiafulvalene]triiodide [(BEDT-TTF)$_2$I$_3$]. Phys. Rev. B **1986**, 34, 117–125.

77. Kajita, K.; Ojiro, T.; Fujii, H.; Nishio, Y.; Kobayashi, H.; Kobayashi, A.; Kato, R. Magnetotransport phenomena of α-Type (BEDT-TTF)$_2$I$_3$ under high pressures. J. Phys. Soc. Jpn. **1992**, 61, 23–26.

78. Katayama, S.; Kobayashi, A.; Suzumura, Y. Pressure-Induced zero-gap semiconducting state in organic conductor α-(BEDT-TTF)$_2$I$_3$ salt. J. Phys. Soc. Jpn. **2006**, 75, 1–6.

79. Takahashi, T.; Nogami, Y.; Yakushi, K. Charge ordering in organic conductors. J. Phys. Soc. Jpn. **2006**, 75.

80. Takano, Y.; Hiraki, K.; Takada, Y.; Yamamoto, H.M.; Takahashi, T. Local spin susceptibility characteristic of zero-gap state of α-(BEDT-TTF)$_2$I$_3$ under pressure. J. Phys. Soc. Jpn. **2010**, 79, 1–7.

81. Montambaux, G.; Piéchon, F.; Fuchs, J.-N.; Goerbig, M.O. Merging of Dirac points in a two-dimensional crystal. Phys. Rev. B **2009**, 80, 1–4.

82. Kino, H. Theoretical electronic band structure of alpha-(BEDT-TTF)$_2$I$_3$. Private Communication, National Research Institute for Metals: Tsukuba, Japan, 2012.

83. Nishikawa, H.; Sekiya, H.; Fujiwara, A.; Kodama, T.; Ikemoto, I.; Kikuchi, K.; Yamada, J.; Oshio, H.; Kobayashi, K.; Yasuzuka, S.; Murata, K. Molecular conductor based on reduced π-system donor: Insulating state of (MeDH-TTP)$_2$AsF$_6$. Chem. Lett. **2006**, 35, 912–913.

84. Yasuzuka, S.; Kobayashi, K.; Nishiwaka, H.; Yoshino, H.; Murata, K. Quantum criticality in reduced pi-donor system (MeDH-TTP)$_2$AsF$_6$. J. Phys. Soc. Jpn. **2006**, 75, 1–4.

85. Zambounis, J.S.; Pfeiffer, J.; Papavassiliou, G.C.; Lagouvardos, D.J.; Terzis, A.; Raptoupoulou, C.P.; Delhaes, P.; Ducasse, L.; Fortune, N.A.; Murata, K. Structural and physical properties of τ-(P-(S,S)-DMEDT-TTF)$_2$(AuBr$_2$)$_1$(AuBr$_2$)$_{0.75}$. Solid State Commun. **1995**, 95, 211–215.

86. Papavassiliou, G.C.; Lagouvardos, D.J.; Zambounis, J.S.; Terzis, A.; Raptopoulou, C.R.; Murata, K.; Shirakawa, N.; Ducasse, L.; Delhaes, P. Structural and physical properties of τ-(EDO-S,S-DMEDT-TTF)$_2$(AuBr$_2$)$_1$(AuBr$_2$)$_y$. Mol. Cryst. Liq. Cryst. **1996**, 285, 83–88.

87. Murata, K.; Yoshino, H.; Tsubaki, Y.; Papavassiliou, G.C. Hysteretic magnetic state in the organic τ-phase conductors. Synthet. Metal. **1998**, 94, 69–72.
88. Konoike, T.; Iwashita, K.; Nakano, I.; Yoshino, H.; Sasaki, T.; Takahashi, T.; Nogami, Y.; Brooks, J.S.; Graf, D.; Mielke, C.H.; Papavassiliou, G.C.; Murata, K. Shubnikov-de Haas oscillations and low temperature electric structure in τ-phase conductors. Synthet. Metal. **2003**, 615-616, 135–136.
89. Konoike, T.; Iwashita, K.; Yoshino, H.; Murata, K.; Sasaki, T.; Papavassiliou, G.C. Shubnikov de Haas oscillations in a 2D organic conductor tau-(EDO-S,S-DMEDT-TTF)$_2$(AuBr$_2$)$_{1+y}$ (y ~ 0.75). Phys. Rev. B **2002**, 66, 1–5.
90. Storr, K.; Balicas, L.; Brooks, J.S.; Graf, D.; Papavassiliou, G.C. Magnetic-Field-Dependent interplay between incoherent and Fermi liquid transport mechanisms in low-dimensional τ-phase organic conductors. Phys. Rev. B **2001**, 64, 1–9.
91. Murata, K.; Yoshino, H.; Nakanishi, T.; Konoike, T.; Brooks, J.; Graf, D.; Mielke, C.; Papavassiliou, G.C. A new quantum hall effect in the two-dimensional organic conductor, tau-(EDO-S,S-DMEDT-TTF)$_2$(AuBr$_2$)$_{1+y}$. Curr. Appl. Phys. **2004**, 4, 488–490.
92. Brooks, J.S.; Graf, D.; Choi, E.S.; Balicas, L.; Storr, K.; Mielk, C.H.; Papavassiliou, G.C. High magnetic field-induced insulating phase in an organic conductor. 2002, arXiv:cond-mat/0209592v1. Available online: http://arxiv.org/abs/physics/0402096 (accessed on 9 October 2012).
93. Yoshino, H.; Aizawa, H.; Kuroki, K.; Anyfantis, G.C.; Papavassiliou, G.C.; Murata, K. Thermoelectric figure of merit of τ-type conductors of several donors. Physica B **2010**, 405, S79–S81.
94. Kikuchi, K.; Isono, T.; Kojima, M.; Yoshimoto, H.; Kodama, T.; Fujita, W.; Yokogawa, K.; Yoshino, H.; Murata, K.; Kaihatsu, T.; Akutsu, H.; Yamada, J. Uniaxial strain orientation dependence of superconducting transition temperature (T_c) and critical superconducting pressure (P_c) in β-(BDA-TTP)$_2$I$_3$. J. Am. Chem. Soc. **2011**, 133, 19590–19593.

95. Nuruzzaman, M.; Yokogawa, K.; Yoshino, H.; Yoshimoto, H.; Kikuchi, K.; Kaihatsu, T.; Yamada, J.; Murata, K. Uniaxial-Strain-Orientation dependence of the competition between Mott and charge ordered phases and their corresponding superconductivity of β-(BDA-TTP)$_2$I$_3$. J. Phys. Soc. Jpn. **2012**. in press.

4

CHAPTER

INFRARED AND RAMAN STUDIES OF CHARGE ORDERING IN ORGANIC CONDUCTORS, BEDT-TTF SALTS WITH QUARTER-FILLED BANDS

Kyuya Yakushi

Toyota Physical and Chemical Research Institute, 41-1, Yokomichi, Nagakute, Aichi 480-1192, Japan;

ABSTRACT

This paper reviews charge ordering in the organic conductors, β″-(BEDT-TTF)(TCNQ), θ-$(BEDT\text{-}TTF)_2X$, and α-$(BEDT\text{-}TTF)_2X$. Here, BEDT-TTF and TCNQ represent bis(ethylenedithio)tetrathiafulvalene and 7,7,8,8-tetracyanoquinodimethane, respectively. These compounds, all of which have a quarter-filled band, were evaluated using infrared and Raman spectroscopy in addition to optical conductivity measurements. It was found that β″-(BEDT-TTF)(TCNQ) changes continuously from a uniform metal to a charge-ordered metal with increasing temperature. Although charge disproportionation was clearly observed, long-range charge order is not realized. Among six θ-type salts, four compounds with a narrow band show the metal-insulator transition. However, they maintain a large amplitude of charge order ($\Delta\rho \sim 0.6$) in both metallic and insulating phases. In the X = $CsZn(SCN)_4$ salt with intermediate bandwidth, the amplitude of charge order is very small ($\Delta\rho < 0.07$) over the whole temperature range. However, fluctuation of charge order is indicated in the Raman spectrum and optical conductivity. No indication of the fluctuation of charge order is found in the wide band X = I_3 salt. In α-(BEDT-TTF)$_2$I$_3$ the amplitude of charge order changes discontinuously from small amplitude at high temperature to large amplitude ($\Delta\rho_{max}$~0.6) at low temperature. The long-range charge-ordered state shows ferroelectric polarization with fast optical response. The fluctuation of multiple stripes occurs in the high-temperature metallic phase. Among α-(BEDT-TTF)$_2$MHg(SCN)$_4$ (X = NH_4, K, Rb, Tl), the fluctuation of charge order is indicated only in the X = NH_4 salt. α′-(BEDT-TTF)$_2$IBr$_2$ shows successive phase transitions to the ferroelectric state keeping a large amplitude of charge order ($\Delta\rho_{max}$~0.8) over the whole temperature range. It was found that the amplitude and fluctuation of charge order in these compounds is enhanced as the kinetic energy (bandwidth) decreases.

Keywords: organic conductor; BEDT-TTF; charge order; infrared and Raman spectroscopy; vibrational spectroscopy; optical conductivity

1. INTRODUCTION

Vibrational spectroscopy is a powerful tool to detect the valence of molecules in organic charge-transfer salts. Among the normal modes of molecular vibrations, several vibrational modes change their eigenfrequencies depending on the valence of the molecule in question. These vibrational modes are called charge-sensitive modes. For example, the C=C stretching mode is utilized as a probe for detecting the valence. As the chemical bond is usually much stronger than intermolecular interactions, the vibrational frequency and symmetry of a free molecule is approximately preserved in the solid state except in the case where a strong vibronic interaction occurs between a vibrational mode and a low-energy electronic state (electron molecular vibration coupling). A vibrational mode that strongly interacts with the electronic state is called structure-sensitive mode. Knowledge of the normal mode of a free molecule is essential, because a deviation from this mode in the solid state tells us which interaction is occurring. In this sense, reliable assignments of vibrational bands in the solid state are the most critical and indispensable. The charge-sensitive mode was applied for the first time to the distribution of charges in the unit cell of Cs_2TCNQ_3, where TCNQ represents 7,7,8,8-tetracyanoquinodimethane, wherein the infrared-active external ring C=C stretching modes (ν_4) of $TCNQ^0$ and $TCNQ^-$ were separately observed [1]. A detailed normal mode analysis of $TCNQ^0$ and $TCNQ^-$ was reported by Bozio et al. [2]. Matsuzaki et al., extended the application to the TCNQ salt with fractional charge [3]. Subsequently, vibrational spectroscopy was applied to mixed-valence compounds of TCNQ [4], where one of the CN stretching modes of TCNQ was utilized as the probe.

The existence of Wigner-type charge order (CO) in organic conductors had been argued based on the finding of $4k_F$ modulation in quasi-one-dimensional tetrathiafulvalene-7,7,8,8-tetracyanoquinodimethane (TTF-TCNQ) [5,6,7,8] because the wavelength of the $4k_F$ modulation corresponds to the distance between localized electrons (holes) which are arranged with equal distance. However, the $4k_F$ modulation found in TTF-TCNQ was related to the $4k_F$ Peierls instability with on-site Coulomb repulsion. Since the amplitude of the charge density wave (CDW) is small, the charge-sensitive mode shows almost no linewidth broadening. The long-range CO of the Wigner-type was reported for quasi-one-dimensional $(DI\text{-}DCNQI)_2Ag$, where DI-DCNQI corresponds to 2,5-diiodo-dicyanoquinodiimine, by Hiraki and Kanoda using ^{13}C-NMR [9]. They showed that the amplitude of CO ($\Delta\rho$ = 0.5) is much larger than that of an incommensurate CDW. At about the same time, Seo theoretically predicted the CO ground state for $(TMTTF)_2X$ (X = Br, SCN), where TMTTF corresponds to tetramethyltetrathiafulvalene, and $(DI\text{-}DCNQI)_2Ag$ using mean-field theory [10]. Later, Chow reported a similar charge order for $(TMTTF)_2X$ (X = PF_6, AsF_6) using ^{13}C-NMR [11]. In the case of a two-dimensional organic compound, Moldenhauer et al., reported that localization of charge occurs in one of the stacks of α-$(BEDT\text{-}TTF)_2I_3$ (BEDT-TTF corresponds to bis(ethylenedithio) tetrathiafulvalene) [12]. After the theoretical prediction of the CO ground state for α-$(BEDT\text{-}TTF)_2I_3$ by Kino and Fukuyama [13], CO in two-dimensional organic compounds was suggested in α-$(BEDT\text{-}TTF)_2I_3$ [14], and more clearly shown in θ-$(BEDT\text{-}TTF)_2RbZn(SCN)_4$ [15]. Stimulated by the experimental findings of CO, theoretical studies were conducted on the role of intersite Coulomb interaction in CO [16,17,18], lattice distortion accompanied by CO [19,20,21], the relationship between CO fluctuation and superconductivity (SC) [22], and quantum criticality at the edge of CO [23]. Subsequently, several experimental and theoretical studies have been conducted [24,25,26,27]. Some of these will be introduced from Section 3 on, in relation to the experimental results. Charge ordering continues to attract experimental and theoretical interest and attention. Some compounds in the CO state are regarded as unconventional ferroelectrics [28]. However, the ferroelectric properties have not been well investigated so far due to experimental difficulties. SC paring mediated by CO fluctuation was proposed by Merino and McKenzie [22]. Spectroscopic studies which may support the theoretical prediction for the SC mechanism have been reported [29,30,31,32,33,34]. It was also predicted that spin fluctuation mediates the SC pairing in a CO system [35]. Recently, it was pointed out that interaction with anions plays an essential role in the CO phase transition [36,37]. These subjects will be further investigated in future.

Along with NMR and x-ray diffraction, vibrational spectroscopy is a powerful tool to investigate the CO state, since the amplitude of CO is large. Raman spectroscopy was first applied to the charge ordering phase transition of θ-(BDT-TTP)$_2$Cu(NCS)$_2$, for which the spectral features drastically changed [38,39]. Although the charge-sensitive mode is expected to show a simple splitting in a CO state, the Raman spectrum showed complicated splitting due to the additional splitting of the structure-sensitive mode. The fluctuation of charge order was obviously detected in the metallic phase. These characteristics were for the most part commonly observed in the compounds discussed here. In this paper, the author reviews infrared, Raman, and optical conductivity measurements of BEDT-TTF salts, which, apart from β″-(BEDT-TTF)(TCNQ), have a herringbone arrangement. This latter compound is included in this paper, because its unique behavior is necessary for understanding the metallic phase of θ- and α-type BEDT-TTF salts.

The paper is organized as follows. Section 2 briefly describes the relation between the site charge and band structure, fluctuation of charge order and time correlation function, charge- and structure-sensitive modes of BEDT-TTF, electron-molecular vibration coupling, and the line shape when the site charge fluctuates. Section 3 describes charge disproportionation in β″-(BEDT-TTF) (TCNQ), wherein a continuous change of the electronic state is discussed combined with optical conductivity. Section 4 describes the charge-ordering phase transition θ-type BEDT-TTF salts; first, how the charge- and structure-sensitive modes split in the CO state; second, the metallic state with large CO amplitude is discussed; third, the metallic state with small CO amplitude is discussed. Section 5 introduces a spectroscopic investigation of α-type BEDT-TTF. Here, the splitting of charge- and structure-sensitive modes in CO phase and ferroelectric properties are described first. Next, the splitting of the charge-sensitive mode and fluctuation of CO in the metallic phase are discussed. Third, the non-uniform site charge distribution in the metallic phase under ambient pressure and hydrostatic pressure is compared with band calculations. Fourth, the optical conductivity is reviewed along with the collective CO excitation Finally, the successive ferroelectric transition of α′-(BEDT-TTF)$_2$IBr$_2$ is briefly described. The final section 6 presents the kinetic energy, which is associated with the bandwidth, of these compounds, and summarizes their electronic states correlated with kinetic energy.

2. DETECTION AND ANALYSIS OF CHARGE ORDER

2.1. Site Charge and Fluctuation

In solid organic charge-transfer salts, valence electrons are delocalized depending upon the magnitude of the overlap integral. Therefore, the valence of the molecule becomes a non-integer value, when the number of valence electrons does not equal the number of molecules. The fractional valence or fractional charge at the jth site (site charge) in the unit cell, ρ_j, is given by the following equation:

$$\rho_j = \sum_{ka,kb}\sum_{m}^{n_M}\langle km|n_j|km\rangle f(\varepsilon) \tag{1}$$

where $|km\rangle$ is the Bloch function of the mth band, n_M is the number of molecules in a unit cell, n_j is the number operator at the jth site, and $f(\varepsilon)$ is the Fermi distribution function for a hole. The site charge is the integrated valence electron at each site below a Fermi level. Therefore, the site-charge distribution is sometimes non-uniform even in a metal, if the site in a unit cell is crystallographically non-equivalent. If the amplitude of charge order $\Delta\rho$ is small, a small portion of the valence electrons near the Fermi level contribute to the charge disproportionation, while in the case of large $\Delta\rho$, most valence electrons participate in the charge disproportionation as the band structure entirely changes

When the site charge dynamically fluctuates as $\rho(t)$, the correlation time τ is defined by the following equation,

$$\tau = \int\frac{\langle\Delta\rho(t')\Delta\rho(t'+t)\rangle}{\Delta^2}dt \tag{2}$$

where $\Delta\rho(t') = \rho(t') - <\rho(t') \geq \rho(t') - 1/2$ the case of quarter filling, $<\Delta\rho(t')\Delta\rho(t'+t)>$ is the time correlation function of ρ, and $\Delta^2 =< \Delta\rho(t')\Delta\rho(t')>$ is the amplitude (variance) of fluctuation. As explained in the preceding paragraph, only the valence electrons near the Fermi level fluctuate, when the fluctuation amplitude is small. The charge fluctuation in a quarter-filled-band system was theoretically discussed in refs. [40,41] using the charge correlation function C(**q**,ω). If the system has CO instability, the charge correlation function has a strong peak at $\mathbf{q}_0$ and $\omega_0 \sim 0$. The wave vector $\mathbf{q}_0$ specifies a charge-order pattern such as checkerboard or stripe. The peak frequency ω_0 is the energy of the excited state which contributes to produce a CO wave in a uniform metal, and it may be related to the amplitude of CO fluctuation. The peak height and width are respectively associated with correlation length and correlation time. The time correlation function of the specified $\mathbf{q}_0$ is the Fourier transform of $C(\mathbf{q}_0,\omega)$.

2.2. Charge-Sensitive Mode of BEDT-TTF

A normal mode analysis of BEDT-TTF^0 and BEDT-TTF^{1+} was reported by Kozlov et al. [42,43]. The normal modes most sensitive to the charge (valence) of the BEDT-TTF molecule are the in-phase (ν_2, Raman active) and out-of-phase (ν_{27} infrared active) stretching modes of the ring C=C bonds, and the stretching mode of the C=C bond bridging two five member rings (ν_3, Raman active). Kozlov et al., analyzed the normal mode assuming the D_{2h} symmetry, and their mode numbering is different from the subsequent numbering using more exact D_2 symmetry [44]. In this paper, Kozlov's numbering shown in Figure 1 will be used to avoid complication. In the solid state, site symmetry is usually lower than D_{2h} or D_2. However, the crystal field in organic conductors is much weaker than the covalent energy within a molecule. Therefore, the symmetry of the free molecule is approximately preserved in a crystal.

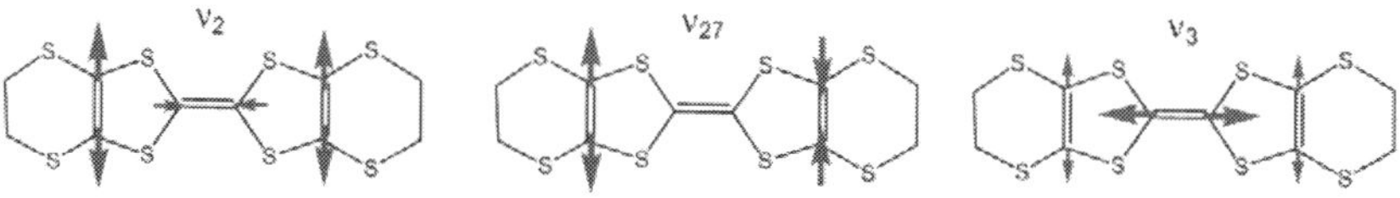

Figure 1. C=C stretching modes of BEDT-TTF. The ν_2 and ν_{27} modes are charge-sensitive, while the ν_3 mode depends upon the network of transfer integrals, and thus it is structure-sensitive.

An attempt to examine the linear relationship between the frequency and fractional charge of BEDT-TTF was made by Wang et al., for the Raman-active ν_2 and ν_3 modes [45]. However, the frequency data were scattered about the linear relationship. In addition, they did not consider the effect of electron-molecular vibration interaction, which will be discussed in the next subsection. The linear relationship for the infrared-active ν_{27} was first utilized by Moldenhauer et al., for estimating the iconicity of BEDT-TTF [12]. However, this relationship is given based on the incorrect assignment of BEDT-TTF$^+$ [46]. Subsequently, the relationship was not examined for long time, because this mode is usually hidden in a huge electronic absorption band when it is measured on the conducting plane or powdered samples. Recently, Yamamoto et al., have examined the assignment of Kozlov et al., and they reassigned ν_2, ν_3, and ν_{27} of BEDT-TTF^+ with the aid of ^{13}C and deuterium substituted compounds [46]. First they examined ν_{27} employing the polarization perpendicular to the conducting layer on the side face of single crystals. They found that the frequency of BEDT-TTF^0 deviated from the linear relation, because it has a non-planar structure. They calculated the frequency of ν_2 and ν_{27} assuming D_2 symmetry employing a DFT method, and found that the frequency of the flat structure BEDT-TTF^0 satisfies the linear relationship. They presented an empirical relationship for the infrared active ν_{27} mode, $\nu_{27}(\rho) = 1398 + 140(1 - \rho)$. Next they examined the Raman-active ν_2 and ν_3 modes. The ν_3 mode is more strongly perturbed by the electron-molecular vibration (EMV) interaction than the effect of charge. As we will explain in the next section, the highest-frequency ν_3 mode does not shift, even if the charge disproportionation occurs, whereas the ν_2 mode at charge-rich site (ν_{2R}) shifts toward lower frequency. Consequently, the line $\nu_{2R}(\rho)$ crosses the line $\nu_3(\rho)$, at $\rho \sim 0.9$ (see Figure 2). As the result of the interaction between ν_{2R} and ν_3, the frequency of ν_{2R} deviates from the linear relationship. They presented the linear relationship for the infrared-active ν_2 as $\nu_2(\rho) = 1447 + 120(1 - \rho)$, which is available only in the range of $0 < \rho < 0.8$. The ρ dependence of the frequency is ascribed to the ρ dependent force constant $F(\rho)$. That is, the linear relationship requires $(dF/d\rho)_{\rho=0} = (dF/d\rho)_{\rho=1}$ [47]. However, the frequency of BEDT-$TTF^{0.5+}$ deviates significantly from the linear relationship toward the low-frequency side, which means that $|(dF/d\rho)_{\rho=0}| > |(dF/d\rho)_{\rho=1}|$ (See Figure 7 of ref. [46]). Other nonlinear relationship at $\rho \sim 0.5$ will be discussed in Section 5.3.1.

2.3. Electron-Molecular Vibration (EMV) Coupling in BEDT-TTF

The EMV interaction was first studied to interpret the strong infrared signal in the reflectivity polarized along the stacking axis in K-TCNQ [48]. Rice et al., reported a symmetric dimer model and presented a comprehensive explanation for the EMV interaction [49]. According to the dimer model, the a_g mode of each molecule splits into in-phase and out-of-phase modes depending upon the magnitude of the coupling constant and transfer integral. The EMV interaction works only for the out-of-phase mode, and the coupling constant is estimated from the frequency difference between the Raman-active in-phase mode and the infrared-active out-of-phase mode. The EMV effect on the Raman-active mode was first pointed out by Girlando et al., in TTF-CA [50].

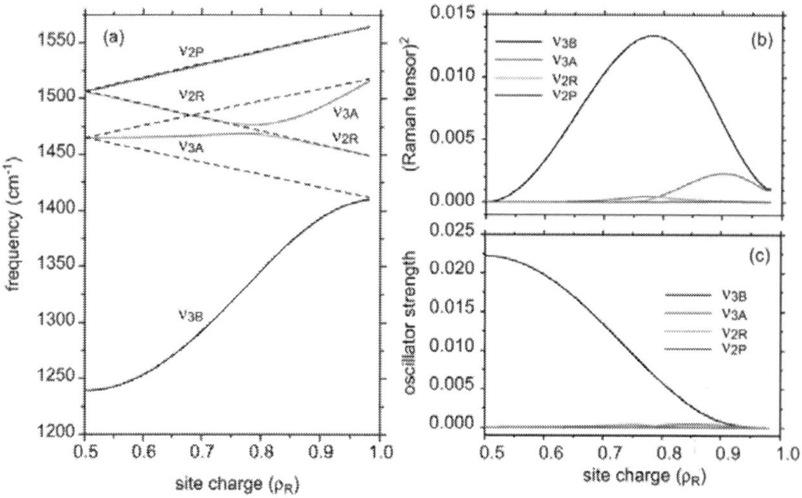

Figure 2. (**a**) Frequency of the vibronic modes plotted against the site charge of charge-rich site ($\rho_P + \rho_R = 1$). The transfer integral is t = 0.2 eV, EMV coupling constants for ν_2 and ν_3 are $g_2 = 0.02$ eV and $g_3 = 0.1$ eV, respectively. The dashed line shows the site-charge dependence when $g_2 = g_3 = 0$, assuming a linear relationship against site charge ρ. The frequencies at ρ = 0 and 1 are $\nu_2(0) = 1567$ cm^{-1}, $\nu_2(1) = 1447$ cm^{-1}, $\nu_3(0) = 1520$ cm^{-1}, and $\nu_3(1) = 1410$ cm^{-1}; (**b**) Contribution of the intermolecular excitation to the Raman tensor of the vibronic modes, which are plotted against the site charge of the charge-rich site; (**c**) Oscillator strength of the vibronic modes plotted against the site charge of the charge-rich site.

Based on the formulation by Painelli and Girlando [51], Yamamoto and Yakushi extended this idea to a disproportionate (asymmetric) dimer with average charge ρ = 0.5, in which each site has a non-equivalent charge like $(BEDT\text{-}TTF)^{\rho+}(BEDT\text{-}TTF)^{(1-\rho)+}$ [52]. Figure 2 shows the numerical calculation of the frequency and intensity of vibronic modes plotted as a function of the site charge of the charge-rich site ($\rho > 0.5$). In this calculation, the transfer integral between two molecules is taken as t = 0.2 eV, and the coupling constants of ν_2 and ν_3 mode are taken as $g_2 = 0.02$ eV and $g_3 = 0.1$ eV. ν_{2P} and ν_{2R}, denote the ν_2 mode of the charge-poor and charge-rich site (molecule), respectively, and ν_{3A}, and ν_{3B} denote the ν_3 mode for the in-phase and out-of-phase oscillation, respectively. The coupling constant in BEDT-TTF is estimated in refs. [43,53]. The coupling constants of ν_2 and ν_3 were respectively estimated to be $g_2 = 0.043$ and $g_3 = 0.071$ eV from an analysis of the infrared spectrum of the isolated dimer $(BEDT\text{-}TTF)_2^{2+}$ [53]. The parameters estimated from the infrared spectra were very scattered ($g_2 \sim 0.007–0.039$ eV and $g_3 = 0.007–0.081$ eV) depending upon the compounds [54]. A more recent estimation gives $g_3 = 0.076$ eV from the analysis of the infrared spectrum of κ-(BEDT-TTF)$_2$Cu[N(CN)$_2$]-Br$_{0.85}$Cl$_{0.15}$ [55]. Using the same data, Girlando interpreted this as $g_2 = 0.075$ eV [56]. The estimation of the coupling constants of ν_2 and ν_3 from the infrared spectrum seems to be unfixed still now. The coupling constants used for the model calculation (Figure 2) are qualitatively consistent with ref. [55] and with the interpretation of the Raman spectra of various compounds appearing in this paper.

As shown in Figure 2a, the weakly coupled modes, ν_{2P} and ν_{2R}, approximately conform to a linear relationship. However, the strongly coupled modes, ν_{3A} and ν_{3B}, deviate remarkably from the linear relation, although the frequency of ν_3 of the free molecule follows a linear relationship as shown by the dashed lines in Figure 2a [56]. The ν_{3A} and ν_{3B} modes are respectively the in-phase and out-of-phase vibrations at $\rho = 0.5$. The noteworthy point here is that the ν_{3A} mode that is Raman-active keeps the frequency corresponding to $\rho = 0.5$ in the range of $0.2 < \rho < 0.8$. This is because the in-phase vibrations at two sites are not perturbed by the EMV interaction [49]. Actually, the frequencies of ν_3 of the 2:1 salts are nearly equal to those of the 3:2 salts in the work by Wang et al. [45] although they claim a linear relationship for ν_3 in their paper. Near $\rho \sim 0.8$, mixing between ν_{2R} and ν_{3A} occurs, and the assignment becomes meaningless. This situation is sometimes seen in the charge-ordered state (See Section 4.1, Section 5.1, and Section 5.3.3). This result means that Raman spectroscopy cannot determine the fractional charge when the molecular vibrations strongly interact with each other through the EMV mechanism. Fortunately, in the case of BEDT-TTF, the ν_2 mode is available to determine the fractional charge, because the EMV coupling is weak enough. However, this mode is available only in the range of $0 < \rho < 0.8$. In this sense, the infrared-active mode, for example, ν_{27} of BEDT-TTF is the best probe to determine the site charge in a charge-ordered system. From the viewpoint of experimental technique, it is extremely difficult to observe ν_{27} in the conducting plane, because the phonon mode is screened by the strong electronic absorption band. Usually the crystal face parallel to a conducting plane is most developed. Therefore, the best way to detect ν_{27} is to measure the reflectivity on the thin side face of the crystal with the polarization perpendicular to the conducting plane.

In the infrared spectrum, we can observe only the out-of-phase mode ν_{3B}, which shows a large downshift due to the large EMV coupling constant, as shown in Figure 2c. As the out-of-phase mode dynamically modulates the orbital energy of the highest occupied molecular orbital (HOMO) of BEDT-TTF in the dimer, it transfers charge back and forth inducing an oscillating dipole along the direction connecting the centers of the molecules [49]. Accordingly, the out-of-phase ν_3 mode is usually found very strongly in the optical conductivity of the conducting plane (See Section 4 and Section 5). The downshift of this mode is sensitive to the transfer integrals between BEDT-TTF molecules (see ref. [54] for example). Yamamoto et al., discussed the relation between the downshift of ν_{3B} and transfer integrals comparing the ν_3 modes among several θ-type of BEDT-TTF salts. [57] Utilizing this property, the symmetry can be determined, for example, by the presence of a glide plane (See Section 4.1) and breaking of inversion symmetry (See Section 5.1). The Raman intensity of the same vibronic mode ν_{3B} is calculated to be strong in the disproportionate dimer (See Figure 2b). In this model, only the intermolecular excitation is considered for the calculation of the Raman tensor in the off-resonance condition. Actually, intramolecular excitations contribute more to the Raman tensor, and the resonance effect is sometimes quite large in the charge-transfer salts of BEDT-TTF. This calculation shows the additional contribution of the intermolecular excitation to the Raman tensor. As shown in Figure 2b, the contribution of this intermolecular excitation is largest in the vibronic ν_3 mode that has a large EMV coupling constant. Actually, the vibronic ν_3 mode is frequently found in the Raman spectrum of charge-ordered system (See Section 4 and Section 5).

2.4. The Line Shape of Charge-Sensitive Mode

The frequency of the charge-sensitive mode strongly depends upon the site charge, for example, $\Delta\omega/\Delta\rho \approx 140\ cm^{-1}/e$ or $\Delta\omega/\Delta\rho \approx -140\ cm^{-1}/h$ for the ν_{27} mode. Let us consider a system in which the unit cell involves two sites (molecules) and one hole. If every site is equivalent, then every site has a uniform site charge (hole) $\rho = 0.5$. When the system undergoes a charge-ordering transition, the site charge is disproportioned into ρ_R and ρ_P. Therefore, the charge-sensitive mode splits into two vibrational bands from a single band. If the site charge fluctuates around $\rho = 1/2$, the frequency also fluctuates around $\nu(\rho)$, and the linewidth is broadened depending on the fluctuation rate. This relation is illustrated in Figure 3.

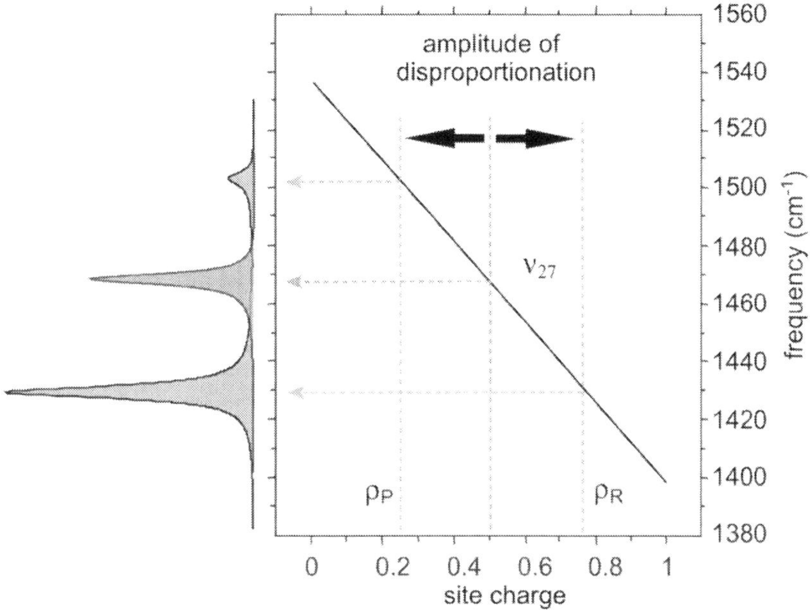

Figure 3. Linear relationship between the site charge and frequency of ν_{27} [46]. The schematic sketch of the spectrum shows that the charge-sensitive mode of uniform site charge (green) is split into two (blue) by charge disproportionation, and the linewidth is broadened by the fluctuation of the site charge.

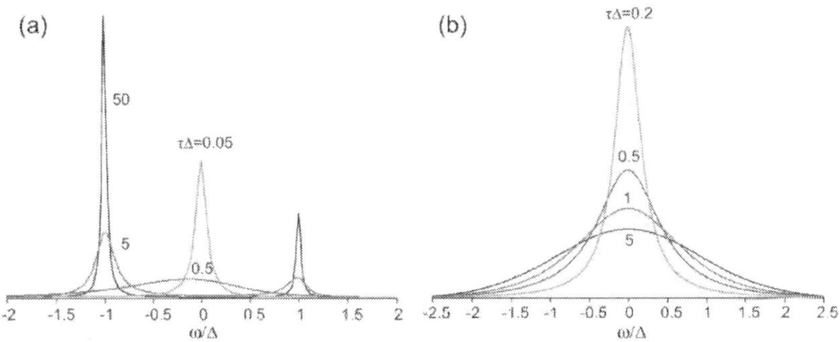

Figure 4. (**a**) Two-states-jump model calculated by Equation (3) with $\omega_{1/2}$ = 0, Γ/Δ = 0.04, a_R = 10, and a_P = 0.3; (**b**) Gaussian model calculated using Equation (4). τ and Δ are the correlation time and amplitude of fluctuation, respectively.

The fluctuation of the site charge is characterized by its amplitude (or variance $\Delta_c = \langle \Delta\rho^2 \rangle^{1/2}$) and correlation time $\tau = \int_0^\infty (\langle \Delta\rho(0)\Delta\rho(t)\rangle / \Delta_c^{\,2} dt$, where $\Delta\rho = \rho(t) - 1/2$ is the time-dependent site-charge variation from the average site charge 1/2. This charge fluctuation modulates the frequencies of ν_{27} and ν_2 with an amplitude Δ_f and correlation time τ. When the site charge fluctuates stochastically between two values such as ρ_R and ρ_P and thus the frequency fluctuates between ω_R and ω_P with a transition rate γ, then the line shape of the charge-sensitive mode is described by the following equation [58].

$$I(\omega) \propto \mathrm{Re}\left[\begin{pmatrix} a_R & a_P \end{pmatrix} \begin{pmatrix} i(\omega - (\omega_{1/2} - \Delta) + \gamma/2 + \Gamma/2 & -\gamma/2 \\ -\gamma/2 & i(\omega - (\omega_{1/2} + \Delta) + \gamma/2 + \Gamma/2 \end{pmatrix}^{-1} \begin{pmatrix} a_R \\ a_P \end{pmatrix} \right] \quad (3)$$

here $\omega_{1/2}$ is the frequency for $\rho = 1/2$, $\Delta = (\omega_P - \omega_R)/2$, Γ is the natural width in the solid state, and a_R and a_P are parameters proportional to the transition dipole moment or Raman tensor of the charge-sensitive mode at charge-rich and charge-poor sites, respectively. In this case, the correlation time is equal to the staying time at a site which is given by $\tau = 1/\gamma$. When $\tau\Delta \ll 1$ or $\tau\Delta \gg 1$, the lineshape approaches a Lorentzian.

When the site charge fluctuates stochastically as a Gaussian process, the correlation function has an exponential decay as, $\langle \Delta\rho(0)\Delta\rho(t)\rangle = \Delta_c^2 \exp(-\gamma t)$, where Δ_c is the amplitude of the fluctuation and γ is the decay rate. In this case, the modulation of the site charge is transformed to a modulation of the frequency of the charge-sensitive mode, and the line shape of the charge-sensitive mode is given by the following equation [59,60],

$$I(\omega) \propto \mathrm{Re} \int_0^\infty \exp\left\{ -\frac{\Delta^2}{\gamma^2}\left(e^{-\gamma t} - 1 + \gamma t\right) - i\omega t \right\} dt \quad (4)$$

where Δ is the amplitude (variance) of the modulation frequency of the charge-sensitive mode. In this case the correlation time is equal to the inverse of the decay rate, $\tau = 1/\gamma$. When $\tau\Delta \gg 1$ and $\tau\Delta \ll 1$, the lineshape approaches a Gaussian and a Lorentzian, respectively. Figure 4 shows the spectral line shape of these two cases plotted as a function of $\omega\Delta$ for various $\tau\Delta$ with $\omega_{1/2} = 0$. In both cases, the line shape is determined by the parameter $\tau\Delta$. When both τ and Δ vary with temperature, it is difficult to determine them from an analysis of the line shape.

3. B″-(BEDT-TTF)(TCNQ)

3.1. Crystal Structure and Superlattice

β″-(BEDT-TTF)(TCNQ) has a segregated stack structure, in which BEDT-TTF and TCNQ are regularly stacked along the c axis. The degree of the charge transfer of $(BEDT\text{-}TTF)^{+\rho}(TCNQ)^{-\rho}$ is estimated to be $\rho = 0.5$ from the frequency shift of ν_{21}, ν_{34}, and ν_4 of TCNQ [61]. Therefore, this compound is regarded as a quarter-filled system. BEDT-TTF has the largest transfer integral along the side-by-side direction (a axis) and TCNQ has the largest transfer integral along the stacking direction [62]. The hopping interaction between BEDT-TTF and TCNQ along the b axis is negligibly small. These theoretical predictions are qualitatively supported by the anisotropic optical conductivity [61]. Very weak superlattice spots are reported in ref. [61]. The observation of vibronic bands of TCNQ only in the E||c optical conductivity suggests that TCNQ is dimerized whereas BEDT-TTF is not. Based on these spectroscopic results and assuming the space group $P\bar{1}$, the proposed model of this superstructure is shown in Figure 5. TCNQ is connected by a center of symmetry ($4k_F$ modulation) and BEDT-TTF is located on the center of symmetry. Therefore, the two TCNQs in the unit cell are crystallographically equivalent, whereas the two BEDT-TTFs are non-equivalent. These superlattice spots disappear at around 170 K. However, the vibronic bands of TCNQ still remain after the superlattice spots vanish. This result implies that the short-range ordered $4k_F$ modulation remains in the TCNQ stack or another superlattice forms. According to the X-ray diffraction study by Nogami, a new superlattice spot for 2a × 2c was found below 170 K. [63]. In any case, the valence electron in the TCNQ stack is localized within the dimer, and the valence electron in the pseudo-one-dimensional BEDT-TTF band is responsible for the charge carriers.

β″-(BEDT-TTF)(TCNQ) is metal-like down to 1.8 K with three anomalies at ~170 K, ~80 K, and ~20 K [62,64]. The magneto-resistance below 20 K was extensively studied by Yasuzuka et al. They found five distinct small Fermi surfaces through the Shubnikov-de-Haas oscillations [65,66,67]. A quasi-one-dimensional periodic orbit resonance was found through the magneto-optical measurement [68,69]. The magneto-resistance and magneto-optical experiments demonstrate the existence of the Fermi surface below 20 K, which implies that the ground state is a Fermi liquid metal. On the other hand, charge disproportionation with an amplitude of $\Delta\rho \sim 0.6$ was reported at room temperature based on the observation of the splitting of charge-sensitive ν_{27} mode of BEDT-TTF [70]. Such a large amplitude at high-temperature indicates a localized nature, so suggesting a phase change between the high- and low-temperature regions.

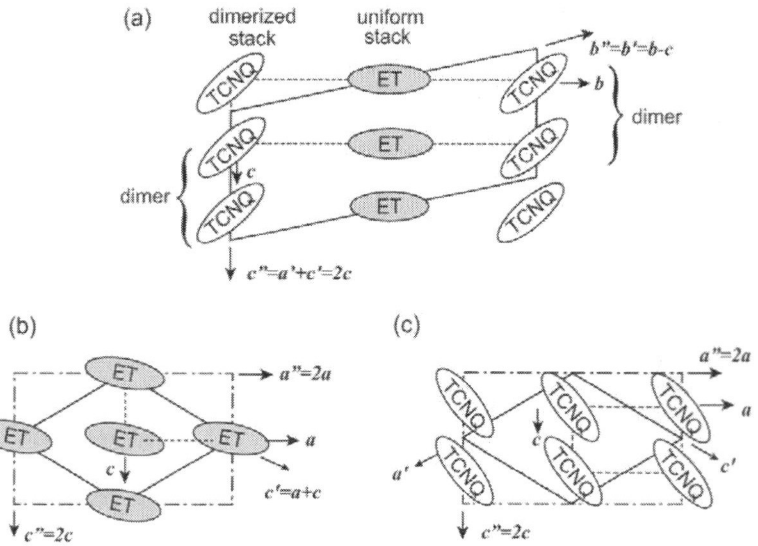

Figure 5. Arrangement of molecules in β''-(BEDT-TTF)(TCNQ). Average lattice (dotted line; crystal axes are **a**, **b**, and **c**), superlattice above 170 K (solid line; crystal axes are **a'**, **b'**, and **c'**), and superlattice 2a × 2c below 170 K (dash dot line; crystal axes are **a''**, **b''**, and **c''**). Arrangement in (**a**) BEDT-TTF (ET) and TCNQ; (**b**) BEDT-TTF; and (**c**) TCNQ.

3.2. Crossover from Uniform Metal to Charge-Ordered Metal

Uruichi et al., have discussed the infrared, Raman, and optical conductivity of β''-(BEDT-TTF) (TCNQ) [61]. Figure 6 shows the infrared-active ν_{27} mode and Raman active ν_2 mode of BEDT-TTF along with the electrical resistivity. Both charge-sensitive modes, ν_{27} and ν_2, appear as a single band at ρ~0.5 below 100 K. Upon increasing the temperature, they split into high-frequency and low-frequency modes. This splitting is comparable with ν_{27} and ν_2 in the CO state of θ-(BEDT-TTF)$_2$RbZn [71,72]. This splitting clearly shows the charge disproportionation from uniform site charge with Δρ~0 to two different site charges with an amplitude of Δρ~0.6. Above 150 K, these two sites appear to correspond to the two non-equivalent sites in the superlattice (center and corner sites in Figure 5b). However, the intensity of the superlattice reflection is extremely weak compared with Bragg reflections. This observation means that the two independent sites are approximately equivalent. Therefore, this charge disproportionation seems not to be long-range ordered, and probably the short-range ordered site charges are dynamically fluctuating. The correlation time was estimated to be $\tau > \sim 0.3$ ps at room temperature assuming a two-states-jump model [61]. This will be further discussed below along with the optical conductivity. On the other hand, the charge-sensitive modes ν_{21}, ν_{34}, and ν_4 of TCNQ show no change. This observation is consistent with the speculation that the valence electrons of TCNQ are localized within the TCNQ dimer over the whole temperature range.

Figure 7 shows the optical conductivity polarized along the a axis, which is the direction of the maximum transfer integral between BEDT-TTF. The temperature dependence is peculiar in that the spectral weight shifts to the low-frequency region on decreasing temperature. At the same time the resistivity also decreases. This characteristic spectral change seems to be interpreted by the theoretical model of Merino et al. [40]. They computed the optical conductivity for a quarter-filled square lattice with various V/t values under the condition of $U/t = 20$, where *U, V,* ant t are the on-site Coulomb energy, intersite Coulomb energy, and intersite hopping energy, respectively. When V/t exceeds the critical value $(V/t)_c$, a CO state is more stable than a metallic state. According to their theory, the density of states in a metallic phase ($V/t < (V/t)_c$) is triply peaked at around the Fermi energy (See the inset of Figure 7). From a low energy, these peaks correspond to the incoherent band associated with the lower Hubbard band, quasiparticle band, and upper incoherent band. When V/t increases, the quasiparticle peak in the density of states decreases, and finally vanishes in the CO state ($V/t > (V/t)_c$). Consequently the optical conductivity consists of three terms, $\sigma(\omega) = \sigma_D(\omega) + \sigma_{IQ}(\omega) + \sigma_{II}(\omega)$, where $\sigma_D(\omega)$ is the Drude term that peaks at $\omega = 0$ which corresponds to the excitation within the quasiparticle band, $\sigma_{IQ}(\omega)$ is the optical transition between the incoherent bands and the quasiparticle band, and $\sigma_{II}(\omega)$ is the optical transition between the incoherent bands (transition between the neighbor sites in real space description) [73]. In the CO state, the gap is open, so that $\sigma_D(\omega)$ and $\sigma_{IQ}(\omega)$ vanish. When V/t decreases, the CO state is melted, and $\sigma_D(\omega)$ and $\sigma_{IQ}(\omega)$ gain the intensity and $\sigma_{II}(\omega)$ loses the intensity. At the same time, the peaks of $\sigma_{IQ}(\omega)$ and $\sigma_{II}(\omega)$ shift toward lower energy. The observation of $\sigma_D(\omega)$ is a sign of the coherent quasiparticle band and the intensity decrease and the low-energy shift of $\sigma_{IQ}(\omega)$ and $\sigma_{II}(\omega)$ corresponds to growth of the coherent band of the quasiparticle. This trend of $\sigma_{IQ}(\omega)$ and $\sigma_{II}(\omega)$ is seen in Figure 8b with decreasing temperature. The Drude term $\sigma_D(\omega)$ appearing below

600 cm^{-1} has not been measured in this compound. Probably $\sigma_D(\omega)$ will quickly grows below 20 K, as the dc conductivity increases remarkably below 20 K as shown in Figure 7.

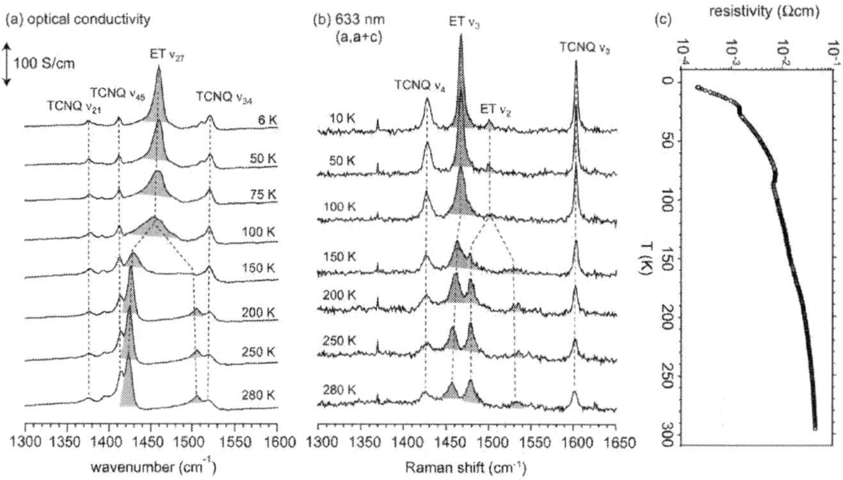

Figure 6. Temperature dependences of the (**a**) ν_{27} mode (blue); (**b**) ν_2 (blue) and ν_3 (green) modes; and (**c**) electrical resistivity of BEDT-TTF in β″-(BEDT-TTF)(TCNQ). Note that the charge-sensitive modes merge into a single band on decreasing the temperature. This figure is modified from Figures 3,4,7 in ref. [61].

According to the charge correlation function C(q, ω) near CO calculated by Merino et al., the excitation energy ω approaches zero and the correlation length diverges [40,41], when the parameters V/t approach the critical value for long-range CO. Accordingly, the amplitude of CO fluctuation Δρ increases and the correlation time τ is expected to diverge. This trend is seen in Figure 8e on increasing the temperature, where the product τΔν is estimated from the two-states-jump model as explained in Section 2.4. Therefore, together with the optical conductivity, β″-(BEDT-TTF)(TCNQ) approaches the long-range CO state with increasing temperature. Above 200 K the increase in the product τΔν levels off, and the optical conductivity appears to have a gap. If this compound attains a long range order above 200 K, the charge-rich and charge-poor molecules are fixed at crystallographically independent sites. In spite of the large amplitude of CO, the two independent sites are approximately equivalent, as described in the first paragraph of this section. This result suggests that short-range charge order fluctuates, and it is not long-range ordered above 200 K. The absence of long-range order as well as the sizable dc conductivity with $d\rho/dT > 0$ in this temperature range, as shown in Figure 8a, suggests the existence of a finite density of states at the Fermi level as schematically drawn in the inset in Figure 7. The apparent optical gap is regarded as a pseudogap. Therefore the high-temperature state is considered to be a charge-ordered metal [74].The amplitude $\Delta\rho \sim 0.6$ and correlation time $\tau > \sim 0.3$ ps may characterize the charge-ordered metal of this compound. The optical conductivity below 600 cm^{-1} is necessary to clarify the finite density of state near the Fermi level.

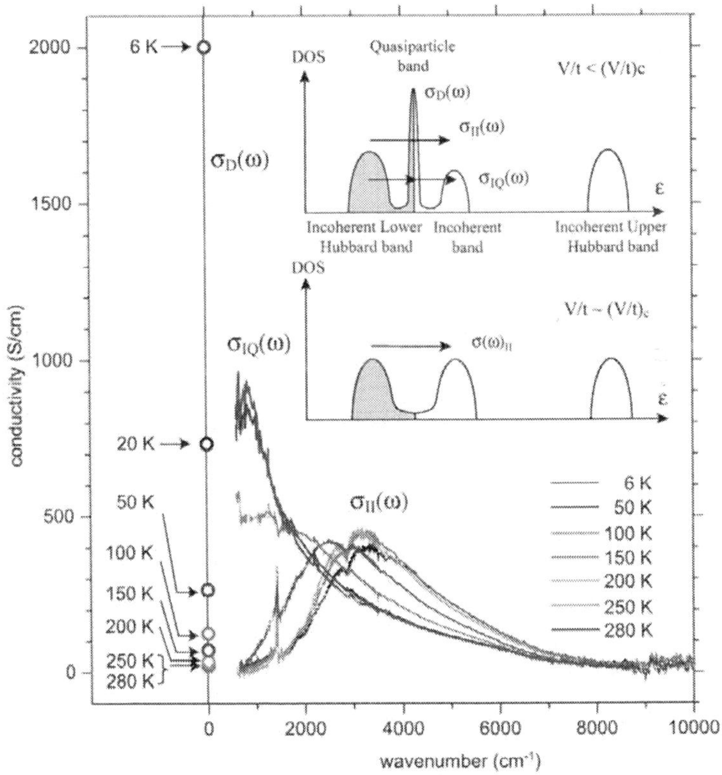

Figure 7. Temperature dependence of the E||a optical conductivity of β″-(BEDT-TTF)(TCNQ). Circles at zero wavenumber denote the dc conductivity at several temperatures. The inset shows the schematic figure of density of states in the quarter-filled band system. See the text for the meaning of the electronic transitions, $\sigma_D(\omega)$, $\sigma_{IQ}(\omega)$, and $\sigma_{II}(\omega)$.

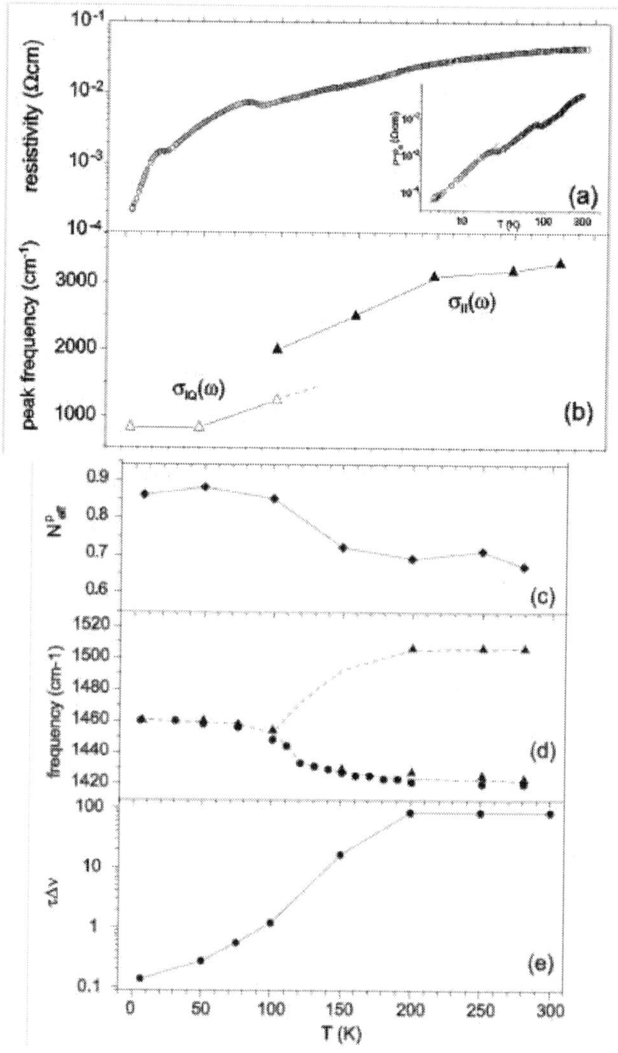

Figure 8. Temperature dependences of the following quantities of β″-(BEDT-TTF)(TCNQ). (**a**) electrical resistivity; (**b**) peak frequency of electronic transition; (**c**) partial effective hole number; (**d**) peak frequency of ν_{27}, where squares and triangles are the data of independent experiment; and (**e**) product of correlation time and splitting width of ν_{27}. Lines are included as guide to the eye. The inset of Figure 8a shows that the resistivity below 20 K follows the relation for a Fermi liquid, $\rho = \rho_0 + AT_2$. Figure 8a,b are taken from Figure 7 of ref. [61].

In contrast, at low temperatures, the CO fluctuation is significantly suppressed, as $\tau\Delta\nu$ is decreased by three orders of magnitude. Merino predicted a transition between a uniform metal and a charge-ordered metal [74]. As shown in Figure 8d, the spectral variation of ν_{27} looks to be continuous at around 100 K. Therefore, this compound shows a crossover from a charge-ordered metal to a uniform metal upon lowering the temperature. Comparing with the theoretical model, it turns out that decreasing temperature corresponds to a decrease of V/t. Although the lattice contraction increases both V and t, the latter is more sensitive to an intermolecular distance. Therefore, lattice contraction is responsible for this crossover behavior. Actually, the increase in t is indicated by an increase in the partial effective number, which is defined by the following equation,

$$N_{eff}^{p} = \frac{2m}{\pi e^2 N} \int_{\omega_{c1}}^{\omega_{c2}} \sigma(\omega) d\omega \tag{5}$$

where m and e are the electron mass and electron charge respectively, N is the hole density of BEDT-TTF, ω_{c1} = 600 cm^{-1} and ω_{c2} = 9000 cm^{-1} are the cut-off frequencies, and $\sigma(\omega)$ are the cut-off frequencies, N^{p}_{eff} increases by about 30% from 280 K to 6 K as shown in Figure 8c, though the spectral weight below 600 cm^{-1} is not taken into account. The role of lattice contraction on the electronic state is also supported by high pressure experiment. When a hydrostatic pressure is applied to this compound at room temperature, the Raman spectrum shows a similar variation as a decrease in the temperature. The split ν_2 modes merge at about 2 kbar [61], that is, the charge-ordered metal changes into a uniform metal due to lattice contraction. Possibly, a quasiparticle band is formed below 20 K because of the high conductivity and Fermi liquid behavior of transport properties [65,66,67,68,69]. To confirm this, the Drude response should be examined in the optical conductivity below 600 cm^{-1}. Finally, the structural change in the superlattice at ~170 K and anomalies in resistivity at ~80 K and ~20 K seem to have no influence on the optical conductivity and the line shape of the charge-sensitive mode. More extensive investigation is necessary for the understanding of this compound.

4. Θ-(BEDT-TTF)$_2$X

The θ-type BEDT-TTF salts, θ-(BEDT-TTF)$_2$X, have a herringbone arrangement of BEDT-TTF, forming a two-dimensional quarter-filled system. This family includes metallic compounds and several compounds which undergo a charge-ordering (CO) phase transition. The electronic phase diagram of this family has been composed employing the dihedral angle φ between the adjacent BEDT-TTF molecules (See Figure 9), which is related to the bandwidth [74]. The dihedral angle was subsequently linked to t_c/t_p to explain the phase diagram, where t_c and t_p are the transfer integrals along the **c** (stacking) direction and **a** ± **c** (diagonal) directions forming a triangular lattice [75,76,77,78]. Watanabe discussed this phase diagram using t_c/t_p as follows [78]. As the dihedral angle φ decreases, the bandwidth increases [79], and the ground state is stabilized in the order of a stripe CO, frustrating CO between the stripe and the threefold lattice, and uniform metal. Accordingly, the temperature range of the highly conductive phase expands to the low-temperature region, as φ is decreased. In the small φ region, X = I_3 (φ = 100°) [80] and X=$CsZn(SCN)_4$ (φ = 104°) [81] salts, the high-temperature phase is metallic, with $d\rho/dT > 0$, whereas it is non-metallic, with $d\rho/dT < 0$ above the CO transition temperature in the large φ region. The activation energy increases with increasing φ as follows: 0.02 eV for X = $RbZn(SCN)_4$ (φ = 111°) [81], 0.08 eV for orthorhombic $TlZn(SCN)_4$ (φ = 117°) [81], 0.11 eV for monoclinic $TlZn(SCN)_4$ (φ = 121°) [81], and 0.17–0.19 eV for $Cu_2(CN)[N(CN)_2]_2$ (φ = 132°) [82]. Hereafter, θ-$(BEDT\text{-}TTF)_2MM'(SCN)_4$ and θ-$(BEDT\text{-}TTF)_2X$ are abbreviated as θ-MM′ and θ-X, respectively.

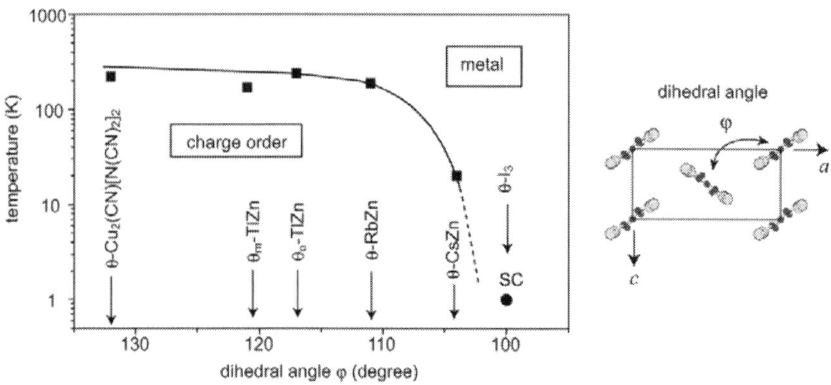

Figure 9. Phase diagram of θ-$(BEDT\text{-}TTF)_2X$ [79]. The squares and the circle represent the charge-ordering phase transition and the superconducting phase transition, respectively.

4.1. Charge-Ordered Phase of X = $RbZn(SCN)_4$, $TlZn(SCN)_4$, $Cu_2(CN)[N(CN)_2]_2$

θ-RbZn undergoes the CO phase transition accompanying a structural change, and this compound has been very extensively studied. As shown in the left panel of Figure 10, Raman-active ν_2 and ν_3 modes show remarkable change in the CO transition. Yamamoto et al., analyzed the spectra comparing them with the spectra for the ^{13}C-substituted compound, in which the carbon atoms in the central bridge C=C bond are substituted by ^{13}C [72]. Since the ν_2 mode is mainly the C=C stretching vibration in the five-member rings and the ν_3 mode is mainly the C=C stretching vibration of the central bridge C=C bond, (See Figure 1) the isotope shift of the latter mode is expected to be significantly larger than the former. Utilizing this property, they assigned all the observed bands as shown in Figure 10. Actually, the isotope shifts of ν_{3A}, ν_{3B}, ν_{3B}, and ν_{3A}, are much larger than those of ν_{2P} and ν_{2R}, while ν_4, and ν_5 show no isotope shift [72]. It becomes clear from this assignment that the ν_2 mode splits into two and the ν_3 mode split into four [83]. As we explained in Section 2.3, the ν_2 mode splits owing to the charge difference and the ν_3 mode splits owing to the EMV mechanism. From the splitting of ν_2, the site-charge difference or amplitude of CO is estimated to be $\Delta\rho \sim 0.6$. This large amplitude is approximately consistent with the site-charge ratio $\rho_P:\rho_R$ = 1:6 estimated from ^{13}C-NMR experiment [15]. From the splitting of ν_3, it is concluded that the single conducting layer in the unit cell contains four molecules. This means that the unit cell is doubled below the CO phase transition.

As shown in the right panel of Figure 10, the four ν_3 modes are classified into two groups by the polarization of the incident and scattered light. The two ν_{3A} modes and two ν_{3B} modes, respectively, have the character of species A and B in the C_2 symmetry. This means that the four molecules in a conducting layer are classified into two groups, and in each group, two equivalent molecules are connected by a screw axis. These groups correspond to charge-rich and charge-poor molecules. Therefore, the charge-rich (charge-poor) molecules are arranged to form a horizontal stripe. This conclusion is consistent with the space group $P2_12_12_1$ which was determined by the low-temperature x-ray diffraction measurements [84]. The factor group of $P2_12_12_1$ is D_2, which is reduced to C_2 for a single conducting layer [85]. The horizontal stripe in the θ-RbZn salt is claimed by Tajima et al., through the analysis of the mid-infrared band in the optical conductivity [86]. Based on the mode of the horizontal stripe, Suzuki et al. [85] and Yamamoto et al. [57] conducted a numerical calculation of the splitting of ν_2 and ν_3. According to their analysis, the four vibronic modes are expressed by the following symmetry coordinates,

$$
\begin{aligned}
Q_{A1} &= \{a(Q_{P1}+Q_{P2})+\sqrt{1-a^2}(Q_{R1}+Q_{R2})\}/\sqrt{2} \\
Q_{A2} &= \{\sqrt{1-a^2}(Q_{P1}+Q_{P2})-a(Q_{R1}+Q_{R2})\}/\sqrt{2} \\
Q_{B1} &= \{b(Q_{P1}-Q_{P2})+\sqrt{1-b^2}(Q_{R1}-Q_{R2})\}/\sqrt{2} \\
Q_{B2} &= \{\sqrt{1-b^2}(Q_{P1}-Q_{P2})-b(Q_{R1}-Q_{R2})\}/\sqrt{2}
\end{aligned}
\tag{6}
$$

where Q_{P1} and Q_{P2} (Q_{R1} and Q_{R2}) are the normal coordinates of charge-poor (charge-rich) site. The different character between ν_2 and ν_3 is ascribed to the difference in EMV coupling constants, which are small for ν_2 and large for ν_3 compared to transfer integrals. When the normal coordinate has a_g symmetry in a free molecule, Q_{A1} and Q_{A2} have A symmetry and Q_{B1} and Q_{B2} have B symmetry. In the case of ν_2 mode, $a \approx 1$ and $b \approx 1$, since the coupling constant is small, in other words, the interaction between charge-rich and charge-poor site is small. Therefore, both ν_{2P} (ν_{A1}, ν_{B1}) and ν_{2R} (ν_{A2}, ν_{B2}) are doubly degenerate. On the other hand, the four ν_3 modes, ν_{3A} (ν_{A1}, ν_{A2}) and ν_{3B} (ν_{B1}, ν_{B2}) are split through the strong EMV interaction.

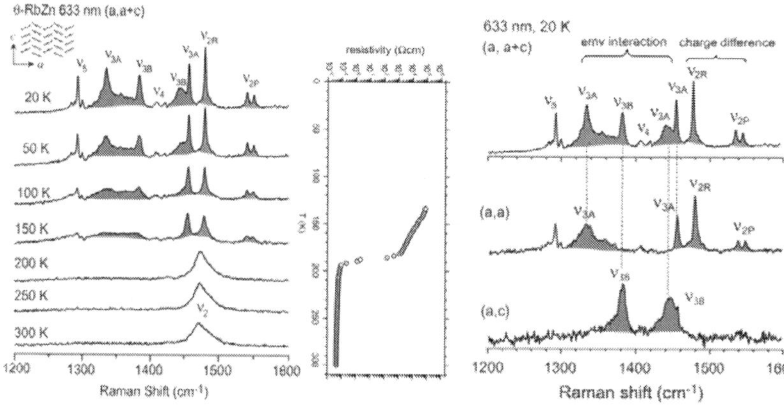

Figure 10. (**left panel**) Temperature dependence of the Raman spectrum and the electrical resistivity of θ-(BEDT-TTF)$_2$RbZn(SCN)$_4$. The ν_2 and ν_3 modes are shown by blue and green colors. The inset shows the herringbone arrangement of BEDT-TTF with respect to the crystallographic axes; (**right panel**) Polarized Raman spectrum of θ-RbZn measured at 20 K. The four ν_3 modes (green) in the top panel are classified into two groups by polarization. The Raman spectra are modified from Figures 14,15 of ref. [24].

Similar analyses were conducted for θ_o-TlZn [85], θ_m-TlZn [85], and θ-$Cu_2(CN)[N(CN)_2]_2$ [57] salts. Since θ_o-TlZn and θ-$Cu_2(CN)[N(CN)_2]_2$ are isostructural to θ-RbZn, the Raman spectra below T_{CO} are exactly the same as that of θ-RbZn as shown in Figure 11. However, θ_m-TlZn belongs to monoclinic system with space group C2. Reflecting this structural difference, the Raman spectrum and electrical resistivity shown in Figure 12 are different from those of the orthorhombic salts [85]. The spectrum at 60 K is interpreted assuming pseudo inversion symmetry in the following way. The ν_2 mode is split into the charge-rich and charge-poor sites, and two in-phase modes are observed among four ν_3 modes. Based on this interpretation a diagonal stripe is claimed to be the CO pattern [85]. A diagonal stripe is theoretically predicted to be almost degenerate with a horizontal stripe. It is theoretically considered that the lattice distortion below T_{CO} stabilizes the horizontal stripe in θ-RbZn [87]. If θ_m-TlZn takes a diagonal stripe, the low-temperature structure is expected to be distorted differently from θ-RbZn.

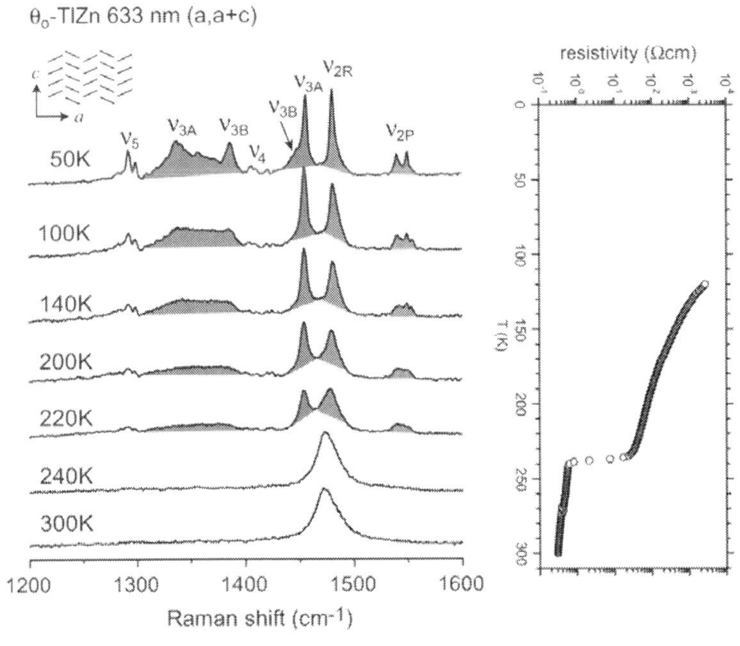

Figure 11. Temperature dependence of the Raman spectrum of θ_o-(BEDT-TTF)$_2$TlZn(SCN)$_4$ along with the electrical resistivity. The inset shows the herringbone arrangement of BEDT-TTF with respect to the crystallographic axes. The spectral change is almost exactly the same as that of θ-(BEDT-TTF)$_2$RbZn(SCN)$_4$. The Raman spectrum is modified from Figure 4 of ref. [85].

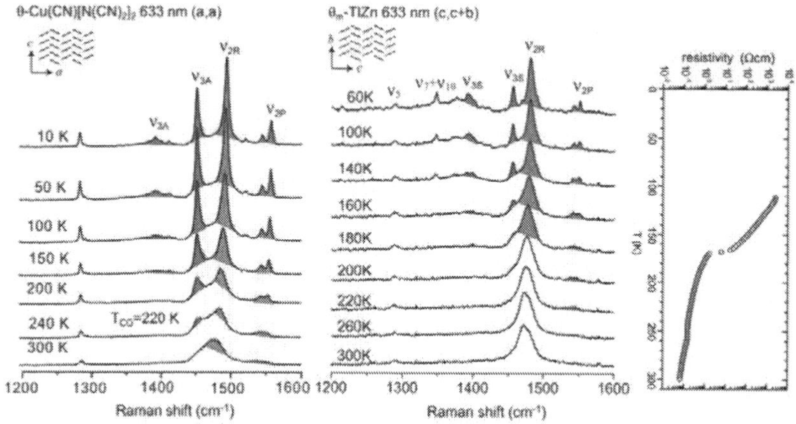

Figure 12. Temperature dependence of the Raman spectrum of θ-(BEDT-TTF)$_2$Cu(CN)[N(CN)$_2$]$_2$ (left) and θ$_m$-(BEDT-TTF)$_2$TlZn(SCN)$_4$ (middle) along with the electrical resistivity. The inset shows the herringbone arrangement of BEDT-TTF with respect to the crystallographic axes. The spectral change and resistivity are rather continuous. The left spectrum is modified from Figure 5 of ref. [57]. The right spectrum is modified from Figure 7 of ref. [85].

4.2. Metallic Phase of X = RbZn(SCN)$_4$, TlZn(SCN)$_4$, Cu$_2$CN[N(CN)$_2$]$_2$

The high-temperature phase of θ-RbZ is regarded as an unconventional metal. There is a finite density of states at the Fermi level, though the temperature derivative of resistivity is negative. This unusual behavior in the high-temperature phase of θ-RbZn was reported in ^{13}C-NMR experiment. The linewidth shows a broadening as the temperature approaches T_{CO} [15]. Chiba analyzed this behavior, and reported the following results [88]. Above T_{CO}, fluctuation of CO arises in space (the fractional charge is distributed from +0.3 to +0.7) and in time (extremely slow on the time scale of ^{13}C-NMR). This short-range ordered CO is related to the x-ray diffuse scattering at $\boldsymbol{q}_1 \approx (1/4, k, 1/3)$, which corresponds to the short-range ordered 4a × 3c superlattice (short-range ordered threefold CO) [84]. When the sample is rapidly cooled, the phase transition is suppressed [89]. This rapid-cooling state is regarded as a frozen state of the metal, and it is in a glassy state [90]. In this frozen state, another diffuse scattering at $\boldsymbol{q}_2 = (0, k, 1/2)$ is found in addition to weak $\boldsymbol{q}_1$ [91]. The diffuse scattering at $\boldsymbol{q}_2$ corresponds to the short-range ordered 2c superlattice (short-range ordered stripe CO). As this metastable state relaxes, the volume fraction of the 2c superlattice decreases, whereas the 4a × 3c superlattice becomes dominant [92]. This result means that threefold CO and stripe CO are competing in the glassy state.

The Raman spectrum shows a single broad band above T_{CO} as shown in Figure 10. This broad band is assigned to the ν_2 mode based on the isotope shift of the ^{13}C-substituted compound [93]. If the high-temperature phase is a metal, the ν_2 mode should be observed between the ν_{2P} and ν_{2R} bands. However, the broad ν_2 band above T_{CO} is located at the ν_{2R} band, and thus ν_{2R} is regarded as being hidden in the background. This result as well as the ^{13}C-NMR data is direct evidence for the fluctuation of charge order in the metallic phase. The large splitting, $\Delta\nu_2 = \nu_{2P} - \nu_{2R}$, means a large amplitude ($\Delta\rho \sim 0.6$) of fluctuation. If the correlation time is very long on the time scale of vibrational spectroscopy as claimed by NMR [88], the line shape is regarded as the case of $\tau\Delta\nu_2 > 1$ (See Figure 4a, $\Delta = \Delta\nu_2/2$), that is, motional narrowing does not work. Therefore, Raman spectroscopy yields a snap shot of the slow fluctuation, and thus the broad linewidth reflects spatially inhomogeneous charge distribution. Accordingly, the line shape of the ν_2 mode in the rapid-cooling state is almost exactly the same as that of the high-temperature spectrum [24,94]. The strong inhomogeneity probably comes from the short-range ordered 3a × 4c superlattice, which involves 48 sites in the unit cell. On the other hand, in the case of ^{13}C-NMR, the line broadening due to the spatial inhomogenity is wiped out at higher temperature due to the motional narrowing, because the time scale of ^{13}C-NMR is slower than that for vibrational spectroscopy. The line broadening comes back at low temperature due to the slower dynamics.

The optical conductivity of θ-RbZn has been reported by Wang et al., down to 50 cm^{-1} [95]. As shown in Figure 13f,g, the optical conductivity above 200 K consists of a broad peak at around 2000 cm^{-1} with no Drude response. Theoretically, the 3a × 4c superlattice is metallic with Fermi surfaces [76]. Nishimoto et al., calculated the optical conductivity of threefold lattice. They show that the Drude weight is significantly suppressed, and the spectral weight shifts to high-frequency excitation (See Figure 13a–e) [96]. The resultant small Drude response may be overdamped in the short-range ordered 3a × 4c superlattice, which results in the pseudogap like optical conductivity. The optical conductivity in the metallic phase exhibits strong vibronic band of ν_3 at ~1300 cm^{-1} [72,95]. As the vibronic band is forbidden in the structure of the metallic phase, this vibronic band is activated due to the local lattice distortion by the fluctuation of threefold and stripe CO. In other words, the appearance of the vibronic band in θ-type BEDT-TTF salts indicates the fluctuation of CO in the metallic phase.

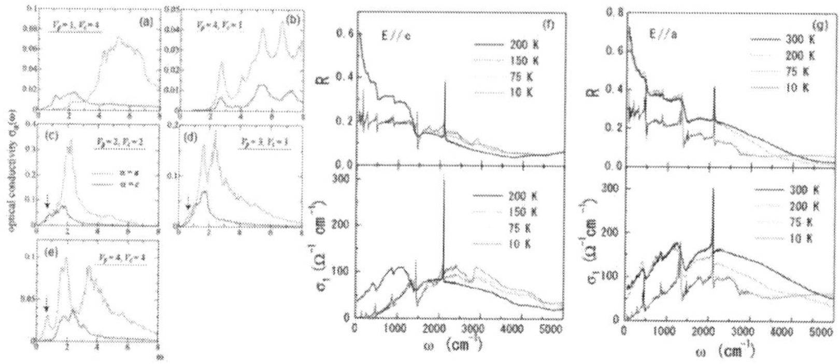

Figure 13. Reflectivity and optical conductivity of θ-(BEDT-TTF)$_2$RbZn(SCN)$_4$. and theoretical calculation of the optical conductivity of a triangular lattice. The arrows in (**c**), (**d**), and (**e**) denote the Drude peaks, which appear at finite energy because the open-end boundary condition is applied for a finite size cluster of L_a = 8 and L_c = 6 [96]. Figure 13a–e are taken from ref. [96], and Figure 13f,g are taken from ref. [95].

Other compounds with $\varphi > 111°$ also show similar $\tau\Delta\nu_2 \gg 1$ spectra with a broad linewidth as shown in Figure 14b,c [97]. As the dihedral angle increases, the ν_{2P} mode, the counterpart of ν_{2R}, appears more clearly. In addition, ν_3 mode in the low-frequency region of the broad ν_{2R} is more enhanced above T_{CO} (See Figure 14b). In particular, in θ-Cu$_2$(CN)[N(CN)$_2$]$_2$, another broad ν_3 mode is found at ~1390 cm^{-1} (not shown) above T_{CO}, when a 780 nm excitation laser is used [57]. The observation of the precursor band ν_3 means that short-range ordered horizontal and diagonal stripes grow more extensively toward T_{CO}. Actually, x-ray diffuse scattering is observed at q_2 = 1/2c* above T_{CO} = 220 K in θ-Cu$_2$(CN)[N(CN)$_2$]$_2$ [57]. Although no diffuse scattering has been found in θ_m-TlZn, the vibronic band of ν_3 is clearly found in the optical conductivity as shown in Figure 15a. Differently from θ-RbZn, the fluctuation of the threefold lattice is absent, whereas the short-range 2c superlattice is dominant. Watanabe explained this trend theoretically as the result of t_c/t_p [78], which decreases from positive value (~0.5) to a negative value (−0.5) according to the increase in the dihedral angle in the phase diagram of the θ-type BEDT-TTF salts [98,99]. In these two compounds, the frustration with threefold CO seems to be reduced more than in θ-RbZn and θ_o-TlZn. This may be related to the continuous transformation of the Raman and optical conductivity spectra and the resistivity of θ_m-TlZn and θ-Cu$_2$(CN)[N(CN)$_2$]$_2$ in contrast to θ-RbZn and θ_o-TlZn (See Figure 12, Figure 15).

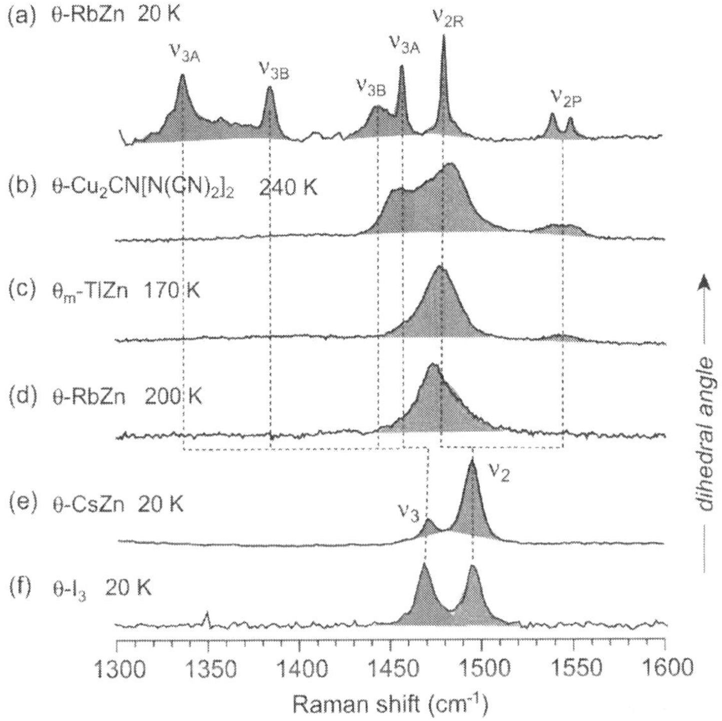

Figure 14. Comparison of the Raman spectra of (**a**) the CO state of θ-(BEDT-TTF)$_2$RbZn(SCN)$_4$; (**b**) high-temperature phase of θ-(BEDT-TTF)$_2$Cu$_2$CN[N(CN)$_2$]$_2$; and the metallic phase of (**c**) θ$_m$-(BEDT-TTF)$_2$TlZn(SCN)$_4$; (**d**) θ-(BEDT-TTF)$_2$RbZn(SCN)$_4$; (**e**) θ-(BEDT-TTF)$_2$CsZn(SCN)$_4$; and (**f**) θ-(BEDT-TTF)$_2$I$_3$. Note that the ν_2 mode is split in (a), (b), (c), and (d), while ν_2 is merged in (e) and (f). The assignments in θ-(BEDT-TTF)$_2$MZn(SCN)$_4$ (M = Rb, Cs) in the metallic phase are conducted with the aid of ^{13}C-substituted compounds [94].

4.3. θ-(BEDT-TTF)$_2$X (X = CsZn(SCN)$_4$, I$_3$)

The θ-CsZn salt is located at the boundary between the metallic θ-I$_3$ salt and the large φ salts that undergo CO transition. The x-ray diffuse rods are found at $\mathbf{q}_1$ = (2/3, k, 1/3) and $\mathbf{q}_2$ = (0, k, 1/2) below 120 K [100,101]. This finding implies that a short-range ordered 3a × 3c superlattice (fluctuation of threefold CO) coexists with the 2c superlattice (fluctuation of stripe CO). The 2c lattice is regarded as the fluctuation of the horizontal stripe which is found in θ-RbZn below T_{CO}. On lowering the temperature, the intensity of $\mathbf{q}_2$ increases as the resistivity increases from about 140 K. Although the correlation length of the 2c lattice grows up on lowering the temperature, it does not attain a long-range order. Interestingly, an external current suppresses the $\mathbf{q}_2$ domain at 12 K keeping the $\mathbf{q}_1$ domain intact [100]. This finding is related to the nonlinear conduction [102] and the thyristor effect [103]. The ^{13}C-NMR study argued the following points. (1) Remarkable charge disproportionation with extremely slow fluctuations is observed in the temperature range of 294 K–140 K; (2) The short-range CO with the amplitude of $\Delta\rho\sim0.4$–0.6 becomes almost static in the region of 140 K–50 K; (3) The amplitude of the CO fluctuation becomes very much reduced below 30 K [104]. The dielectric permittivity study proposes a glass-like state at low-temperature which consists of conglomerate domains of short-range ordered threefold and stripe CO [105]. Several theoretical studies have been conducted to explain the coexistent and/or frustrating CO fluctuation using a triangular lattice model [76,78,96,106]. The stripes and non-stripe threefold CO are competing near an isotropic triangular lattice ($V_c\sim V_p$), forming a charge-ordered metal [78,96]. The origin of the threefold CO is the nesting of the Fermi surface, while the origin of the stripe CO is the off-site Coulomb interaction [106], or both instabilities come from the Fermi surface nesting [76]. In all cases, there is a wide metallic region due to the frustration between the stripe and threefold CO fluctuation near the isotropic triangular lattice.

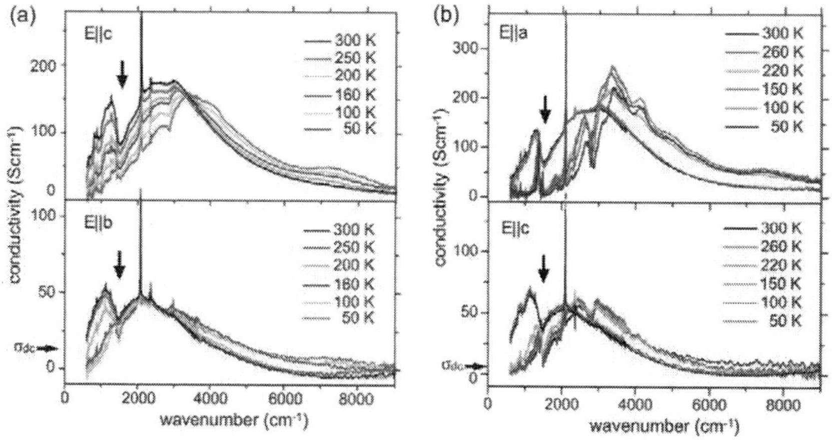

Figure 15. Optical conductivity of (**a**) θ_m-(BEDT-TTF)$_2$TlZn(SCN)$_4$ and (**b**) θ_o-(BEDT-TTF)$_2$TlZn(SCN)$_4$. Note that the spectral variation is discontinuous in θ_o-(BEDT-TTF)$_2$TlZn(SCN)$_4$, whereas it is continuous in θ_m-(BEDT-TTF)$_2$TlZn(SCN)$_4$. The arrow denotes the vibronic band of ν_3.

Figure 16 shows the Raman spectra of θ-CsZn and θ-I_3 along with the electrical resistivity of θ-CsZn [94]. θ-$(ET)_2I_3$ is metallic down to low temperature and undergoes a superconducting transition at 3.6 K [107,108]. In both compounds, ν_2 appears as a single peak near the frequency of $\rho = 0.5$ [46] over the whole temperature range. Different from θ-I_3, a very broad vibronic mode is observed at 1200–1300 cm^{-1} in θ-CsZn as shown in Figure 16. The Raman cross section of this vibronic mode is enhanced when charge disproportionation is enlarged as shown in Figure 2b (Section 2). This observation suggests that the correlation length of the CO fluctuation grows upon decreasing the temperature, since the intensity of the vibronic band increases. From the polarization dependence, ν_{3A} and ν_{3B} are respectively classified into A and B species in the factor group C_2 [94]. This result means that the short-range CO domain has a screw axis symmetry. Therefore, the 2c superlattice with the horizontal stripe seems to contribute to this vibronic mode [109]. This observation is consistent with x-ray diffuse scattering that shows the evolution of the short-range ordered 2c superlattice. Although threefold CO fluctuation is observed in both θ-CsZn and θ-RbZn, the features of ν_2 of θ-CsZn are very different from that of θ-RbZn. The ν_2 mode of θ-CsZn appears as a slightly broad single band at the frequency of $\rho \sim 0.5$, whereas the ν_2 mode of θ-RbZn appears as a much broader band at the frequency of the charge-rich site. The infrared-active mode ν_{27} also appears in the same way as ν_2 [94]. This result means that the product of the correlation time and splitting width is smaller than unity, $\tau\Delta\nu_2 < 1$, in θ-CsZn (See Figure 4a, $\Delta = \Delta\nu_2/2$), whereas $\tau\Delta\nu_2 > 1$ in θ-RbZn as described in Section 4.2. Therefore, the correlation time and/or CO amplitude of θ-CsZn is much smaller than that of θ-RbZn. If the fluctuation is almost static on the time scale of vibrational spectroscopy as suggested by NMR [104], $\Delta\nu_2$ is dominated by spatial inhomogeneity. The linewidth is about 10 cm^{-1} below 100 K. If we assume a static Gaussian distribution for the site charge, the amplitude is roughly estimated as $\Delta\rho < 0.08$. This small amplitude is inconsistent with the argument of ^{13}C-NMR [104]. However, the small amplitude is consistent with the much weaker intensity (diffuse/Bragg ~ 10^{-5}) of the diffuse scattering in θ-CsZn [100] compared with that (diffuse/Bragg ~ 10^{-3}) in θ-RbZn [91]. The small amplitude seems to be attributed mainly to the strong frustration between the threefold and stripe fluctuation. Seo demonstrated that in such a case, both peaks in the charge correlation function $C(q,\omega)$ corresponding to stripe and threefold fluctuation have higher excitation energy compared to the non-frustrate system [41]. This result implies that the amplitude of CO cannot grow when they are competing.

The optical conductivity of θ-CsZn is shown in Figure 17 [94]. Compared with the optical conductivity of θ-RbZn [95], the spectral weight of θ-CsZn is located in the low-frequency region, which is similar to the optical conductivity of β″-(ET)(TCNQ) shown in Figure 8. Since the conductivity should approach the dc conductivity shown in Figure 17, the peak which corresponds to σ_{IQ} is expected below 600 cm^{-1}, and the Drude response is likely to be missing. This speculation is consistent with the optical conductivity reported by Wang [110], wherein a strong peak accompanied by several vibronic bands appears below 600 cm^{-1} without Drude response. The V/t dependence of the optical conductivity similar to β″-(ET)(TCNQ) has been theoretically calculated for a triangular lattice [96]. The dips at 1300 cm^{-1} and 877 cm^{-1} are the vibronic modes respectively corresponding to ν_3 and ν_7, which arise from the EMV interaction. These vibronic modes are forbidden in the average structure of θ-CsZn. The appearance of these vibronic modes also strongly indicates structural fluctuations, which probably corresponds to the 3a × 3c and 2c short-range superlattices.

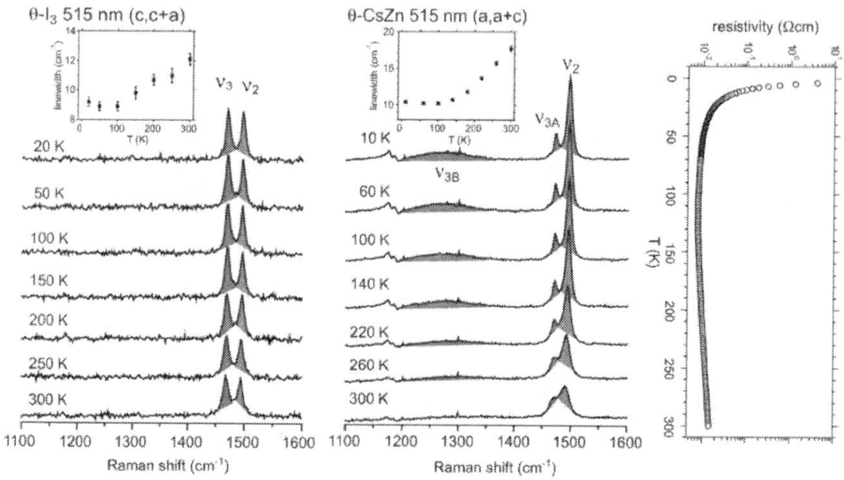

Figure 16. Temperature dependence of the Raman spectra of θ-(BEDT-TTF)$_2$I$_3$ and θ-(BEDT-TTF)$_2$CsZn(SCN)$_4$ along with the electrical resistivity of θ-(BEDT-TTF)$_2$CsZn(SCN)$_4$. The inset shows the temperature dependence of the linewidth of the ν_2 mode. The Raman spectra of θ-(BEDT-TTF)$_2$I$_3$ and θ-(BEDT-TTF)$_2$CsZn(SCN)$_4$ are modified from Figure 2c of ref. [111] and Figure 3 of ref. [94], respectively.

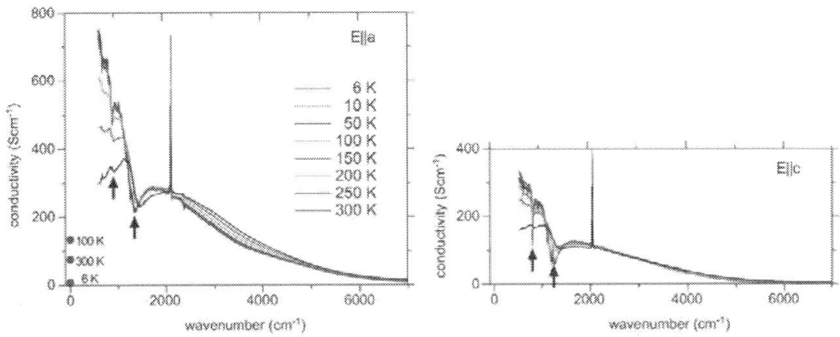

Figure 17. Optical conductivity of θ-(BEDT-TTF)$_2$CsZn(SCN)$_4$. Blue colored circles in the bottom panel show the dc conductivity values. The arrows show the vibronic modes appearing as dips. The color of the E||c spectrum denotes the same temperature as that of the E||a spectrum.

The dihedral angle of θ-I$_3$ is the largest among the θ-type BEDT-TTF salts, and the fluctuation of CO is suppressed in θ-I$_3$ [78]. If we estimate the amplitude in the same way as for θ-CsZn, the amplitude is estimated as $\Delta\rho < 0.07$. Actually the ν_2 mode appears as a single band and the vibronic band is not found in the Raman spectrum of θ-I$_3$ as shown in the left panel of Figure 16. Takenaka et al., reported the optical conductivity in the frequency range of 80–48000 cm^{-1} [112]. As the bandwidth is widest among the θ-type BEDT-TTF salts, the spectral weight of the optical conductivity in θ-I$_3$ is located in the frequency region lower than that of θ-CsZn. As is the case of β″-(BEDT-TTF)(TCNQ), the collapse of the coherent quasiparticle state accompanying the high-frequency shift of the spectral weight is observed upon increasing the temperature. The sublattice of BEDT-TTF in θ-I$_3$ is described by the orthorhombic cell which is the same as the average unit cell of θ-CsZn [113]. However, a weak vibronic mode is found at ~1410 cm^{-1} as a dip [112,114]. This vibronic band might suggest structural fluctuation perhaps associated with charge order.

5. A-(BEDT-TTF)$_2$I$_3$ AND RELATED COMPOUNDS

Among the various known organic conductors [115,116,117,118], α-(BEDT-TTF)$_2$I$_3$ (abbreviated as α-I$_3$) shows rich electronic properties such as charge ordering [119], superconductivity [120], zero-gap state (ZGS) [121,122], photo-induced phase transition [123], and nonlinear optical response [124].The crystal structure of α-I$_3$ consists of alternating layers of the anion and donor [125]. The donor layer has a herringbone arrangement of BEDT-TTF molecules. The unit cell accommodates two holes, which are distributed in four BEDT-TTFs named A, A′, B, and C. At room temperature, A and A′ are connected by inversion symmetry, and B and C are located on the inversion center. α-I_3 exhibits a first-order metal-insulator (MI) phase transition at T_{MI} = 135 K [125]. Kino and Fukuyama examined the effect of on-site Coulomb interaction within the Hartree-Fock approximation, and they predicted that a charge localization with a vertical stripe is the cause of the MI transition [13,126]. Before the theoretical prediction, Moldenhauer et al., reported that the localization of electrons occurs at the B and C sites whereas the electrons at the A and A′ sites are delocalized below T_{MI} [12]. The ^{13}C-NMR study suggested the existence of two differently charged molecules below T_{MI} [119,127]. Seo introduced intersite Coulomb interaction V, and he proposed that the localized charge would form a horizontal stripe [16]. These two models can be distinguished by the symmetry consideration, since the vertical stripe is compatible with the space group $P\bar{1}$, whereas the horizontal stripe requires P1. Raman spectroscopy was used to examine the selection rule of the charge-sensitive mode and the symmetry reduction from $P\bar{1}$ to P1 below T_{MI} was claimed [111]. An x-ray diffraction study supported P1, and showed a horizontal stripe structure [128]. Decisive evidence for the breaking of inversion symmetry was given by the experiment of second harmonic generation (SHG) [124]. Recently, Alemany et al., pointed out that the hydrogen bond between the ethylene groups of BEDT-TTF and I_3^- plays an essential role in the CO transition of α-I_3 [129].

In the MI transition of α-I_3, Raman [111] and infrared [130] spectra showed drastic changes as seen in Figure 18. The assignment of these bands along with vibronic bands was conducted with the aid of ^{13}C and deuterium substituted compounds. The CO and metallic phases will be discussed respectively in Section 5.1 and Section 5.2 along with the detailed assignment. Unlike θ-type BEDT-TTF salts, it has been reported for α-I_3 that the site-charge distribution is non-uniform in the metallic phase [13,111,128], although the average hole number per site is 0.5 in the case of uniform distribution. The non-uniform site charge distribution of metallic α-I_3 will be discussed along with the isostructural metallic α-(BEDT-TTF)$_2$$NH_4(SCN)_4$ in Section 5.3. One of the most attractive properties of α-I_3 is the ferroelectric properties in the CO state. In this context, another herringbone type compound α′-IBr_2 will discussed in Section 5.4.

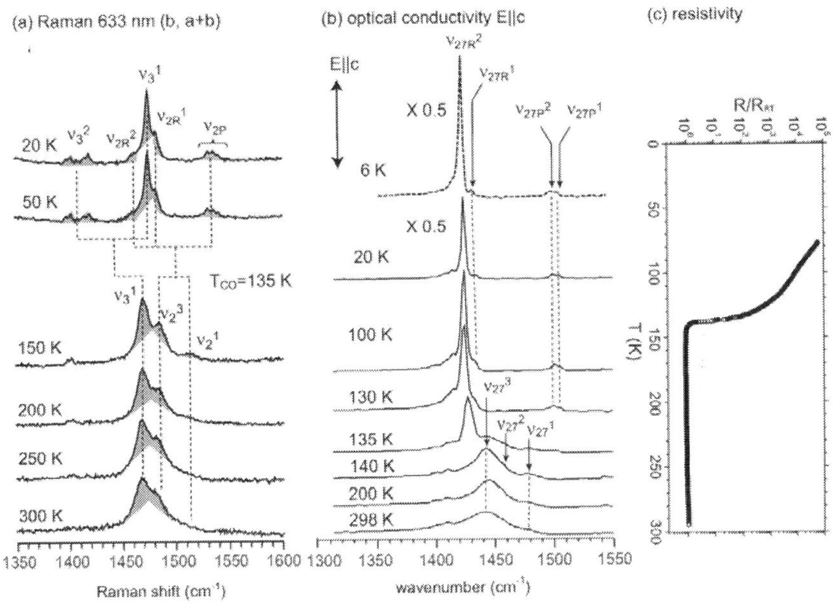

Figure 18. (**a**) Temperature dependence of the Raman spectrum of α-(BEDT-TTF)$_2$I$_3$. The ν$_2$ and ν$_3$ modes are shown as blue and green, respectively. A drastic spectral change is observed below 120 K. The difference in the transition temperature comes from the laser heating effect; (**b**) Temperature dependence of the infrared spectrum of α-(BEDT-TTF)$_2$I$_3$, which was obtained by the Kramer-Kronig transformation of the reflectivity polarized along the c-axis (perpendicular to the conducting plane). The 6 K spectrum drawn by the dashed line is the optical conductivity of the deuterium substituted compound, α-(d$_8$-BEDT-TTF)$_2$I$_3$; (**c**) Electrical resistivity normalized by room temperature value. The optical conductivity is taken from Figure 1 of ref. [130].

5.1. Charge-Ordered Phase of α-(BEDT-TTF)$_2$I$_3$

Figure 19 shows infrared and Raman spectra of α-I$_3$, α-(d$_8$-BEDT-TTF)$_2$I$_3$, and α-(^{13}C-BEDT-TTF)$_2$I$_3$, all in the CO state. In this spectral region, BEDT-TTF has three C=C stretching modes, ν$_2$, ν$_3$, and ν$_{27}$ which are shown in Figure 1, and bending modes of ethylene groups. The assignment of the Raman-active modes is the same as in ref. [111], except for the Raman bands at 1476 cm^{-1} and 1462 cm^{-1}, which were assigned to ν_3^1 and ν_{2R}^2, respectively. These are interchanged in Table 1, as the isotope shift of ν$_2$ and ν$_3$ became closer to each other within each mode. Strictly speaking, however, this kind of assignment makes no sense, as these two modes are strongly mixed with each other. The vibronic modes in the conducting plane are assigned with the aid of isotope shift and comparison with the Raman spectrum. The very broad band, wherein the peak was different between the E||a and E||b spectra, was assigned to ν_3^4, since the frequency of the vibronic mode depended upon the electronic excitation spectrum. The vibrational ν_{27} mode is observable in the optical conductivity spectrum polarized along the c-axis. Some bending modes of ethylene group are overlapped in the region of the ν_{27R} modes (See Table 1 and Figure 19a). Therefore, the spectrum of α-(d_8-BEDT-TTF)$_2$I$_3$ is necessary for the assignment of ν_{27R}. The large dip at ~2690 cm^{-1} appearing in the high-frequency region of the optical conductivity (not shown) is interpreted by Yamamoto as the Fano anti-resonance due to the vibronic overtone of ν_3 [131]. Recently, Yamamoto et al., have discussed the relationship between the vibronic overtone and optical nonlinearity and ferroelectricity [132].

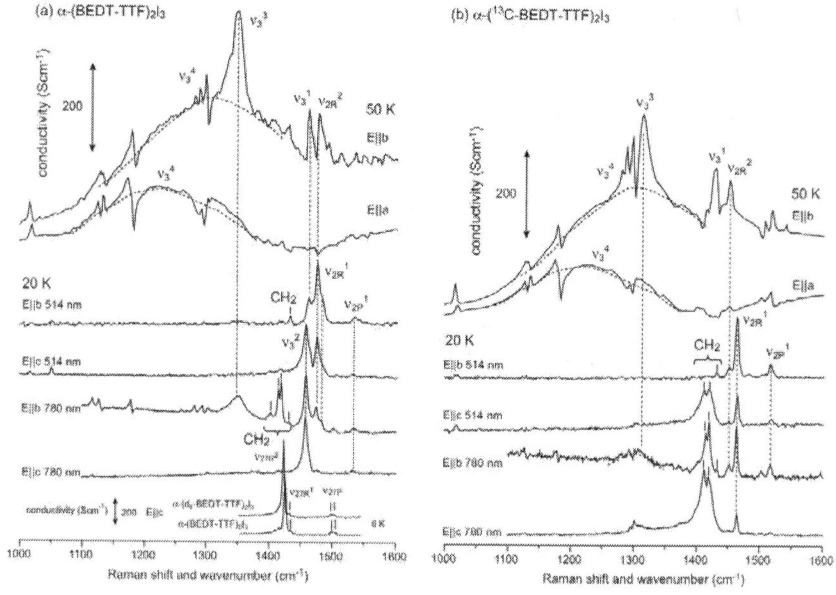

Figure 19. (**a**) Optical conductivity polarized along the b and a directions, Raman spectra of α-(BEDT-TTF)$_2$I$_3$, and optical conductivity polarized along the c direction; (**b**) Optical conductivity polarized along the b and a directions and Raman spectra of α-(^{13}C-BEDT-TTF)$_2$I$_3$.

Table 1. Assignment of the infrared and Raman modes of α-(BEDT-TTF)$_2$I$_3$. The Δν values show the isotope shift, $\Delta\nu_j = \nu_j^{12} - \nu_j^{13}$.

	Raman (20 K)			Infrared (50 K)		
	α-I$_3$	α-(^{13}C)$_2$I$_3$	Δν	α-I$_3$	α-(^{13}C)$_2$I$_3$	Δν
dip ($2\nu_3^2$)	-	-	-	2688	2628	60
ν_{2P}^1	1536	1517	19		(1518?)	-
ν_{2P}^2	-	-	-	-	(1508?)	-
ν_{2R}^1	1483	1465	18	-	-	-
ν_{2R}^2	1476	1452	24	1477	1454	23
ν_3^1	1462	-	-	1462	1431	31
ν_3^2	1458	-	-	-	-	-
ν_3^3	1349	1315	34	1346	1315	31
ν_3^4	-	-	-	~1300 (E‖b) ~1230 (E‖a)	~1300(E‖b) ~1230(E‖a)	-
CH$_2$-Bending	1431	1431	-	-	-	-
	1419	1420	-	-	-	-
	1414	1414	-	-	-	-
	1402	1402	-	-	-	-
ν_{27P}^1	-	-	-	1506 (E‖c)	1505 (E‖c)	-
ν_{27P}^2	-	-	-	1501 (E‖c)	1500 (E‖c)	-
ν_{27R}^1	-	-	-	1435 (E‖c)	1432 (E‖c)	-
ν_{27R}^2	-	-	-	1425 (E‖c)	1423 (E‖c)	-

If the unit cell has inversion symmetry, Raman-active band cannot be observed in an infrared spectrum and vice versa. As shown in Table 1, ν_{2R}^2, ν_3^1, and ν_3^3 are found in both the Raman and infrared spectra, which violate the mutual exclusion rule. As describe above, comparison of the Raman spectrum with the infrared spectrum is a good method to judge the breaking of inversion symmetry. Based on the assignment, it becomes clear that the ν_2 and ν_{27} modes are split into two groups, ν_{2P} and (ν_{2R1}, ν_{2R2}) in the Raman spectrum and (ν_{27P}^1, ν_{27P}^2) and (ν_{27R}^1, ν_{27R}^2) in the infrared spectrum as shown in Figure 18. Applying the linear relationship [46] to the ν_{27} mode (50 K), the site charge distribution is estimated as (0.8, 0.7, 0.3, 0.2) [130], which are approximately equal to the values (0.85, 0.80, 0.20, 0.15) estimated by Ivek et al. [133]. The CO amplitude ($\Delta\rho_{max}$ = 0.6) is comparable to that of θ-RbZn. Kakiuchi et al., examined the breakdown of Friedel's law by anomalous scattering effect, and they concluded a symmetry reduction from $P\bar{1}$ to P1 [128]. Using the space group P1, they determined the molecular geometry of each site, and provided consistent result for the site charge distribution as ρ_A = 0.82(9), ρ_B = 0.73(9), $\rho_{A'}$ = 0.29(9), ρ_C = 0.26(9) at 20 K. Recently, Alemany et al., reported the site charges as ρ_A = 0.638, ρ_B = 0.577, $\rho_{A'}$ = 0.438, ρ_C = 0.359 using a numerical atomic orbital density functional theory (DFT) approach [129]. This CO amplitude ρ_{max} = 0.28 is significantly smaller than the experimentally estimated value of ρ_{max} = 0.56–0.70. The site charge distribution has been estimated from the ^{13}C-NMR shift, which consists of Knight shift and chemical shift [15,134]. The chemical shift reflects the local charges, whereas the Knight shift is proportional to the local spin susceptibility which reflects the density of states at the Fermi level [135]. Therefore, the chemical shift should be analyzed to estimate the site charges. Kawai and Kawamoto conducted a ^{13}C-NMR study using a single side ^{13}C-enriched molecule [135]. To avoid the effect of the Knight shift and ring current within the molecule, they applied a magnetic field nearly parallel to the long axis of the molecule in the non-magnetic state (at 60 K). They determined the chemical shift tensor of the four sites, and estimated the site charge to be $\rho_R \sim 0.74$ and $\rho_P \sim 0.19$. If we simply apply the relation between the chemical shift and site charge which they present, site charge distribution is estimated as $\rho_A = 0.74$, $\rho_B = 0.76$, $\rho_{A'} = 0.18$, and $\rho_C = 0.20$. All of these results studied by vibrational spectroscopy, x-ray diffraction, and ^{13}C-NMR approximately consistent with each other.

Yamamoto et al., found strong second harmonic generation (SHG) in the CO state, using 4 ps pulse laser with a wavelength of 1.4 μm as the fundamental light [124,136]. The evolution of the second harmonic light just below T_{CO} is direct evidence for the breakdown of the inversion symmetry. They determined the nonlinear electric susceptibility $\chi^{(2)}_{ij}(2\omega; \omega, \omega)$, where i and j denote the polarization of the second harmonic light and fundamental light, respectively. The relative values against BBO (β-barium borate), which is a representative crystal for nonlinear optics, are 21, 8.5, 44, 31 for the polarization aa, ab, ba, and bb. Such a strong susceptibility value suggests that the generated electric dipole in a unit cell is coherently extended over a macroscopic domain, that is, a ferroelectric domain. As in conventional ferroelectrics, the crystal is expected to form a multiple domain structure. Ferroelectric domains with opposite polarization are observed utilizing the experimental technique of SHG interferometry [137]. The domain size is found to be more than 0.2 mm × 0.2 mm, which is almost a single domain. Although polarization reversal by an external electric field has not been realized due to high conductivity and low breakdown field, this compound in the CO state is classified as an unconventional ferroelectrics called electronic ferroelectrics [138]. In conventional ferroelectrics, ionic polarization coming from displacement or ordering of ions plays a role in creating a polar domain. On the other hand, in electronic ferroelectrics, ions are replaced by electrons, and electronic polarization by charge order causes a polar domain. To show that the polar domain mainly originated from electrons, Yamamoto demonstrated fast photoresponse where the second harmonics diminish by 50% within 100 fs and recovers within several tenths of a picosecond after the irradiation of a 100 fs laser. This fast photoresponse was studied by Iwai et al., as a photoinduced insulator-to-metal phase transition. Using mid-infrared pump-probe spectroscopy, they observed the transient reflectivity change in the mid-infrared region [123,139]. The decay process depends on the intensity of the light pulse. In the case of weak excitation, one photon produces a microscopic metallic domain consisting of about 100 molecules, which decays quickly. In the case of strong excitation, on the other hand, the microscopic domains aggregate with each other forming a quasi-macroscopic metallic domain, which decays slowly. The recovery of the second harmonics corresponds to this slow decay. Recently, Kawakami et al., uncovered the early dynamic process of the photoinduced phase transition using a 12 fs laser [140].

5.2. Metallic Phase of α-(BEDT-TTF)$_2$I$_3$ and Isostructural Metallic Compounds

5.2.1. Assignment in Metallic Phase

We have shown that charge order is the origin of the splitting of the ν_2 and ν_{27} modes. If a metallic compound involves crystallographically non-equivalent sites in the unit cell, the charge-sensitive modes split reflecting the non-equivalent sites. This splitting is also described by Equation (1). Figure 20 shows the Raman spectra of metallic α-I$_3$, α-(BEDT-TTF)$_2$NH$_4$Hg(SCN)$_4$, (abbreviated as α-NH$_4$Hg) and θ-I$_3$. The first two compounds are isostructural with each other, while the latter one takes herringbone arrangement with higher symmetry. α-NH$_4$Hg is metallic down to low temperature, and shows superconductivity at 1 K [141]. Similar to α-I$_3$, the unit cell of α-NH$_4$Hg involves four ET molecules (see Figure 1 of ref. [142] for the definition of A, A', B, and C). As described in the last section, the unit of the α-type compound involves three Raman-active modes and one infrared-active mode for ν_2 and ν_3. In the case of α-I$_3$, one ν_2 mode is missing, whereas all of the three appear to be present in ^{13}C-substitued compound as shown in Figure 20a,a',b,b'. From the isotope shift, the three bands in Figure 20a can be safely assigned to ν_2^1, ν_2^2, and ν_3^3. In the case of α-NH$_4$Hg, three ν_2 and two ν_3 modes are found and they are assigned based on the isotope shift shown in Figure 20c,c'.The remaining Raman-active ν_3 mode is missing from this Raman spectrum. Similar splitting of ν_{27} is described in ref. [130]. The crystal of θ-I$_3$ belongs to the monoclinic system with space group P2$_1$/c, and the unit cell contains four BEDT-TTFs. However, the arrangement of BEDT-TTF can be approximately described by the orthorhombic system with space group Pnma with the unit cell accommodating two crystallographically equivalent BEDT-TTFs [113,143]. Due to the high symmetry of the unit cell, ν_2 and ν_3 appear as a single band at $\rho \sim 0.5$ as shown in Figure 20d. The frequencies of the ν_2 and ν_3 modes of these compounds are shown in Table 2.

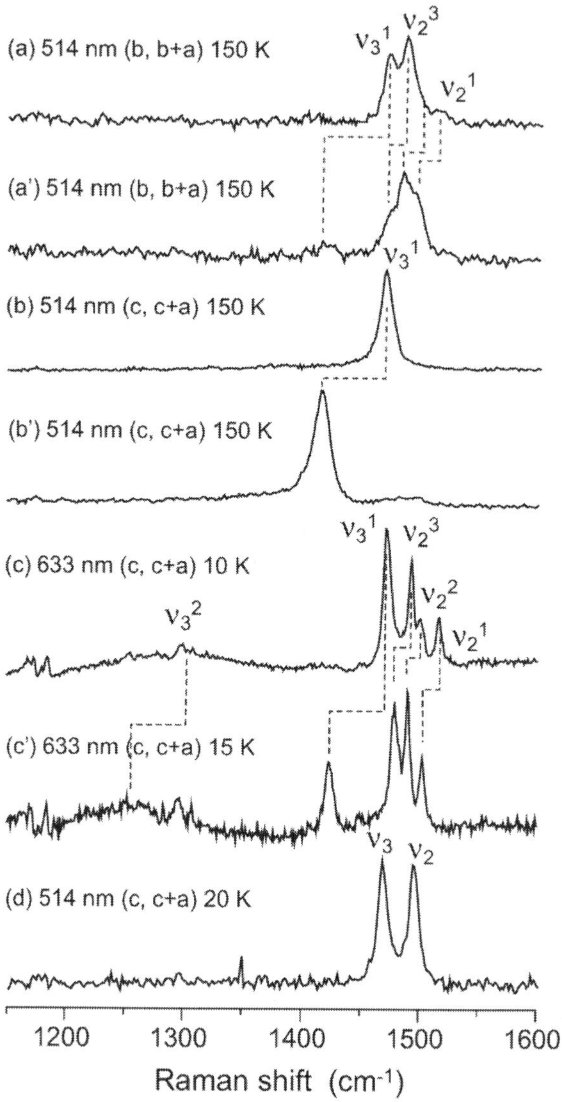

Figure 20. Raman spectra of α-(BEDT-TTF)$_2$I$_3$ and α-(BEDT-TTF)$_2$NH$_4$Hg(SCN)$_4$, and the ^{13}C-substitued compound, in which two carbon atoms of the central C=C bond of BEDT-TTF are substituted by ^{13}C. (**a**) and (**b**) α-(BEDT-TTF)$_2$I$_3$; (**a′**) and (**b′**) α-(^{13}C-BEDT-TTF)$_2$I$_3$; (**c**) α-(BEDT-TTF)$_2$NH$_4$Hg(SCN)$_4$; (**c′**) α-(^{13}C-BEDT-TTF)$_2$NH$_4$Hg-(SCN)$_4$, and (**d**) θ-(BEDT-TTF)$_2$I$_3$.

Table 2. Assignment of the Raman modes of α-$(BEDT\text{-}TTF)_2I_3$, α-$(BEDT\text{-}TTF)_2NH_4Hg(SCN)_4$, and θ-$(BEDT\text{-}TTF)_2I_3$.

	α-$(^{12}C\text{-}ET)_2I_3$	α-$(^{13}C\text{-}ET)_2I_3$		α-$(^{12}C\text{-}ET)_2$ $NH_4Hg(SCN)_4$	α-$(^{13}C\text{-}ET)_2$ $NH_4Hg(SCN)_4$		θ-$(ET)_2I_3$
	Raman (150 K)	Raman (150 K)	Isotope shift ν^{12}-ν^{13}	Raman (10 K)	Raman (15 K)	Isotope shift ν^{12}-ν^{13}	Raman (20 K)
ν_2^1	1516	1496	~20	1515	1503	13	
ν_2^2	-	~1484	-	1500	1490	10	1496
ν_2^3	1487	1470	17	1492	1479	14	
ν_3^1	1471	~1415	~56	1471	1423	48	
ν_3^2	-	-	-	~1300	~1250	~50	1469

5.2.2. Fluctuation of Charge Order in Metallic Phase

Although ν_2 and ν_{27} of α-I_3 and α-NH_4Hg are split into three with similar split widths, the linewidth of each mode is very different. Yue et al., showed that the linewidth of metallic α-I_3 is about two to three times as large as that of α-NH_4Hg [130]. As shown in Figure 21a, the linewidth of ν_{27} of α-I_3 is much broader than the charge-insensitive mode such as the CH_2 bending mode. As the frequencies of ν_{27} and ν_2 strongly depend upon the site charge (140 cm^{-1}/e for ν_{27} and 120 cm^{-1}/e for ν_2) [46], such a broad linewidth should be associated with fluctuation of the site charge. Assuming that the site charge fluctuates stochastically as a Gaussian process, Yue et al., analyzed the line shape of the somewhat isolated ν_2^1 mode, and obtained a width of $\Delta\nu \sim 9\text{–}17$ cm^{-1}, which corresponds to the CO amplitude of $\Delta\rho \approx 0.08\text{–}0.14$. Comparing with the theoretical curve, they estimated the parameter to be $\tau\Delta\nu \sim 0.7\text{–}10$ and the correlation time for the fluctuation to be $\tau^{-1} \sim 1\text{–}25$ cm^{-1}. This fluctuation rate is much faster than that of θ-RbZn, but it is sufficiently slow for the electron dynamics. This slow fluctuation rate suggests collective motion of site charge. In contrast to the θ-type BEDT-TTF salts, x-ray diffuse scattering has not been reported. Tanaka and Yonemitsu calculated the finite temperature free energy for the horizontal, vertical, and diagonal stripes, and threefold CO of α-I_3 within the framework of the mean-field approximation, using realistic parameters and taking electron-lattice coupling into account [87]. At low temperatures, these CO states and the paramagnetic metallic state are distributed within a narrow energy range. Therefore, the free energy in the metallic phase above T_{MI} is approximately multiply degenerate. Yue et al., speculated that the metallic state of α-I_3 above T_{MI} involves short-range correlation of these multiple stripes to increase entropy. This situation at finite temperature resembles the case of θ-CsZn in which stripe and threefold CO are competing. The slow fluctuation of the site charge is regarded as dynamical fluctuation of various types of CO domains. Such CO fluctuation modulates the charge density at each site and broadens the linewidth of the charge-sensitive mode, when the fluctuation rate is sufficiently slow.

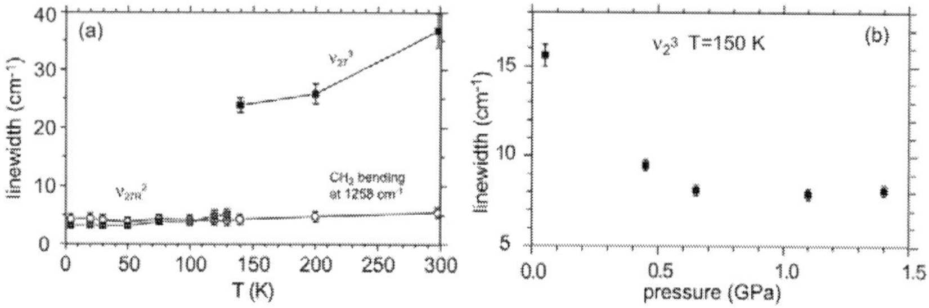

Figure 21. The linewidth of ν_{27} and $\nu_2{}^3$ of α-(BEDT-TTF)$_2$I$_3$. (**a**) Temperature dependence of the linewidth of infrared-active ν_{27}, which becomes very broad in the metallic phase; (**b**) Pressure dependence of the linewidth of Raman active $\nu_2{}^3$, which decreases on increasing pressure. These figures are modified from parts of Figures 2,4 of ref. [130].

As shown in Figure 21b, the linewidth of $\nu_2{}^3$ of α-I$_3$ decreases by almost half at 0.65 GPa and it becomes comparable to the linewidth of $\nu_2{}^3$ of α-NH$_4$Hg at 100 K. It is known that the application of hydrostatic pressure suppresses the long-range CO state [144]. This result implies that the paramagnetic metal is stabilized more than other stripe COs, and thus the energy difference between them is enlarged under hydrostatic pressure. Therefore, the fluctuation of stripe COs is suppressed under hydrostatic pressure, and the amplitude and correlation length may be reduced, leading to a higher fluctuation rate. This fast modulation narrows the linewidth of the charge-sensitive mode in addition to the small amplitude under hydrostatic pressure. The same speculation is applicable to stable metal α-NH$_4$Hg. The narrow linewidth of the charge-sensitive mode of α-NH$_4$Hg is consistent with this speculation.

The existence of CO fluctuation in α-NH$_4$Hg is indicated by the vibronic band $\nu_3{}^2$ in the Raman spectrum (See Figure 20c,c′) as in the case of θ-CsZn. On the other hand, the vibronic band is not found in the Raman spectrum of the isostructural α-KHg, α-RbHg and α-TlHg salts. This results suggests that the fluctuation in α-NH$_4$Hg is largest in the group of α-MHg (M = NH$_4$, K, Rb, Tl). Hiejima et al., observed the infrared-active ν_{27} modes on the side face of the crystal with the polarization perpendicular to the conducting plane [145]. In addition to the three ν_{27} modes, they found a weak new band growing below 200 K in α-MHg (M = K, Rb, Tl). They interpreted this band as the forbidden mode among the four ν_{27} modes, which means breaking of inversion symmetry. They speculated that fluctuation of stripe CO locally breaks the inversion symmetry as in α-I_3. If this is the case, the CO fluctuation in α-NH_4Hg is negligibly small, since the weak band is not found in α-NH_4Hg.

To reconcile these contradictory arguments, the charge density wave (CDW) should be reconsidered. Foury-Leylekian reported an incommensurate CDW for α-MHg (M = K, Rb) using x-ray diffraction [146]. The satellite reflections are already detected at high temperature (250–300 K). Recently, they pointed out the anion sublattice instability leading to deformation of the SCN tetrahedral coordination of Hg^{2+} and SCN tetragonal antiprismatic coordination of K^+ in α-KHg [147]. They claimed that the Fermi-surface-nesting instability of the BEDT-TTF layer is coupled with the instability of the anion sublattice. This sublattice instability is common in K, Rb, and Tl. As this deformation breaks inversion symmetry, the deformation of the anion sublattice brings about the non-centrosymmetric displacement of BEDT-TTF. This mechanism seems to be more reasonable to account for the breaking of the selection rule of ν_{27}. Since this CDW is incommensurate, the ν_{27} modes will become broad and unresolved, if the amplitude of CDW is large enough $\Delta\rho > 0.05$ [148]. Actually, the ν_{27} modes of α-NH_4Hg are more resolved than those of α-MHg (M = K, Rb, Tl) [145]. As the ν_{27} modes of the latter compounds are barely resolved, the CDW amplitude is considered to be $\Delta\rho < 0.05$. The above discussion leads to the conclusion that the fluctuation of CO is indicated only in α-NH_4Hg which shows superconductivity.

5.3. Non-Uniform Site-Charge Distribution in Metallic Phase

Non-uniform site charges in α-I_3 have been reported using Raman spectroscopy [111] and ^{13}C-NMR [134] experiments. Wojciechowski et al., interpreted them simply by non-uniform (anisotropic) network of transfer integrals in the unit cell using Equation (1) [111]. Yamamoto et al., conducted a systematic study of β″-type BEDT-TTF salts [149]. They qualitatively interpreted the site-charge distribution from non-uniform (anisotropic) intersite Coulomb interaction especially along the stacking direction. Band calculation that takes Coulomb interaction into account will incorporate the effect of both anisotropic networks on some level. Moroto et al., claimed that the non-uniform site charges are a precursor of charge order [134]. Bangura et al., attributed the splitting of ν_2 in metallic β″-(BEDT-TTF)$_4$[(H_3O)Ga(C_2O_4)]·$C_6H_5NO_2$ to the signature of charge order [150]. Recently, Kaiser et al., reported the splitting of ν_{27} in metallic β″-

$(BEDT\text{-}TTF)_2SF_5CH_2CF_2SO_3$, and they interpreted the growth of splitting as a fluctuation of CO [31]. All of these compounds involve non-equivalent sites in the unit cell: three sites in α-I_3 and two sites in the latter two compounds. Therefore, the splitting of the charge-sensitive mode in a metallic state is not always caused by charge order. The splitting of the charge-sensitive mode in the metallic phase should be carefully interpreted, considering the splitting width, line broadening, indication of symmetry breaking, etc. In this section, site-charges in metallic phase are more precisely discussed to compare the band calculation.

5.3.1. Relation between Frequency and Site Charge Near ρ = 0.5

The relationship between the frequency of ν_2 and the valence of the molecule, ρ, was examined by Yamamoto [46]. They assumed a linear relationship in the range of $0 \leq \rho \leq 0.8$ [151]. The ρ dependence of the frequency of ν_2 mode is ascribed to the ρ dependent force constant F(ρ). That is, the linear relationship requires $(dF/d\rho)_{\rho=0} = (dF/d\rho)_{\rho=1}$ [152]. However, the frequency of $BEDT\text{-}TTF^{0.5+}$significantly deviates from the linear relationship toward the low-frequency side, which implies the relation $|(dF/d\rho)_{\rho=0}| > |(dF/d\rho)_{\rho=1}|$. To estimate the site charge more reliably at ρ~0.5, the empirical equation, $\nu_2(\rho) = 1570 - 180\rho + 57\rho^2$ (ρ~0.5) is proposed, using $\nu_2(0)$ = 1570 cm^{-1} for the calculated frequency of the flat neutral BEDT-TTF molecule, [46] $\nu_2(0.5)$ = 1494 cm^{-1}, 1494 cm^{-1}, 1495 cm^{-1} respectively for β'-$(BEDT\text{-}TTF)_2ICl_2$, κ-$(BEDT\text{-}TTF)_2Cu_2(CN)_3$, and κ-$(BEDT\text{-}TTF)_2Cu[N(CN)_2]Cl$ at room temperature, and $\nu_2(1)$ = 1447 cm^{-1} for $(BEDT\text{-}TTF)ClO_4$ and $(BEDT\text{-}TTF)AuBr_2Cl_2$ at room temperature [153]. All of the compounds with ρ = 0.5 shows a frequency upshift by ~6 cm^{-1} on cooling from room temperature to 10 K due to hardening of the crystal lattice. The frequency shift due to lattice expansion is described by the equation, $\frac{1}{\omega}\frac{d\omega}{dT} = -\gamma\beta(T)$, where γ is the optical Gruneisen parameter and β(T) is the coefficient of volume thermal expansion. Therefore, this temperature dependence is given by the following equation, $\nu_2(0.5,T) = \omega_0 \exp\{-\gamma \int_0^T \beta(T)dT\} = \omega_0 \exp\{-\gamma(\beta_0 T + \beta_1 T^2/2)\}$, if β(T) is approximated by the equation, $\beta(T) = \beta_0 + \beta_1 T$ [154]. The examples of for the ν_2 modes of κ-$Cu[N(CN)_2]Cl$ and β'-ICl_2 are shown in Figure 22. Although the function form of β(T) is too crude to express actual one, the frequency shift is well reproduced. If we assume that $\nu_2(0)$ and $\nu_2(1)$ also show the same temperature dependence, the quadratic expression can be corrected as

$$\nu_2(\rho) = 1502\exp(-2.8\times10^{-6}T - 4.3\times10^{-8}T^2) - 180\rho + 57\rho^2 \quad (7)$$

for T < 300 K. However, the temperature dependence of molecular vibration varies depending on the compound. For example, the ν_2 mode of θ-I_3 shifts less than 3 cm^{-1} at 10 K. Therefore the estimated site charge has an ambiguity of about ±0.03.

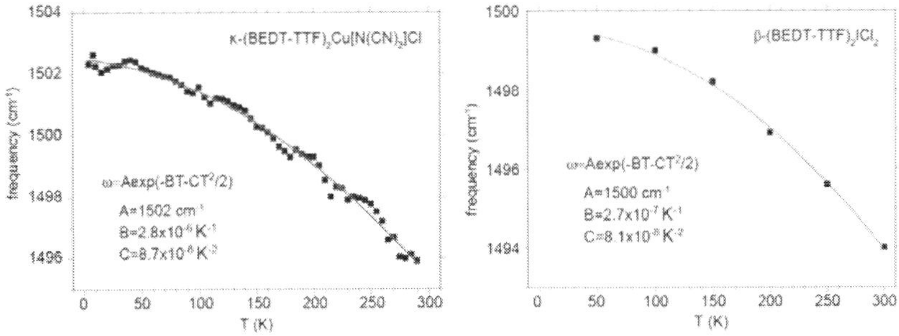

Figure 22. Temperature dependence of the frequency of ν_2 of κ-(BEDT-TTF)$_2$-Cu[N(CN)$_2$]Cl and β'-(BEDT-TTF)$_2$ICl$_2$. The red line is the best fit equation, $\omega = A\exp(-BT-CT^2/2)$, where $A = \omega_0$, $B = \gamma\beta_0$, and $C = \gamma\beta_1$.

5.3.2. Site Charge Distribution in Metallic Phase at Ambient Pressure

The second column of Table 3 shows the site charge of α-I_3 in the metallic phase estimated from the frequency of ν_2 using Equation (7). According to an x-ray diffraction study, the site charge follows the order, $\rho_B > \rho_A = \rho_{A'} > \rho_C$, in the metallic state [128]. $\rho_A = \rho_{A'}$ shown in parentheses in Table 3 is calculated assuming this order. This assumption is consistent with the assignment that ν_2^2 is missing in α-I_3 (See Figure 20a,a'). These hole numbers are compared with the corresponding values estimated from the infrared-active ν_{27} mode [130], assuming a linear relationship [155].The site charge was calculated by Wojciechowski et al., neglecting Coulomb interaction [111]. Katayama et al., examined the effect of the anion potential, and they showed that anion potential has a small influence on the site charge [156]. Kobayashi et al., calculated the temperature dependence of the site hole number within the mean-field approximation for extended Hubbard model [157]. Ishibashi et al., made Mulliken charge analyses in their ab initio calculation of the band structure of α-I_3 [158]. Recently, the site hole numbers were reported by Alemany using the DFT approach considering only the valence electrons [129]. In Table 3, their results are compared with the experimental results. The mean-field calculation overestimates the site-charge difference between the sites B and C. The simplified Coulomb energy parameters may be one of the reasons for the overestimation. In the ab initio and DFT calculations, this site-charge difference is more suppressed, and the agreement with experimental values is good taking the experimental errors into account.

Table 3. Site hole numbers of metallic α-(BEDT-TTF)$_2$I$_3$. MF denotes mean-field calculation, and DFT denotes density functional theory.

	Raman (150 K)	Infrared [130] (140 K)	X-ray [128] (150 K)	MF [157] (150 K)	*ab initio* [158] (298 K)	*ab initio* [158] (8 K)	DFT [129] (RT)
$\rho_A = \rho_{A'}$	(0.52(3))	~0.6	0.59(3)	0.52	0.53	0.54	0.521
ρ_B	0.60(3)	0.68	0.67(2)	0.71	0.55	0.57	0.546
ρ_C	0.37(3)	0.44	0.42(2)	0.25	0.40	0.37	0.382

In the same way, the site charges of metallic α-NH$_4$Hg at 10 K are estimated to be 0.57(3), 0.50(3), and 0.39(3) using the frequencies of ν_2 shown in Table 2. Although the corresponding site cannot be decided by vibrational spectroscopy, Hiejima et al., assigned the three infrared-active ν_{27} modes with the aid of ^{13}C-NMR, and they interpreted the site A to be most charge-poor [145]. Apart from the result of ^{13}C-NMR, the following argument also supports the above interpretation. In Section 5.2.2. we discussed the breaking of inversion symmetry in α-MHg (M = K, Rb, Tl), based on the growth of the forbidden mode. The forbidden mode is the out-of-phase mode between ν_{27A} and $\nu_{27A'}$ which are coupled through dipole-dipole interaction [145]. As the dipole is induced parallel to the long axis of molecule, the frequency of the out-of-phase mode is lower than that of the in-phase mode. The out-of-phase mode grows at a slightly lower frequency of the highest frequency mode among the three ν_{27} modes, in-phase mode, ν_{27B} and ν_{27C}. This means that the highest frequency mode is the in-phase mode of ν_{27A}, that is, the A site is the most charge-poor. As the other sites cannot be decided, the assignment of ρ_B and ρ_C in Table 4 is interchangeable. These hole numbers are compared with the corresponding values estimated from the infrared-active ν_{27} mode [130], assuming a linear relationship [155]. Table 4 shows the site-charge distribution of α-NH$_4$Hg calculated by the mean-field approximation with the transfer integrals given by Mori et al. [142] and Coulomb interaction parameters, U = 0.4, V_c = 0.17, V_p = 0.05 eV [157]. Foury-Leylekian et al., calculated the band structure of isostructural α-KHg at 104 K using a numerical atomic orbital DFT approach, and presented the site charges [147]. Their values are shown in Table 4 along with the hole numbers of isostructural compounds, α-MHg (M = K, Rb, Rl). The mean-field calculation underestimates the non-uniformity among site charges, while the first-principles calculation enhances the non-uniformity and approaches the experimental values. However, the most charge-poor site is the B site in both theoretical calculations. This result is different from the interpretation of the experimental result of the ν_{27} mode (See Figure 1 of ref. [145]).

Table 4. Site hole numbers of α-(BEDT-TTF)$_2$MHg(SCN)$_4$ (M = NH_4, K, Rb, Tl). MF denotes mean-field calculation.

	α-NH_4Hg			α-KHg		α-RbHg	α-TlHg
	Raman (10 K)	Infrared [130] (150 K)	MF (10 K)	Raman (10 K)	*ab initio* [147] (104 K)	Raman (10 K)	Raman (10 K)
$\rho_A = \rho_{A'}$	0.39(3)	0.40	0.51	0.42(3)	0.520	0.42(3)	0.42(3)
ρ_B	[0.50(3)	[0.56	0.47	[0.55(3)	0.424	[0.55(3)	[0.55(3)
ρ_C	0.57(3)]	0.62]	0.50	0.61(3)]	0.527	0.61(3)]	0.61(3)]

From the symmetry of the unit cell of θ-I_3, every site is approximately equivalent. Therefore, the site charge should be ρ = 0.5. Applying the empirical Equation (7), the site charge is calculated as ρ = 0.51(3) at 298 K (1493 cm^{-1}) and ρ = 0.54(3) (1496 cm^{-1}) at 20 K. The deviation from 0.5 at 20 K comes from the temperature dependence of the ν_2 mode that is different from that for κ-type ET salts which are used to obtain the first term of Equation (7).

5.3.3. Site-Charge Distribution under Hydrostatic Pressure

Tamura et al., showed that metallic θ-I_3 undergoes a first order phase transition showing a resistivity jump at around 0.5 GPa [159]. Tajima et al., further investigated the transport properties under hydrostatic pressure [160]. They found that the electrical resistance under 1 GPa behaves similar to that of α-I_3 at the same pressure. In addition, the Hall coefficient showed the same temperature dependence as that of α-I_3 at the same pressure. These two transport properties suggest a zero-gap state (ZGS) in θ-I_3 as well as in α-I_3 [120,121]. Recently, Miyagawa et al., investigated θ-I_3 under hydrostatic pressure using ^{13}C-NMR [161]. The behavior of the spin shift and spin-lattice relaxation rate is also consistent with ZGS. They suggested a structural transformation to the α-type structure at 0.5 GPa. Figure 23a shows the pressure dependence of the Raman spectrum of ν_2 and ν_3 measured at 50 K. The Raman spectrum shows a frequency upshift under hydrostatic pressure. However, no distinct change was observed, and the spectrum is different from that of α-I_3 under hydrostatic pressure (See

Figure 23a). This result suggests that high symmetry of the unit cell of θ-I_3 is preserved at least approximately up to 3.1 GPa. If this compound is in a ZGS above 0.5 GPa, there should be a contact point at the Fermi level. However, it is concluded from Figure 23a that the band structure of θ-I_3 at 0.5 GPa is quite different from that of α-I_3. The conservation of the pseudo-symmetry provides a restriction when we speculate the crystal and band structure under hydrostatic pressure.

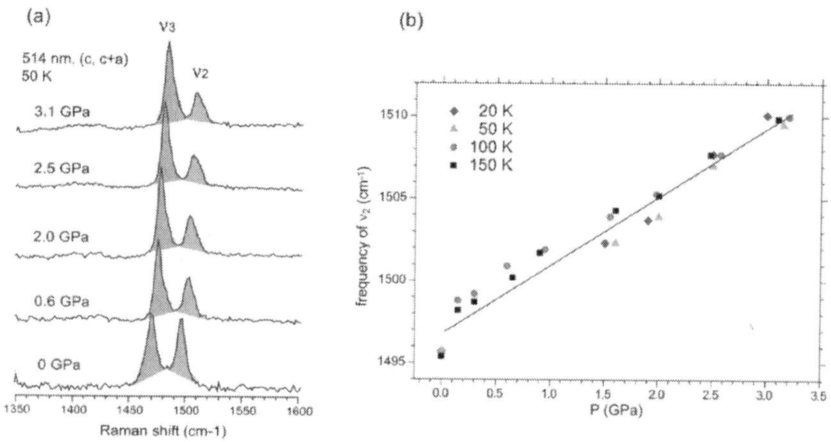

Figure 23. (**a**) Pressure dependence of the Raman spectrum of θ-(BEDT-TTF)$_2$I$_3$ measured at 50 K; (**b**) Pressure dependence of the frequency of ν_2 in θ-(BEDT-TTF)$_2$I$_3$.

When hydrostatic pressure is applied, the molecular vibration shows a frequency upshift ($\omega + \Delta\omega$) due to hardening of the lattice. In addition to the lattice phonon mode, the frequency shift against pressure is described by $\Delta\omega/\Delta P = \omega\gamma\kappa$, where γ and κ are the Gruneisen constant and compressibility, respectively. Therefore, the pressure dependent frequency is given by the equation, $\omega(P) = \omega(1 + \gamma\kappa P)$. It is found that $\gamma\kappa$ is approximately temperature independent within an error of 10% in the temperature range from 20 K to 150 K as shown in Figure 23b. Since the hydrostatic pressure induces no structural change in this compound, the parameter was estimated to be $\gamma\kappa \approx 3 \times 10^{-3}$ GPa^{-1} from Figure 23b.

Figure 24 shows the pressure dependence of the Raman spectrum of α-I_3 at 150 K, 100 K, and 20 K. At 150 K, the compound is metallic at ambient pressure. On increasing pressure, the Raman spectrum does not show a drastic change. At 100 K and 20 K, the compound is in a CO phase at ambient pressure. In this case, remarkable spectral changes are found to occur between 0.1 GPa and 0.45 GPa at 100 K and between 1.2 GP and 1.5 GPa at 20 K. These spectral changes correspond to the phase transition from the CO phase to the metallic phase, which is consistent with the suppression of the MI transition by hydrostatic pressure. In the metallic phase, the ν_2^1 and ν_2^2 modes do not show a parallel shift, rather they tend to merge on increasing the pressure. This trend is much more remarkable in the CO phase as shown in Figure 24c. Hydrostatic pressure not only hardens the lattice but also modulates the transfer integrals non-uniformly, which changes the site-charge distribution. After eliminating the lattice hardening effect assuming $\gamma\kappa \approx 3 \times 10^{-3}$ GPa^{-1} for α-I_3, the site charge is estimated using Equation (7). The results are shown in the bottom panel of each figure. In a CO phase, the amplitude of charge order decreases on increasing pressure [162]. This trend is probably associated with the increase of transfer integrals under hydrostatic pressure.

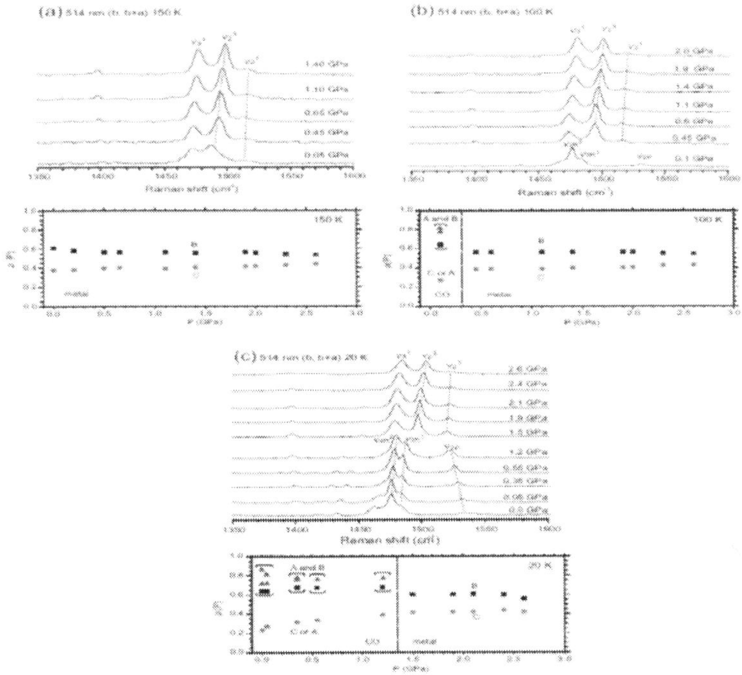

Figure 24. Pressure dependence of the Raman spectra and site charges of α-(BEDT-TTF)$_2$I$_3$ analyzed at (**a**) 150 K; (**b**) 100 K; and (**c**) 20 K. Ambiguity of the hole numbers at charge-rich site is very large, because ν_{2R} is mixed with ν_3 in the CO state. The symbols, ▲ and ■, in parentheses show the nominal site charge calculated by Equation (7). Figure 24 is modified from Figure 5 of ref. [111].

In a metallic phase, the site charges approach uniform charge distribution as in θ-I_3 due to a non-uniform increase in transfer integrals on increasing hydrostatic pressure. This result suggests that the crystal structure of the zero-gap state of α-I_3 is slightly modulated so as to equalize sites B and C. However, there still remains a site-charge difference in ZGS. Alemany et al., reported the band calculation of α-I_3 [129] using the atomic parameters at 1.76 GPa that is determined by Kondo et al. [163]. They used the generalized gradient approximation to DFT, and found the contact point at the Fermi level. According to their calculation, the site hole numbers changes from $\rho_A = \rho_{A'} = 0.521$, $\rho_B = 0.546$, $\rho_C = 0.382$ at ambient pressure to $\rho_A = \rho_{A'} = 0.576$, $\rho_B = 0.457$, $\rho_C = 0.361$ at 1.76 GPa, where the most charge-rich site is interchanged between A=A′ and B sites and the maximum amplitude is enlarged from $\Delta\rho_{max} = 0.164$ at ambient pressure to $\Delta\rho_{max} = 0.215$. This change is opposite to the above mentioned trend shown in Figure 24a.

5.4. Optical Conductivity of *α-(BEDT-TTF)$_2$I$_3$*

The optical conductivity of α-I_3 has been published by several groups. Yakushi et al., reported optical conductivity from 720 cm^{-1} to 25000 cm^{-1} [164]. They found a drastic change in reflectivity at T_{MI}, and presented a calculation of the optical conductivity in the metallic phase based on a tight-binding approximation neglecting Coulomb interaction. The polarization (anisotropic) ratio of the optical weight was qualitatively reproduced. Zelezny et al., reported temperature dependent reflectivity in the range of 10–700 cm^{-1}, and transmittance in the 10–33 cm^{-1} range [165]. They found a gap-like decrease in optical conductivity below 100 cm^{-1} in the CO phase, and they found a lattice phonon at 31.7 cm^{-1} in the E||b spectrum at 20 K. Dressel et al., reported optical conductivity in the range of 15–5500 cm^{-1} along with temperature-dependent microwave conductivity (10, 12, 35, 60, and 100 GHz) [166]. They determined the optical gap to be 400 cm^{-1}, and found a strong phonon at 35 cm^{-1} below T_{MI}. The similarity to the CDW system was pointed out from the experimental findings of non-linear transport, high dielectric permittivity, and 35 cm^{-1} band, although α-I_3 is not a CDW system. More recently, Clauss et al., reported the optical conductivity in the range of 10–10000 cm^{-1} [133,167]. They found no Drude response above T_{MI}, and they determined the CO gap to be 600 cm^{-1} below T_{MI}. They estimated the effective transfer integrals comparing with theoretical calculation of the extended Hubbard model for a quarter-filled square lattice [40]. Figure 25 shows the reflectivity and optical conductivity of α-I_3 that is reported by Yue et al. [130]. The optical conductivity agrees well with that reported by Clauss et al. [133,167]. According to the theoretical model by Merino et al. [40], the spectral weight of the electronic excitation at ~6t (t is the transfer integral of square lattice) in the CO phase shifts to ~2t–3t with Drude response in the metallic phase. The corresponding peak shift is not clear in the E||b spectrum. As Clauss et al., pointed out, the spectral weight in the infrared region shifts to the gap region in the metallic phase. This behavior is typically found in the E||a spectrum. No Drude response was observed in the metallic phase. It is difficult to specify the cause of non-Drude behavior, because the metallic phase exists only at high temperature, at which thermal excitations collapse the quasi-particle band as described for the optical conductivity of θ-I_3in Section 4.3, and/or they blur the small semimetallic Fermi surface. Incidentally, a well-defined Drude response is observed in α-MHg (M = NH_4, K, Rb, Tl), because these compounds are a metallic down to low temperature [29,30].

The collective excitation in the CO state has been investigated in analogy with CDW. The experimental findings of large permittivity in the range of 50–50000 Hz and nonlinear conductivity of α-I_3 in the CO phase recalled the sliding of CDW [166]. The observation of the voltage oscillation in the 7–28 kHz region suggests collective transport [168]. A recent permittivity experiment from 1 Hz to 1 MHz indicates the existence of a large dielectric relaxation mode [133,169]. Ivek et al., interpreted this low-frequency dielectric relaxation mode to be a phason-like behavior. Dressel et al., reported an attempt to find a collective CO mode in the low-frequency region, and they assigned the strong 35 cm^{-1} band to a collective CO excitation [170]. However, the corresponding strong band is not indicated in ref. [133,167]. Yue et al., reported the transmittance of the single crystal of α-I_3.down to 28 cm^{-1} [130]. They confirmed the 31 cm^{-1} band which Zelezny et al., reported [165] but did not find such a strong band at 35 cm^{-1}. More extensive experimental and theoretical studies will be necessary to establish collective CO excitation.

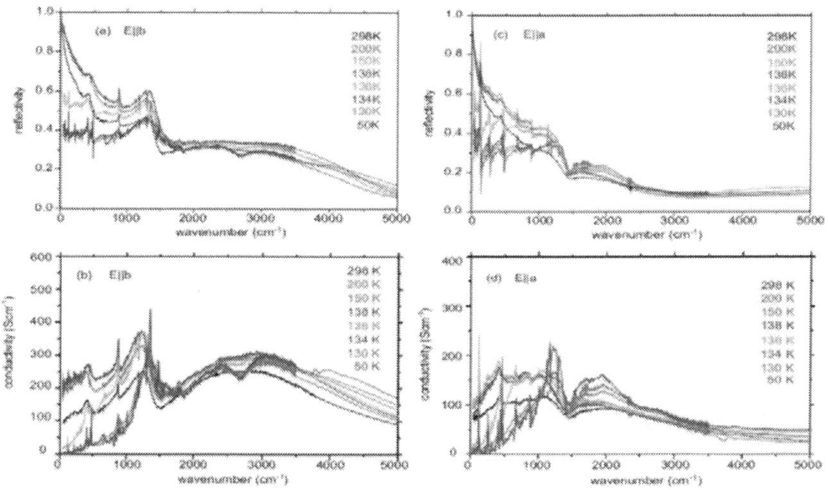

Figure 25. Temperature dependent reflectivity and optical conductivity of α-$(BEDT\text{-}TTF)_2I_3$. The optical conductivities are taken from ref. [130].

5.5. α′-$(BEDT\text{-}TTF)_2IBr_2$

α′-$(BEDT\text{-}TTF)_2IBr_2$ (abbreviated as α′-IBr_2) has a layered structure similar to α-I_3, but it is not isostructural with the α-type BEDT-TTF salts [171]. Similar to α-I_3, the unit cell contains four molecules with a herringbone arrangement. Unlike α-I_3, on the other hand, the unit cell contains two crystallographically independent groups (A, A′) and (B, B′), in which A and A′ (B and B′) are connected by inversion symmetry within each group. The band structure is calculated to be semimetallic, and the bandwidth of α′-IBr_2 is the narrowest among the α-type BEDT-TTF salts [171]. Reflecting the narrow bandwidth, α′-IBr_2 is semiconducting (ρ_{RT}~0.3 Ωcm) with an activation energy of ~0.16 eV at around room temperature. The electrical resistivity increases steeply by an order of magnitude at around 200 K, and this compound moves into a high-resistivity state with the same activation energy [172]. The phase transition of α′-IBr_2 was investigated by means of NMR [173] and x-ray diffraction [171,174] experiments. The NMR study suggested that the thermal motion of the ethylene groups of BEDT-TTF begins to freeze gradually at around 250 K [173]. X-ray diffraction study confirmed this suggestion analyzing the temperature factors of the carbon atoms in one of the ethylene groups [171]. The ordering of the ethylene group occurs continuously from room temperature down to 70 K. Except for this ordering, no distinct structural change was reported above and below 200 K. Neither superlattice spots nor diffuse x-ray scattering has been found down to 20 K [171]. The magnetic susceptibility also shows no clear anomaly at around 200 K [172]. Inokuchi et al., reported the optical conductivity and Raman spectrum of this compound, and they suggested that the unit cell involves two non-equivalent molecules with different valence [175]. X-ray diffraction study suggested that the electron-phonon interaction is extremely weak in this system.

Yue et al., investigated the infrared and Raman spectra of α'-IBr_2, and assigned the Raman spectrum with the aid of polarization dependence and ^{13}C-substituted compounds [176]. Based on the assignment, they showed clear evidence that the low-temperature high-resistivity phase is a charge-ordered state. Combined with the ν_{27} mode, they estimated the site charge distribution to be (0.9, 0.8, 0.2, 0.1), and thus the maximum CO amplitude is very large, $\Delta\rho_{max}$ = 0.8. Although the side band of ν_{27R}^2 (indicated by an arrow in Figure 26) is assigned to CH_2 bending mode, it may be assigned to ν_{27R}^1 instead of another side band, because the low-frequency side band is observed in the deuterium-substituted compound. In this case, the site charge distribution and $\Delta\rho_{max}$ are estimated to be (1.0, 0.8, 0.2, 0.1) and $\Delta\rho_{max}$ = 0.9. Furthermore, they found that the inversion symmetry is broken in the CO state. If the unit cell has inversion symmetry, the ν_2 mode is classified into two Raman-active (in-phase) modes and two infrared-active (out-of-phase) modes. In the CO phase, they found four ν_2 modes in the Raman spectrum of the ^{13}C-substituted α-IBr_2. They took notice of the stretching vibration of IBr_2^- as a marker band for the inversion symmetry. However, the splitting of this band seems not to be decisive to judge the breaking of inversion symmetry, as the frequency of this mode is comparable with the lattice phonon. Since IBr_2^- is not located on the center of symmetry in the unit cell, the intramolecular vibration is influenced by the non-centrosymmetric crystal field. The ν_2 or ν_3 mode is much safer as a marker band. Taking the symmetry into account, they proposed two possible arrangements, diagonal stripes along the a-b and b axes, as the CO pattern. These two stripes are probably competing and generate frustration in this compound.

Yue et al., analyzed the resistivity and magnetic susceptibility [58]. They found that the resistivity curve is approximately expressed by a single exponential curve below ~210 K, whereas it cannot be expressed by a single exponential curve above ~210 K, where the activation energy exponentially decreases against 1/T. The magnetic susceptibility is expressed by an alternating Heisenberg model down to 30 K. At around 30 K this compound abruptly loses susceptibility and becomes non-magnetic with hysteresis [177]. They pointed out that every crystal cracked when the crystal was cooled across this transition temperature. As the deuterium substituted α'-IBr_2 does not show the 30 K phase transition, the interaction with IBr_2^- may be associated with this magnetic phase transition. Yamamoto et al., found an onset of SHG at 160 K using an SHG microscope, and they argued for the growth of ferroelectric polarization below 160 K [136,178]. Therefore, this compound has three phase transitions at ~210 K, ~160 K, and ~30 K. This behaviour is different from that of α-I_3, in which all of these transitions occur at the same temperature (~135 K).

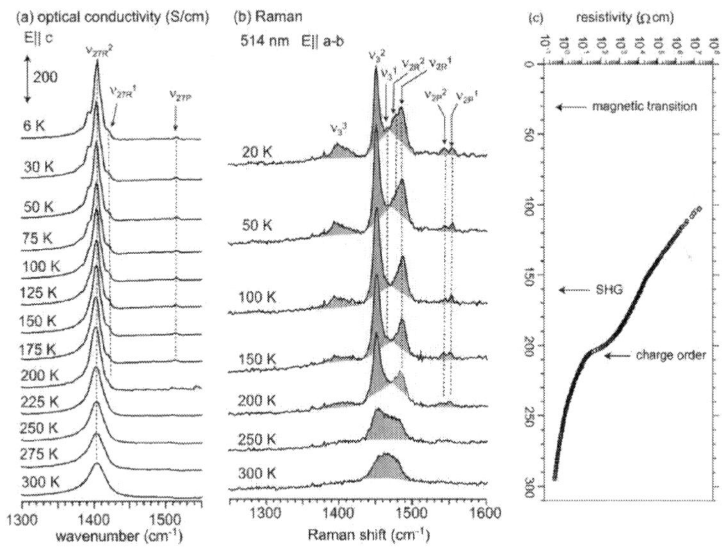

Figure 26. Temperature dependences of the (**a**) Infrared-active ν_{27} mode; (**b**) Raman-active ν_2 and ν_3 modes; and (**c**) electrical resistivity in α'-(BEDT-TTF)$_2$IBr$_2$. See ref. [176] for the argument of assignment. Figure 26a,c are taken from ref. [58].

Figure 26 shows the infrared-active ν_{27} mode and Raman-active ν_2 and ν_3 modes along with the electrical resistivity [58]. The charge-sensitive ν_{27} and ν_2 modes are already split at room temperature, and the spectral change against temperature is continuous as in θ-$Cu_2(CN)[N(CN)_2]_2$ and θ_m-TlZn. It is therefore obvious that the line shapes of the ν_{27} and ν_2 modes reflect the case of $\tau\Delta\nu >> 1$ over the whole temperature range. On increasing temperature, as shown in Figure 26b, the ν_3 mode is weakened toward the transition temperature at ~210 K. This spectral change suggests shortening of the coherence length ζ of the CO domain [58]. The optical conductivity has a clear gap below 200 K, whereas it has finite density of states above

210 K. The temperature dependence is continuous as for θ_m-TlZn. In the limit of $\zeta = 0$, the individual holes, perhaps polaron, are randomly hopping at high temperature. Assuming this extreme case, Yue et al., interpreted the phase transition at 210 K as an order-disorder transition [58]. They qualitatively explained the optical conductivity change based on the order-disorder model of the atomic limit. According to this simple model, the finite density of states above 210 K is generated by the disordered configurations of localized holes, the energy of which is given by the coordination number around the hole [58,177]. The electrical and magnetic properties and spectroscopic feature closely resemble those of θ-$Cu_2(CN)[N(CN)_2]_2$, hence the high-temperature phase might be understood in the same way as θ-$Cu_2(CN)[N(CN)_2]_2$, in which the short-range CO domain grows extensively. However, strong frustration perhaps suppresses the growth of short-range order in this compound.

This simple model cannot explain the difference in temperature between the order-disorder (~210 K) and ferroelectric (~160 K) transitions. In the atomic limit, the Coulomb repulsion along two diagonal stripes is nearly the same. These nearly degenerate short-range CO domains compete with each other and generate frustration, when the coherence length of CO increases. This frustration may cause the above successive ferroelectric transition. Recently, weak superlattice spots for a × b × 2c were found in an x-ray study conducted at BL-8A beamline of the Photon Factory Facility, KEK [179]. Based on the structural modulation, an antiferroelectric state is proposed as the intermediate phase (160–210 K) before the ferroelectric transition.

Yue et al., estimated the kinetic energy from the integration of optical conductivity [177], which is 0.1, 0.3, and 0.4 eV per unit cell for α′-IBr_2, α-I_3, and α-NH_4Hg, respectively. This result indicates that the transfer integrals of α′-IBr_2 are very small compared with those of α-I_3, and α-NH_4Hg. Therefore, in the case of α′-IBr_2, the spin degree of freedom survives after the CO transition due to the small spin gap (~82 K) coming from the small exchange energy along the stripe. On the other hand, in α-I_3, the exchange energy is larger and the dimerization along the stripe is larger than that in α′-IBr_2, and furthermore, the dimerization is enhanced below the phase transition of α-I_3 [128,171]. These two factors result in a large spin gap (440 K) as reported by Sugano et al. [180]. The difference in the magnitude of spin gap is the cause of the simultaneous transition of charge order and non-magnetic state. The difference in ferroelectric transition is not clear at the moment. Possibly, the frustration plays a role in the successive transition to the ferroelectric state in α′-IBr_2.

6. SUMMARY

The charge-ordering phase transition arises from the competition between the kinetic energy gain and the cost of on-site and nearest-neighbor inter-site Coulomb energy. In some compounds such as θ-type BEDT-TTF salts, geometrical frustration stabilizes the metallic state, and lattice distortion sometimes plays a dominant role for dissolving the frustration. Even in such a case, the transfer integral plays an essential role for the properties of the materials [78]. Transfer integrals depend considerably on the counter anions and temperature, whereas Coulomb interaction is almost insensitive to such a modification. The rough values of the kinetic energy per unit formula (=hole) are shown in Figure 27, which are estimated from the room-temperature optical conductivity [181]. The kinetic energy is a stabilization energy due to the delocalization of the valence electron over the crystal. The kinetic energy is associated with the bandwidth: the wider the bandwidth, the larger the kinetic energy. The kinetic energy including the correlation effect is experimentally estimated from the optical conductivity [182]. The kinetic energy $-E_K$ was calculated in the same way as reported by Yue et al. [177].

The kinetic energy of β″-(BEDT-TTF)(TCNQ) is comparable with those of θ_m-TlZn and θ_o-TlZn, which undergo a metal-insulator transition while maintaining a large CO amplitude ($\Delta\rho\sim0.6$) over the whole temperature range. In contrast to the θ-type salts, this compound continuously changes the spectral feature from a large CO amplitude ($\Delta\rho\sim0.6$) to a small amplitude ($\Delta\rho\sim0$) on lowering the temperature. Experimentally, the split charge-sensitive modes merge into a single mode in the low-temperature region. The transport properties below 20 K indicate Fermi-liquid behavior. It is speculated from the optical conductivity and charge-sensitive modes that the coherent state below 20 K collapses above 20 K and continuously changes into a charge-ordered metal with a pseudogap. As the hydrostatic pressure reproduces the same spectral change for the Raman-active mode, the variation of the transfer integral due to lattice contraction has a big influence on the electronic state of this compound.

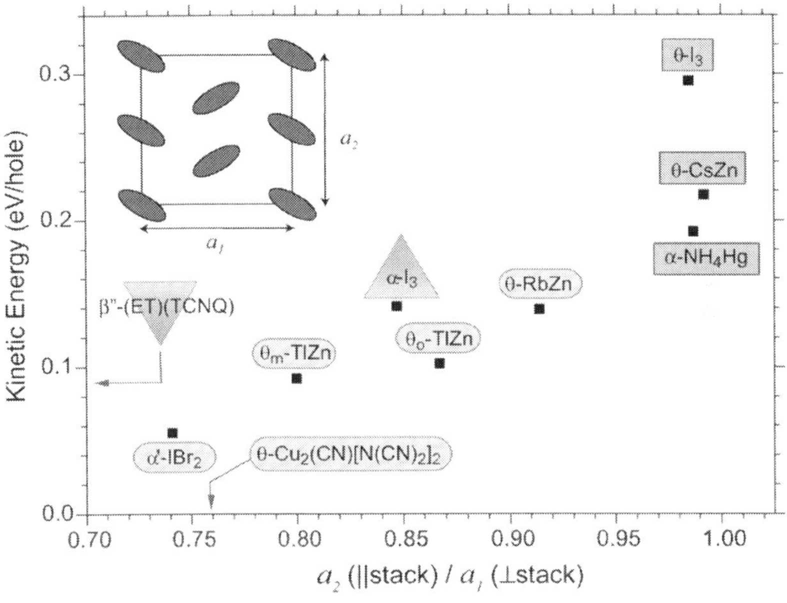

Figure 27. Kinetic energy per a formula unit (■) plotted as a function of the lattice parameter ratio. In the case of the θ-type crystal, the crystal axis along the stacking direction is doubled. The inset shows the schematic molecular arrangement and crystal axes. In the compounds framed by the green rectangle, the CO amplitude is small ($\Delta\rho \sim 0.1$) over the whole temperature range. In the compounds framed by the yellow rounded rectangle, the CO amplitude is large ($\Delta\rho \sim 0.6$) over the whole temperature range. In the compound framed by the triangle, the CO amplitude changes at the phase transition, small amplitude ($\Delta\rho \sim 0.2$) at high temperature and large amplitude ($\Delta\rho_{max} \sim 0.6$) at low temperature. In the compound framed by the inverted triangle, the amplitude changes continuously, large amplitude at high temperature ($\Delta\rho \sim 0.6$) and small amplitude ($\Delta\rho \sim 0$) at low temperature.

The four θ-type BEDT-TTF salts, θ-RbZn, θ$_o$-TlZn, θ$_m$-TlZn, and θ-$Cu_2(CN)[N(CN)_2]_2$, (yellow color in Figure 27) show an MI transition. However, they maintain a large CO amplitude ($\Delta\rho \sim 0.6$) in both metallic and insulating phases. In the CO phase of the θ-type BEDT-TTF salts, the CO pattern is a horizontal stripe in θ-RbZn, θ$_o$-TlZn, and θ-$Cu_2(CN)[N(CN)_2]_2$, as concluded by the selection rule of the ν_3 modes. On the other hand, the diagonal stripe is claimed in θ$_m$-TlZn. In contrast to the α-type BEDT-TTF salts, the θ-type compounds except θ$_m$-TlZn have a non-polar space group $P2_12_12_1$ in the CO phase. Probably, the θ-type compounds except for θ$_m$-

TlZn do not have ferroelectric polarization in the CO phase in spite of the non-centrosymmetric space group. As the CO amplitude is large even in the metallic phase, this phase is regarded as a CO metal. However, the compound can be classified into two groups, [θ-RbZn, θ_o-TlZn] and [θ_m-TlZn, θ-$Cu_2(CN)[N(CN)_2]_2$]. The phase transition in the former is discontinuous, whereas it is continuous in the latter. The activation energy of the resistivity in the latter is significantly larger, and the precursor ν_3 bands are found in the latter. In the metallic phase of the latter group, short-range stripe CO probably grows more extensively than in the former group.

The kinetic energy of θ-CsZn and θ-I_3 is significantly larger than the others among the θ-type BEDT-TTF salts (green color in Figure 27). The CO amplitude of these compounds is much smaller than in the others. The amplitude of the former is $\Delta\rho < 0.08$, and $\Delta\rho < 0.07$ in the latter. An indication of CO fluctuation is found in the Raman spectrum and optical conductivity of θ-CsZn, in which the frustration between the stripe and threefold COs is dominant. θ-I_3 has a zero-energy (Drude) peak in the optical conductivity at 4 K, while the coherent state collapses at 100 K.

The kinetic energy of α-I_3 is comparable with that of θ-RbZn. The amplitude of CO discontinuously changes from a small amplitude (at high temperature) to a large amplitude (at low temperature) at the MI transition temperature. Below the MI transition temperature, the localized charge is arranged in a horizontal stripe involving two charge-rich sites and two charge-poor sites with an amplitude of $\Delta\rho \approx 0.4$–0.6. In the CO phase, the inversion symmetry is broken, and strong SHG is found. Observation of a macroscopic polar (ferroelectric) domain is demonstrated by SHG interferometry. A fast optical response is also reported. This unconventional ferroelectricity is called electronic ferroelectricity. In the metallic phase, the broad linewidth of the charge-sensitive mode indicates fluctuation of the CO. This fluctuation is well suppressed by hydrostatic pressure.

The kinetic energy of α-NH_4Hg is significantly larger than that of α-I_3. The linewidth of the charge-sensitive mode is narrow in α-MHg (M = NH_4, K, Rb, Tl). Among these four compounds, the inversion symmetry is locally broken in α-MHg (M = K, Rb, Tl). This phenomenon is attributed to the anion-sublattice coupled incommensurate CDW. The amplitude of the CDW is estimated to be $\Delta\rho < 0.05$. On the other hand, the fluctuation of CO is indicated in the Raman spectrum of α-NH_4Hg which undergoes superconducting state at 1 K.

α-I_3 in the metallic phase and α-NH_4Hg involve three non-equivalent sites in the unit cell. Reflecting the non-uniform site, the charge-sensitive mode split into three. The maximum site-charge difference is estimated to be $\Delta\rho_{max}$~0.2 both in metallic α-I_3 and α-NH_4Hg. The most charge-poor site is the C site in α-I_3 and the A site in α-NH_4Hg. First principle calculation agrees with the experimental results for α-I_3. In α-NH_4Hg, however, there is disagreement in the most charge-poor site, between the band calculation and the interpretation of the charge-sensitive mode. The site charge distribution under hydrostatic pressure is estimated for α-I_3 and θ-I_3. In the metallic phase, the site charge distribution approaches a uniform charge distribution on increasing pressure. However, a non-uniform site charge distribution remains in the zero-gap state of α-I_3. The zero-gap state of θ-I_3 under hydrostatic pressure keeps uniform charge distribution unlike in α-I_3.

The kinetic energy of α'-IBr_2 is smallest among the compounds discussed in this paper, and probably comparable with that of θ-$Cu_2(CN)[N(CN)_2]_2$. The electrical and magnetic properties and spectroscopic feature closely resemble each other. The ground state is a non-magnetic CO state with large amplitude ($\Delta\rho \approx 0.6$–0.8). α'-IBr_2 shows successive transitions at ~210 K and ~160 K and magnetic transition at ~30 K. The large CO amplitude is maintained over the whole temperature range. The ordering of charge sets in at ~210 K, the ferroelectric domain begins to grow at ~160 K, and spin singlet state is formed at ~30 K accompanied by structural distortion.

ACKNOWLEDGMENTS

The author deeply thanks all the collaborators for their invaluable contribution to this work. In particular, collaborations and discussions with K. Yamamoto, M. Uruichi, and T. Yamamoto of Institute for molecular science, A. Kawamoto of Hokkaido University, and H. M. Yamamoto and R. Kato of RIKEN are highly appreciated. This work was partly supported by a Grant-in-Aid for Scientific Research (Grant No. 19350074) from MEXT Japan.

REFERENCES AND NOTES

1. Girlando, A.; Bozio, R.; Pecile, C. Infrared and Raman spectroscopic evidence of ground state charge densities at TCNQ sites in crystalline Cs2TCNQ3. Chem. Phys. Lett. 1974, 25, 409–412.

2. Bozio, R.; Girlando, A.; Pecile, C. Vibrational analysis of spectra of quinoid molecular ions. J. Chem. Soc. Faraday Trans. 1975, 71, 1237–1254.
3. Matsuzaki, S.; Kuwata, R.; Toyoda, K. Raman spectra of conducting TCNQ salts; estimation of the degree of charge transfer from vibrational frequencies. Solid State Commun. 1980, 33, 403–405.
4. Chappell, J.S.; Bloch, A.N.; Bryden, W.A.; Maxfield, M.; Poehler, T.O.; Cowan, D.O. Degree of charge transfer in organic conductors by infrared absorption spectroscopy. J. Am. Chem. Soc. 1980, 103, 2442–2443.
5. Kagoshima, S.; Ishiguro, T.; Anzai, H. X-ray scattering study of phonon anomalies and superstructures in TTF-TCNQ. J. Phys. Soc. Jpn. 1976, 41, 2061–2071.
6. Kondo, J.; Yamaji, K. Density correlation of classical 1-d electron gas with reference to the 4κF anomaly in TTF-TCNQ. J. Phys. Soc. Jpn. 1977, 43, 424–436.
7. Torrance, J.B. Spin waves, scattering at 4κF, and spin-Peierls fluctuations in organic metal: Tetrathiafulvalene-Tetracyanoquinodimethane (TTF-TCNQ). Phys. Rev. 1978, 17, 3099–3103.
8. Hubbard, J. Generalized wigner lattices in one dimension and some applications to tetracyanoquinodimethane (TCNQ) salts. Phys. Rev. 1978, 17, 494–505.
9. Hiraki, K.; Kanoda, K. Wigner crystal type of charge ordring in an organic conductor with a quarter-filled band: (DI-DCNQI)2Ag. Phys. Rev. Lett. 1998, 80, 4737–4740.
10. Seo, H.; Fukuyama, H. Antiferromagntic phases of one-dimensional quarter-filled organic conductors. J. Phys. Soc. Jpn. 1997, 66, 1249–1252.
11. Chow, D.S.; Zamborszky, F.; Alavi, B.; Tantillo, D.J.; Baur, A.; Merlic, C.A.; Brown, S.E. Charge ordering in the TMTTF family of molecular conductors. Phys. Rev. Lett. 2000, 85, 1698–1701.
12. Moldenhauer, J.; Horn, C.; Pokhondnia, K.I.; Schweitzer, D.; Heinen, I.; Keller, H.J. FT-IR absorption spectroscopy of BEDT-TTF radical salts: Charge transfer and donor-anion interaction. Synth. Metal. 1993, 60, 31–38.
13. Kino, H.; Fukuyama, H. On the phase transition of α-(ET)2I3. J. Phys. Soc. Jpn. 1995, 64, 1877–1880.

14. Takano, Y.; Hiraki, K.; Yamamoto, H.M.; Nakamura, T.; Takahashi, T. Charge disproportionation in the organic conductor, α-(BEDT-TTF)2I3. J. Phys. Chem. Solid 2001, 62, 393–395.
15. Miyagawa, K.; Kawamoto, A.; Kanoda, K. Charge ordering in a quasi-two-dimensional organic conductor. Phys. Rev. 2000, 62, R7679–R7682.
16. Seo, H. Charge ordering in organic ET compounds. J. Phys. Soc. Jpn. 2000, 69, 805–820.
17. McKenzie, R.H.; Merino, J.; Marston, J.B.; Sushkov, O.P. Charge ordering and antiferromagnetic exchange in layered molecular crystals of the θ-type. Phys. Rev. 2001, 64, 085109:1–085109:11.
18. Kuroki, K. Theoretical aspects of charge correlations in θ-(BEDT-TTF)2X. Sci. Technol. Adv. Mater. 2009, 10, 024312:1–024312:11.
19. Mazumdar, S.; Clay, R.T.; Campbell, D.K. Bond-order and charge-density waves in the isotropic interacting two-dimensional quarter-filled band and the insulating state proximate to organic superconductivity. Phys. Rev. 2000, 62, 13400–13425.
20. Clay, R.T.; Mazumdar, S.; Campbell, D.K. Charge ordering in θ-(BEDT-TTF)2X materials. J. Phys. Soc. Jpn. 2002, 71, 1816–1819.
21. Dayal, S.; Clay, R.T.; Li, H.; Mazumdar, S. Paired-electron crystal: Order from frustration in the quarter-filled band. Phys. Rev. 2011, 83, 245106:1–245106:12.
22. Merino, J.; McKenzie, R.H. Superconductivity mediated by charge fluctuations in layered molecular crystals. Phys. Rev. Lett. 2001, 87, 237002:1–237002:4.
23. Cano-Cortes, L.; Merino, J.; Fratini, S. Quantum critical behavior of electrons at the edge of charge order. Phys. Rev. Lett. 2010, 105, 036405:1–036405:4.
24. Takahashi, T.; Nogami, Y.; Yakushi, K. Charge ordering in organic conductors. J. Phys. Soc. Jpn. 2006, 75, 051008:1–051008:17.
25. Seo, H.; Merino, J.; Yoshioka, H.; Ogata, M. Theoretical aspects of charge ordering in molecular conductors. J. Phys. Soc. Jpn. 2006, 75, 051009:1–051009:20.
26. Dressel, M.; Drichko, N. Optical properties of two-dimensional organic conductors: Signatures of charge ordering and correlation effects. Chem. Rev. 2004, 104, 5689–5715.

27. Dressel, M. Quantum criticality in organic conductors? Fermi liquid versus non-Fermi-liquid behavior. J. Phys. Condens. Matter 2011, 23, 293201:1–293201:21.
28. Nad, F.; Monceau, P. Dielectric response of the charge ordered state in the quasi-one-dimensional conductors. J. Phys. Soc. Jpn. 2006, 75, 051005:1–051005:12.
29. Dressel, M.; Drichko, N.; Schlueter, J.; Merino, J. Proximity of the layered organic conductors α-(BEDT-TTF)2MHg(SCN)4 (M = K, NH4) to a charge-ordering transition. Phys. Rev. Lett. 2003, 90, 167002:1–167002:4.
30. Drichko, N.; Dressel, M.; Kuntscher, C.A.; Pashkin, A.; Greco, A.; Merino, J.; Schlueter, J. Electronic properties of correlated metals in the vicinity of a charge-order transition: Optical spectroscopy of α-(BEDT-TTF)2MHg(SCN)4 (M = NH4, Rb, Tl). Phys. Rev. 2006, 74, 235121:1–235121:11.
31. Kaiser, S.; Dressel, M.; Sun, Y.; Greco, A.; Schlueter, J.A.; Gard, G.L.; Drichko, N. Bandwidth tuning triggers interplay of charge order and superconductivity in two-dimensional organic materials. Phys. Rev. Lett. 2010, 105, 206402:1–206402:4.
32. Tanaka, M.; Yamamoto, K.; Uruichi, M.; Yamamoto, T.; Yakushi, K.; Kimura, S.; Mori, H. Infrared and Raman study of the charge-ordered state in the vicinity of the superconducting state in the organic conductor β-(meso-DMBEDT-TTF)2PF6. J. Phys. Soc. Jpn. 2008, 77, 024714:1–024714:8.
33. Morinaka, N.; Takahashi, K.; Chiba, R.; Yoshikane, F.; Niizeki, S.; Tanaka, M.; Yakushi, K.; Koeda, M.; Hedo, K.; Fujiwara, T.; et al. Superconductivity competitive with checkerboard-type charge ordering in the organic conductor β-(meso-DMBEDT-TTF)2PF6. Phys. Rev. 2009, 80, 092508:1–092508:4.
34. Girlando, A.; Masino, M.; Kaiser, S.; Sun, Y.; Drichiko, N.; Dressel, M.; Mori, H. Spectroscopic characterization of charge order fluctuations in BEDT-TTF metals and superconductors. Phys. Status Solidi 2012, 249, 953–956.
35. Kobayashi, A.; Katayama, S.; Suzumura, Y. Superconductivity in charge ordered metal for quasi-two-dimensional organic conductor. J. Phys. Soc. Jpn. 2005, 74, 2897–2900.
36. Pouget, J.-P. Bond and charge ordering in low-dimensional organic conductors. Physica B 2012, 407, 1762–1770.

37. Pouget, J.-P.; Foury-Leylekian, P.; Alemany, P.; Canadell, E. Charge ordering in low dimensional organic conductors: Structural aspects. Phys. Status Solidi 2012, 249, 937–942.
38. Ouyang, J.; Yakushi, K.; Misaki, Y.; Tanaka, K. Raman spectroscopic evidence for the charge disproportionation in a quasi-two-dimensional organic conductor θ-(BDT-TTP)2Cu(NCS)2. Phys. Rev. 2001, 63, 054301:1–054301:6.
39. Yakushi, K.; Yamamoto, K.; Simonyan, M.; Ouyang, J.; Nakano, C.; Misaki, Y.; Tanaka, K. Charge-ordering and magnetic phase transitions in θ-(BDT-TTP)2Cu(NCS)2. Phys. Rev. 2002, 66, 235102:1–235102:5.
40. Merino, J.; Greco, A.; McKenzie, R.H.; Calandra, M. Dynamical properties of a strongly correlated model for quarter-filled layered organic molecular crystals. Phys. Rev. 2003, 68, 245121:1–245121:15.
41. Seo, H.; Tsutsumi, K.; Ogata, M.; Merino, J. Charge fluctuation in geometrically frustrated charge ordering system. J. Phys. Soc. Jpn. 2006, 75, 114707:1–114707:8.
42. Kozlov, M.E.; Pokhondnia, K.I.; Yurchenko, A.A. The assignment of fundamental vibrations of BEDT-TTF and BEDT-TTF-d8. Spectrochim. Acta 1987, 43A, 323–329.
43. Kozlov, M.E.; Pokhondnia, K.I.; Yurchenko, A.A. Electron molecular vibration coupling in vibrational spectra of BEDT-TTF based radical cation salts. Spectrochim. Acta 1989, 45A, 437–444.
44. Eldridge, J.E.; Homes, C.C.; Williams, J.M.; Kini, A.M.; Wang, H.H. The assignment of the normal modes of the BEDT-TTF electron-donor molecule using the infrared and Raman spectra of several isotopic analogs. Spectrochim. Acta 1995, 51A, 947–960.
45. Wang, H.H.; Ferraro, J.R.; Williams, J.M.; Geiser, U.; Schlueter, J.A. Rapid Raman spectroscopic determination of the stoichiometry of microscopic quantities of BEDT-TTF-based organic conducors and superconductors. J. Chem. Soc. Chem. Commun. 1994, 1893–1894.
46. Yamamoto, T.; Uruichi, M.; Yamamoto, K.; Yakushi, K.; Kawamoto, A.; Taniguchi, H. Examination of the charge-sensitive vibrational modes in bis(ethylenedithio)tetrathiafulvalene. J. Phys. Chem. 2005, 109, 15226–15235.

47. Although frequency is proportional to the square root of force constant, the frequency practically linearly depends upon the force constant in a narrow frequency range, $\Delta\omega = \omega(0) - \omega(1)$
48. Rice, M.J.; Lipari, N.O.; Strassler, S. Dimerized organic linear-chain conductors and the unambiguous experimental determination of electron-molecular-vibration coupling constants. Phys. Rev. Lett. 1977, 21, 1357–1362.
49. Rice, M.J. Towards the experimental determiation of the fundamental microscopic parameters of organic ion-radical compounds. Solid State Commun. 1979, 31, 93–98.
50. Girlando, A.; Bozio, R.; Pecile, C.; Torrance, J.B. Discovery of vibronic effects in the Raman spectra of mixed-stack charge-transfer crystals. Phys. Rev. 1982, 26, 2306–2309.
51. Painelli, A.; Girlando, A. Electron-molecular vibration (e-mv) coupling in charge-transfer compounds and its consequences on the optical spectra: A theoretical framework. J. Chem. Phys. 1986, 84, 5655–5671.
52. Yamamoto, K.; Yakushi, K. Electron-molecular vibration coupling effect on the Raman spectrum of organic charge transfer salts. J. Phys. (France) 2004, 114, 153–155.
53. Visentini, G.; Masino, M.; Bellitto, C.; Girlando, A. Experimental determination of BEDT-TTF+ electron-molecular vibration coupling constants through optical microreflectance. Phys. Rev. 1998, 58, 9460–9467.
54. Yartsev, V.M.; Graja, A. Electron-intramolecular vibration coupling in charge-transfer salts studied by infrared spectroscopy. Int. J. Mod. Phys. 1998, 12, 1643–1672.
55. Faltermeier, D.; Barz, J.; Dumm, M.; Dressel, M.; Drichko, N.; Petrov, B.; Semkin, V.; Vlasova, R.; Meziere, C.; Batil, P. Bandwidth-controlled Mott transition in κ-(BEDT-TTF)2Cu[N(CN)2]-BrxCl1−x: Optical studies of localized charge excitations. Phys. Rev. 2007, 76, 165113:1–165113:12.
56. Girlando, A. Charge sensitive vibrations and electron-molecular vibration coupling in bis(ethylenethio)-thetrathiafulvalene (BEDT-TTF). J. Phys. Chem. 2011, 115, 19371–19378.
57. Yamamoto, T.; Yakushi, K.; Shimizu, Y.; Saito, G. Infrared and Raman study of the charge-ordered state of θ-(ET)2Cu(CN)[N(CN)2]2. J. Phys. Soc. Jpn. 2004, 73, 2326–2332.

58. Yue, Y.; Nakano, C.; Yamamoto, K.; Uruichi, M.; Wojciechowski, R.; Inokuchi, M.; Yakushi, K.; Kawamoto, A. Charge order-disorder phase transition in α′-(Bis(ethylenedithio)tetrathiafulvalene)2 IBr2 [α′-(BEDT-TTF)2IBr2]. J. Phys. Soc. Jpn. 2009, 78, 044701:1-044701:10.
59. Kubo, R. A stochastic theory of line shape. Adv. Chem. Phys. 1969, 15, 101–127.
60. Sue, J.; Yan, Y.J.; Mukamel, S. Raman excitation profiles of polyatomic molecules in condensed phases. A stochastic theory. J. Chem. Phys. 1986, 85, 462–474.
61. Uruichi, M.; Yakushi, K.; Yamamoto, H.M.; Kato, R. Infrared and Raman studies of the charge-ordering phase transition at ~170 K in the quarter-filled organic conductor, β″-(ET) (TCNQ). J. Phys. Soc. Jpn. 2006, 75, 074720:1-074720:10.
62. Yamamoto, H.M.; Hagiwara, M.; Kato, R. New phase of (BEDT-TTF)(TCNQ). Synth. Metal. 2003, 133, 449–451.
63. Nogami, Y. Private communication. 2009; Okayama University.
64. Yamamoto, H.M.; Tajima, N.; Hagiwara, M.; Kato, R.; Yamaura, J.I. Strange electric magnetic behaviour of new (BEDT-TTF)(TCNQ). Synth. Metal. 2003, 135, 623–624.
65. Yasuzuka, S.; Terakura, C.; Terashima, T.; Yakabe, T.; Terai, Y.; Yamamoto, H.M.; Kato, R.; Uji, S. Fermi surface and resistance anomalies in ET-TCNQ. Synth. Metal. 2003, 135, 647–648.
66. Yasuzuka, S.; Graf, D.; Choi, E.S.; Brooks, J.S.; Terashima, T.; Konoike, T.; Enomoto, K.; Nishimura, M.; Yamamoto, H.M.; Kato, R.; et al. Pressure-induced Fermi surface change in quasi-one-dimensional conductor β″-(ET)(TCNQ). J. Phys. IV Fr. 2004, 114, 157–158.
67. Yasuzuka, S.; Uji, S.; Enomoto, K.; Konoike, T.; Nishimura, M.; Terashima, T.; Graf, D.; Choi, E.S.; Brooks, J.S.; Yamamoto, H.M.; et al. Pressure effect on fermi surface in β″-(ET)(TCNQ). Synth. Metal. 2005, 152, 437–440.
68. Kimata, M.; Ohta, H.; Koyama, K.; Oshima, Y.; Motokawa, M.; Yamamoto, H.M.; Kato, R. Development of the high field magneto-optical measurement system with a rotational cavity for the study of organic conductors. Int. J. Mod. Phys. 2004, 18, 3803–3806.

69. Kimata, M.; Oshima, Y.; Ohta, H.; Koyama, K.; Motokawa, M.; Yamamoto, H.M.; Kato, R. Magnetooptical measurements of β″-(BEDT-TTF)(TCNQ). Physica B 2004, 346, 382–386.
70. Yakushi, K.; Uruichi, M.; Yamamoto, H.M.; Kato, R. Dynamical fluctuation of the site-charge density in metallic β″-(BEDT-TTF)(TCNQ). J. Phys. IV Fr. 2004, 114, 149–151.
71. Drichko, N.; Kaiser, S.; Sun, Y.; Clauss, C.; Dressel, M.; Mori, H.; Schlueter, J.; Zhyliaeva, E.I.; Torunova, S.A.; Lyubovskaya, R.N. Evidence for charge order in organic superconductors obtained by vibrational spectroscopy. Physica B 2009, 404, 490–493.
72. Yamamoto, K.; Yakushi, K.; Miyagawa, K.; Kanota, K.; Kawamoto, A. Charge ordering in θ-(BEDT-TTF)2RbZn(SCN)4 studied by vibrational spectroscopy. Phys. Rev. 2002, 65, 085110:1–085110:8.
73. There is another incoherent optical excitation at ~U with very weak intensity, which corresponds to the optical transition from the lower Hubbard to the upper Hubbard band.
74. Merino, J. Nonlocal Coulomb Correlation in metals close to a charge order insulator transition. Phys. Rev. Lett. 2007, 99, 036404:1–036404:4.
75. Kondo, R.; Higa, M.; Kagoshima, S.; Hoshino, H.; Mori, T.; Mori, H. Electrical and structural properties of θ-type BEDT-TTF organic conductors under uniaxial strain. J. Phys. Soc. Jpn. 2006, 75, 044716:1–044716:7.
76. Kuroki, K. The origin of the charge ordering and its relevance to superconductivity in θ-(BEDT-TTF)2X: The effect of the fermi surface nesting and the distant electron-electron interactions. J. Phys. Soc. Jpn. 2006, 75, 114716:1–114716:14.
77. Hotta, C. Classification of quasi-two dimensional organic conductors based on a new minimal model. J. Phys. Soc. Jpn. 2003, 72, 840–853.
78. Watanabe, H.; Ogata, M. Novel charge order and superconductivity in two-dimensional frustrated lattice at quarter filling. J. Phys. Soc. Jpn. 2006, 75, 063702:1–063702:4.
79. Mori, H.; Tanaka, S.; Mori, T. Systematic study of the electronic state of θ-type BEDT-TTF organic conductors by changing electronic correlation. Phys. Rev. 1998, 57, 12023–12029.
80. Kobayashi, H.; Kato, R.; Kobayashi, A.; Nishio, Y.; Kajita, K.; Sasaki, W. New molecular superconductor (BEDT-TTF)2(I3)x(AuI2)1−x (x < 0.02). Chem. Lett. 1986, 789–792.

81. Mori, H.; Tanaka, S.; Mori, T.; Kobayashi, A.; Kobayashi, H. Crystal structure and physical properties of M = Rb and Tl salts of (BEDT-TTF)2MM′(SCN)4, [M′ = Co, Zn]. Bull. Chem. Soc. Jpn. 1998, 71, 797–806.
82. Komatsu, T.; Sato, H.; Nakamura, T.; Matsukawa, N.; Yamochi, H.; Saito, G.; Kusunoki, M.; Sakagushi, K.; Kagoshima, S. The electrical and magnetic properties of a novel two-dimensional antiferromagnet based on BEDT-TTF, θ-(BEDT-TTF)2Cu(CN)[N(CN)2]2. Bull Chem. Soc. Jpn. 1995, 68, 2233–2244.
83. The doublet of ν2P is interpreted as the Fermi resonance [72].
84. Watanabe, M.; Noda, Y.; Nogami, Y.; Mori, H. Transfer integrals and spatial pattern of charge ordering in θ-(BEDT-TTF)2RbZn(SCN)4 at 90 K. J. Phys. Soc. Jpn. 2004, 73, 116–122.
85. Suzuki, K.; Yamamoto, K.; Yakushi, K. Charge-ordering transition in orthorhombic and monoclinic single crystals of θ-(BEDT-TTF)2TlZn(SCN)4 studied by vibrational spectroscopy. Phys. Rev. 2004, 69, 085114:1–085114:11.
86. Tajima, H.; Kyoden, S.; Mori, H.; Tanaka, S. Estimation of charge-ordering patterns in θ-ET2MM′(SCN)4(MM′ = RbCo, RbZn, CsZn) by reflection spectroscopy. Phys. Rev. 2000, 62, 9378–9385.
87. Tanaka, Y.; Yonemitsu, K. Effects of electron-lattice coupling on charge order in θ-(ET)2X. J. Phys. Soc. Jpn. 2008, 77, 034708:1–034708:9.
88. Chiba, R.; Hiraki, K.; Takahashi, T.; Yamamoto, H.M.; Nakamura, T. Extremely slow charge fluctuations in the metallic state of the two-dimensional molecular conductor θ-(BEDT-TTF)2RbZn(SCN)4. Phys. Rev. Lett. 2004, 93, 216405:1–216405:4.
89. Nakamura, T.; Miyagawa, W.; Kinami, R.; Konishi, Y.; Takahashi, T. Low-temperature electronic states in θ-(BEDT-TTF)2RbZn(SCN)4: Competition of different ground states. Synth. Metal. 1999, 103, 1898–1899.
90. Kanoda, K.; Ohnou, K.; Kodama, M.; Miyagawa, K.; Itou, T.; Hiraki, K. Charge ordering and frustration in organic conductors. J. Phys. IV Fr. 2005, 131, 21–26.
91. Watanabe, M.; Noda, Y.; Nogami, Y.; Mori, H. Investigation of X-ray diffuse scattering in θ-(BEDT-TTF)2RbM′(SCN)4. Synth. Metal. 2003, 135, 665–666.

92. Nogami, Y.; Hanasaki, N.; Watanabe, M.; Yamamoto, K.; Ito, T.; Ikeda, N.; Ohsumi, H.; Tokawa, H.; Noda, Y.; Terasaki, I.; et al. Charge order competition leading to nonlinearity in organic thyristor family. J. Phys. Soc. Jpn. 2010, 79.
93. A weak broad ν3 band is probably involved in the low-frequency region of the broad ν2 band. The assignment of the broad ν2 band with the aid of isotope shift is given in ref. [94].
94. Suzuki, K.; Yamamoto, K.; Yakushi, K.; Kawamoto, A. Infrared and Raman studies of θ-(BEDT-TTF)2CsZn(SCN)4: Comparison with the frozen state of θ-(BEDT-TTF)2RbZn(SCN)4. J. Phys. Soc. Jpn. 2005, 74, 2631–2639.
95. Wang, N.L.; Mori, H.; Tanaka, S.; Dong, J.; Clayman, B.P. Far infrared study of insulator-metal transition in θ-(BEDT-TTF)2RbZn(SCN)4. J. Phys. Condens. Matter 2001, 13, 5463–5470.
96. Nishimoto, S.; Shingai, M.; Ohta, Y. Coexistence of distinct charge fluctuations in θ-(BEDT-TTF)2X. Phys. Rev. 2008, 78, 035113:1–035113:9.
97. The reason why the ν2 mode of metallic I3 salt is significantly deviated from the center of ν2P and ν2R is ascribed to the interaction between ν2R and the neighbor ν3A. [46].
98. Mori, T. Non-Stripe charge order in the θ-phase organic conductors. J. Phys. Soc. Jpn. 2003, 72, 1469–1475.
99. Mori, T. Estimation of off-site coulomb integrals and phase diagrams of charge ordered states in the θ-phase organic conductor. Bull. Chem. Soc. Jpn. 2000, 73, 2243–2253.
100. Watanabe, M.; Nogami, Y.; Oshima, K.; Mori, H.; Tanaka, S. Novel pressure-induced 2κF CDW state in organic low-dimensional compound θ-(BEDT-TTF)2CsCo(SCN)4. J. Phys. Soc. Jpn. 1999, 68, 2654–2663.
101. Watanabe, M.; Yamamoto, K.; Ito, T.; Nakashima, Y.; Tanabe, M.; Hanasaki, N.; Ikeda, N.; Nogami, Y.; Ohsumi, H.; Toyokawa, H.; et al. Non-thermal evidence for current-induced melting of charge order in θ-(BEDT-TTF)2CsZn(SCN)4. J. Phys. Soc. Jpn. 2008, 77, 065004:1–065004:2.
102. Inagaki, K.; Terasaki, I.; Mori, H.; Mori, T. Large dielectric constant and giant nonlinear conduction in organic conductor θ-(BEDT-TTF)2CsZn(SCN)4. J. Phys. Soc. Jpn. 2004, 73, 3364–3369.

103. Sawano, F.; Terasaki, I.; Mori, H.; Mori, T.; Watanabe, M.; Ikeda, N.; Nogami, Y.; Noda, Y. An organic thyristor. Nature 2005, 437, 522–524.

104. Chiba, R.; Hiraki, K.; Takahashi, T.; Yamamoto, H.M.; Nakamura, T. Charge disproportionation and dynamics in θ-(BEDT-TTF)2CsZn(SCN)4. Phys. Rev. 2008, 77, 115113:1–115113:10.

105. Nad, F.; Monceau, P.; Yamamoto, H.M. A possible glass-like state in θ-(BEDT-TTF)2CsZn(SCN)4 at low temperature. J. Phys. Condens. Matter 2008, 20, 485211:1–485211:6.

106. Udagawa, M.; Motome, Y. Charge ordering and coexistence of charge fluctuations in quasi-two-dimensional organic conductors θ-(BEDT-TTF)2X. Phys. Rev. Lett. 2007, 98, 206405:1–206405:4.

107. Kajita, K.; Nishio, Y.; Moriyama, S.; Sasaki, W.; Kato, R.; Kobayashi, H.; Kobayashi, A. ew organic suprconductors κ- and θ-(BEDT-TTF)2I3; Transport property. Solid State Commun. 1987, 64, 1279–1284.

108. Salameh, B.; Nothardt, A.; Balthes, E.; Schmidt, W.; Schweitzer, D.; Strempfer, J.; Hinrichsen, B.; Jansen, M.; Maude, D.K. Electronic properties of the organic metals θ-(BEDT-TTF)2I3 and θT-(BEDT-TTF)2I3. Phys. Rev. 2007, 75, 054509:1–054509:13.

109. The symmetry of 3a × 3c superlattice has not been clarified.

110. Wang, N.L.; Feng, T.; Chen, Z.J.; Mori, H. Infrared properties of θ-type ET charge-transfer salts: θ-(ET)2RbZn(SCN)4 vs. θ-(ET)2CsZn(SCN)4. Synth. Metal. 2003, 135, 701–702.

111. Wojciechowski, R.; Yamamoto, K.; Yakushi, K.; Inokuchi, M.; Kawamoto, A. High-pressure Raman study of the charge ordering in α-(BEDT-TTF)2I3. Phys. Rev. 2003, 67, 224105:1–224105:11.

112. Takenaka, K.; Tamura, M.; Tajima, N. Collapse of coherence quasiparticle states in θ-(bedt-ttf)2i3 observed by optical spectroscopy. Phys. Rev. Lett. 2005, 95, 227801:1–227801:4.

113. Kobayashi, A.; Kato, R.; Kobayashi, H.; Moriyama, S.; Nishio, Y.; Kajita, K.; Sasaki, W. Anion arrangement in a new molecular superconductor, θ-(BEDT-TTF)2(I3)1−x(AuI2)x, (x < 0.02). Chem. Lett. 1986, 2017–2020.

114. Tamura, M.; Yakushi, K.; Kuroda, H. Temperature dependence of the polarized reflectance spectra of the θ-Type of bis(ethylenedithio)tetrathiafulvalenium) triiodide θ-(BEDT-TTF)2I3: Estimation of band parameters. J. Phys. Soc. Jpn. 1988, 57, 3239–3247.

115. ACS. Molecular Conductors: Thematic Issue (Special issue); ACS Publication: Washington, DC, USA, 2004; pp. 4887–5781.
116. Mori, T. Structural genealogy of BEDT-TTF-based organic conductors I. Parallel molecules: β and β″ phases. Bull. Chem. Soc. Jpn. 1998, 71, 2509–2526.
117. Mori, T. Structural genealogy of BEDT-TTF-based organic conductors II. Inclined molecules: θ- and α-, and κ-phases. Bull. Chem. Soc. Jpn. 1999, 72, 179–197.
118. Mori, T. Structural genealogy of BEDT-TTF-based organic conductors III. Twisted molecules: δ- and α'-phases. Bull. Chem. Soc. Jpn. 1998, 72, 2011–2027.
119. Takano, Y.; Hiraki, K.; Yamamoto, H.M.; Nakamura, T.; Takahashi, T. Charge ordering in α-(BEDT-TTF)2I3. Synth. Metal. 2001, 120, 1081–1082.
120. Tajima, N.; Ebina-Tajima, A.; Tamura, M.; Nishio, Y.; Kajita, K. Effects of uniaxial strain on transport properties of organic conductor α-(BEDT-TTF)2I3 and discovery of superconductivity. J. Phys. Soc. Jpn. 2002, 71, 1832–1835.
121. Tajima, N.; Kajita, K. Experimental study organic zero-gap conductor α-(BEDT-TTF)2I3. Sci. Technol. Adv. Mater. 2009, 10, 024308:1–024308:7.
122. Kobayashi, A.; Katayama, S.; Suzumura, Y. Theoretical study of the zero-gap organic conductor α-(BEDT-TTF)2I3. Sci. Technol. Adv. Mater. 2009, 10, 024309:1–024309:15.
123. Iwai, S.; Yamamoto, K.; Kashiwazaki, A.; Hiramatsu, F.; Nakaya, H.; Kawakami, Y.; Yakushi, K.; Okamoto, H.; Mori, H.; Nishio, Y. Photoinduced melting of a stripe-type charge-order and metallic domain formation in a layered BEDT-TTF-based organic salt. Phys. Rev. Lett. 2007, 98, 097402:1–097402:4.
124. Yamamoto, K.; Iwai, S.; Boyko, S.; Kashiwazaki, A.; Hiramatsu, F.; Okabe, C.; Nishi, N.; Yakushi, K. Strong optical nonlinearity and its ultrafast response associated with electron ferroelectricity in an organic conductor. J. Phys. Soc. Jpn. 2008, 77, 074709:1–074709:6.
125. Bender, K.; Henning, I.; Schweitzer, D.; Dietz, K.; Endres, H.; Keller, H.J. Synthesis, structure, and physical properties of a two-dimensional organic metal, di[bis(ethylenedithiolo)tetrathio-fulvalene] triiodide, (BEDT-TTF)2+I3-. Mol. Cryst. Liq. Cryst. 1984, 108, 359–371.

126. Kino, H.; Fukuyama, H. Phase diagram of two-dimensional organic conductors: (BEDT-TTF)2X. J. Phys. Soc. Jpn. 1996, 65, 2158–2169.
127. Takahashi, T.; Hiraki, K.; Moroto, S.; Tajima, N.; Takano, Y.; Kubo, Y.; Satsukawa, H.; Chiba, R.; Yamamoto, H.M.; Kato, R.; et al. Charge disproportionation, everywhere! J. Phys. IV Fr. 2004, 114, 3–8.
128. Kakiuchi, T.; Wakabayashi, Y.; Sawa, H.; Takahashi, T.; Nakamura, T. Charge ordering in α-(BEDT-TTF)2I3 by synchrotron x-ray diffraction. J. Phys. Soc. Jpn. 2007, 76, 113702:1–113702:10.
129. Alemany, P.; Pouget, J.-P.; Canadell, E. Essential role of anions in the charge ordering transition of α-(BEDT-TTF)2I3. Phys. Rev. B 2012, 85, 195118:1–195118:10.
130. Yue, Y.; Yamamoto, K.; Uruichi, M.; Nakano, C.; Yakushi, K.; Yamada, S.; Hiejima, T.; Kawamoto, A. Nonuniform site-charge distribution and fluctuations of charge order in the metallic state of α-(BEDT-TTF)2I3. Phys. Rev. 2010, 82, 075134:1–075134:8.
131. Yamamoto, K.; Kowalska, A.A.; Yue, Y.; Yakushi, K. Vibronic activation of molecular vibrational overtones in the infrared spectra of charge-ordered organic conductors. Phys. Rev. 2011, 84, 064306:1–064306:13.
132. Yamamoto, K.; Kowalska, A.A.; Yue, Y.; Yakushi, K. E-mv coupling of vibrational overtone in organic conductors: Relationship to optical nonlinearity and ferroelectricity. Physica B 2012, 407, 1775–1778.
133. Ivek, T.; Korin-Hamzic, N.; Milat, O.; Tomic, S.; Clauss, C.; Drichko, N.; Schweitzer, D.; Dressel, M. Electrodynamic response of the charge ordering phase: Dielectric and optical studies of α-(BEDT-TTF)2I3. Phys. Rev. 2011, 83, 165128:1–165128:13.
134. Moroto, S.; Hiraki, K.-I.; Takano, Y.; Kubo, Y.; Takahashi, T.; Yamamoto, H.M.; Nakmura, T. Charge disproportionation in the metallic state of α-(BEDT-TTF)2I3. J. Phys. IV Fr. 2004, 114, 339–340.
135. Kawai, T.; Kawamoto, A. 13C-NMR study of charge ordering state in the organic conductor, α-(BEDT-TTF)2I3. J. Phys. Soc. Jpn. 2009, 78, 074711:1–074711:6.
136. Yamamoto, K.; Yakushi, K. Second-Harmonic Generation Study Of Ferroelectric Organic Conductors α-(BEDT-TTF)2X (X = I3 and I2Br). In Molecular Electronic and Related Materials—Control and Probe with Light; Naito, T., Ed.; Transworld Research Network: Kerala, Indian, 2010; pp. 185–201.

137. Yamamoto, K.; Kowalska, A.A.; Yakushi, K. Direct observation of ferroelectric domains by Wigner crystallization of electrons in α-[bis(ethylenedithio)tetetrathiafulvalene]2I3. App. Phys. Lett. 2010, 96, 122901:1–122901:3.

138. Ikeda, N.; Ohsumi, H.; Ohwada, K.; Ishii, K.; Inami, T.; Kakura, K.; Murakami, Y.; Yoshii, K.; Mori, S.; Horibe, Y.; et al. Ferroelectricity from iron valence ordering in the charge-frustrated system LuFe2O4. Nature 2005, 436, 1136–1138.

139. Iwai, S.; Yamamoto, K.; Hiramatsu, F.; Nakaya, H.; Kawakami, Y.; Yakushi, K. Hydrostatic pressure effect on photoinduced insulator-to-metal transition in the layered organic salt α-(BEDT-TTF)2I3. Phys. Rev. 2008, 77, 125131:1–125131:5.

140. Kawakami, Y.; Fukatsu, T.; Sakurai, Y.; Unno, H.; Itoh, H.; Iwai, S.; Sasaki, T.; Yamamoto, K.; Yakushi, K.; Yonemitsu, K. Early-State dynamics of light-matter interaction leading to the insulator-to-metal transition in a charge ordered organic crystal. Phys. Rev. Lett. 2010, 105, 246402:1–246402:4.

141. Mori, H.; Tanaka, S.; Oshima, K.; Saito, G.; Mori, T.; Maruyama, Y.; Inokuchi, H. Electrical properties and crystal structure of mercury (II) thiocyanate salts based upon BEDT-TTF with Li+, K+, NH4+, Rb+, and Cs+. Synth. Metal. 1991, 42, 2013–2018.

142. Mori, H.; Tanaka, S.; Oshima, M.; Saito, G.; Mori, T.; Maruyama, Y.; Inokuchi, H. Crystal and electronic structures of (BEDT-TTF)2[MHg(SCN)4] (M = K and NH4). Bull. Chem. Soc. Jpn. 1990, 63, 2183–2190.

143. Kobayashi, H.; Kato, R.; Kobayashi, A.; Nishio, Y.; Kajita, K.; Sasaki, W. Crystal and electronic structures of layered molecular superconductor, θ-(BEDT-TTF)2(I3)x(AuI2)1−x (x < 0.02). Chem. Lett. 1986, 833–836.

144. Tajima, N.; Tamura, M.; Nishio, Y.; Kajita, K.; Iye, Y. Transport property of an organic conductor α-(BEDT-TTF)2I3 under hydrostatic pressure—Discovery of a novel type of conductor. J. Phys. Soc. Jpn. 2000, 69, 543–551.

145. Hiejima, T.; Yamada, S.; Uruichi, M.; Yakushi, K. Infrared and Raman studies of α-(BEDT-TTF)2 MHg(SCN)4 with M = NH4 and K at low temperature: Breaking of inversion symmetry due to charge-ordering fluctuation. Physica B 2010, 405, 5153–5156.

146. Foury-Leylekian, P.; Ravy, S.; Pouget, J.-P.; Muller, H. X-ray study of the density wave instability of α-(BEDT-TTF)2MHg(SCN)4 with M = K and Rb. Synth. Metal. 2003, 137, 1271–1272.

147. Foury-Leylekian, P.; Pouget, J.-P.; Lee, Y.-J.; Nieminen, R.M.; Ordejon, P.; Canadell, E. Density-wave instability in α-(BEDT-TTF)2KHg(SCN)4 studied by x-ray diffuse scattering and by first-principle calculations. Phys. Rev. 2010, 82, 134116:1–134116:14.

148. This value was estimated from the separation of ν27j mode, (~7 cm−1)/(140 cm−1/e).

149. Yamamoto, T.; Yamamoto, H.M.; Kato, R.; Uruichi, M.; Yakushi, K.; Akutsu, H.; Sato-Akutsu, A.; Kawamoto, A.; Turner, S.S.; Day, P. Inhomogeneous site charges at the boundary between the insulating, superconducting, and metallic phases of β″-type bis-ethylenedithio-tetrathiafulvalene molecular charge-transfer salts. Phys. Rev. 2008, 77, 205120:1–205120:13.

150. Bangura, A.F.; Coldea, A.I.; Singleton, J.; Ardavan, A.; Akutsu-Sato, A.; Akutsu, H.; Turner, S.S.; Day, P.; Yamamoto, T.; Yakushi, K. Robust superconducting state in the low-quasiparticle-density due to proximity to a charge-ordered state. Phys. Rev. 2005, 72, 014543:1–014543:13.

151. See ref. [46] about the reason why this equation is applicable in the range of $0 \leq \rho \leq 0.8$.

152. Although frequency is proportional to the square root of force constant, the frequency practically linearly depends upon the force constant in a narrow frequency range, $\Delta\omega = \omega(0) - \omega(1)$

153. The κ- and β′-ET salts with $\rho = 0.5$ have a dimer unit with large transfer integral. However, the blue shift of ν2 is not attributed to the emv (electron-molecular-vibration) coupling, because similar blue shift is found in non-dimeric θ-(ET)2I3 which will be shown in the next subsection.

154. Hess, L.A.; Prasad, P.N. Vibrational dephasing in organic solids: Temperature dependence of Raman active internal mode of naphthalene. J. Phys. Chem. 1980, 72, 573–579.

155. The frequency of the infrared-active ν27 mode at $\rho = 0.5$ is also deviated from the linear relationship. The examination similar to ν2 is necessary including temperature dependence.

156. Kayatama, S.; Kobayashi, A.; Suzumura, Y. Effect of anion potential on the zero-gap state in the two-dimensional organic conductor α-(BEDT-TTF)2I3. J. Phys. Conf. Ser. 2008, 132, 012003:1–012003:7.
157. Kobayashi, A.; Katayama, S.; Suzumura, Y.; Fukuyama, H. Massless fermions in organic conductor. J. Phys. Soc. Jpn. 2007, 76, 034711:1–034711:6.
158. Ishibashi, S.; Tamura, T.; Kohyama, M.; Terakura, K. Ab initio electronic-structure calculation. J. Phys. Soc. Jpn. 2006, 75, 015005:1–015005:2.
159. Tamura, M.; Matsuzaki, F.; Tajima, N.; Nishio, Y.; Kajita, K. Transport study of an organic conductor, θ-(BEDT-TTF)2I3. Synth. Metal. 1997, 86, 2007–2008.
160. Tajima, N.; Tajima, A.; Tamura, M.; Kato, R.; Nishio, Y.; Kajita, K. Pressure control of transport property of organic conductors; α-, θ-(BEDT-TTF)2I3 and θ-(DIETS)2[Au(CN)4]. J. Phys. IV Fr. 2004, 114, 263–267.
161. Miyagawa, K.; Hirayama, M.; Tamura, M.; Kanoda, K. 13C NMR study on zero-gap state in the organic conductor θ-(BEDT-TTF)2I3 under pressure. J. Phys. Soc. Jpn. 2010, 79, 063703:1–063703:3.
162. As ν2R is mixed with ν31, the hole numbers shown in the CO state shown in Figure 5b,c are inaccurate.
163. Kondo, R.; Kagoshima, S.; Tajima, N.; Kato, R. Crystal and electronic structures of the quasi-two-dimensional organic conductor α-(BEDT-TTF)2I3 and its selenium analogue α-(BEDT-TSeF)2I3 under hydrostatic pressure at room temperature. J. Phys. Soc. Jpn. 2009, 78, 114714:1–114714:7.
164. Yakushi, K.; Kanbara, H.; Tajima, H.; Kuroda, H.; Saito, G.; Mori, T. Temperature dependence of the reflectance spectra of the single crystals of bis(ethylenedithio)tetrathiafulvalene salts. α-(BEDT-TTF)3(ReO4)2 and α-(BEDT-TTF)2I3. Bull. Chem. Soc. Jpn. 1987, 60, 4251–4257.
165. Zelezny, V.; Petzelt, J.; Swietlik, R.; Gorshunov, B.P.; Volkov, A.A.; Kozlov, G.V.; Schweitzer, D.; Keller, H.J. Far infrared response of α- and αt-(BEDT-TTF)2I3. J. Phys. Fr. 1990, 51, 869–881.
166. Dressel, M.; Gruner, G.; Pouget, J.P.; Breining, A.; Schweitzer, D. Field and frequency dependent transport in the two-dimensional organic conductor α-(BEDT-TTF)2I3. J. Phys. I Fr. 1994, 4, 579–594.

167. Clauss, C.; Drichko, N.; Schweitzer, D.; Dressel, M. Charge-order gap in α-(BEDT-TTF)2I3. Physica B 2010, 405, S144–S146.

168. Tamura, K.; Ozawa, T.; Bando, Y.; Kawamoto, T.; Mori, T. Voltage oscillation associated with nonlinear conductivity in the organic conductor α-(BEDT-TTF)2I3. J. Appl. Phys. 2010, 107, 103716:1–103716:5.

169. Ivek, T.; Korin-Hamzic, B.; Milat, O.; Tomic, S.; Clauss, C.; Drichko, N.; Schweitzer, D.; Dressel, M. Collective excitation in the charge-ordered phase of α-(BEDT-TTF)2I3. Phys. Rev. Lett. 2010, 104, 206406:1–206406:4.

170. Dressel, M.; Drichko, N.; Kaiser, S. Collective charge-order excitations. Physica C 2010, 470, S589–S591.

171. Watanabe, M.; Nishikawa, M.; Nogami, Y.; Oshima, K.; Saito, G. Low temperature structure of the α'-(BEDT-TTF)2IBr2 organic magnetic semiconductor. J. Korean Phys. Soc. 1997, 31, 95–98.

172. Tokumoto, M.; Anzai, H.; Ishiguro, T.; Saito, G.; Kobayashi, H.; Kato, R.; Kobayashi, A. Electrical and magnetic properties of organic semiconductors, (BEDT-TTF)2X (X = IBr2, IBrCl, and ICl2). Synth. Metal. 1987, 19, 215–220.

173. Maniwa, Y.; Takahashi, T.; Saito, G. 1H NMR in organic superconductor β-(BEDT-TTF)2I3. J. Phys. Soc. Jpn. 1986, 55, 47–50.

174. Nogami, Y.; Kagoshima, S.; Sugano, T.; Saito, G. X-ray evidence for structural changes in the organic conductors, α-(BEDT-TTF)2I3, α-(BEDT-TTF)2IBr2,and β-(BEDT-TTF)2I3. Synth. Metal. 1986, 16, 367–377.

175. Inokuchi, M.; Yakushi, K.; Kinoshita, M.; Saito, G. Optical properties of α'-(BEDT-TTF)2IBr2. Synth. Metal. 1992, 103, 2102–2103.

176. Yue, Y.; Nakano, C.; Yamamoto, K.; Uruichi, M.; Yakushi, K.; Kawamoto, A. Charge-ordering phase transition in α'-(BEDT-TTF)2IBr2. J. Phys. Conf. Ser. 2008, 132, 012007:1–012007:7.

177. Yue, Y.; Yamamoto, K.; Nakano, C.; Uruichi, M.; Yakushi, K.; Inokuchi, M.; Hiejima, T.; Kawamoto, A. Order-disorder type of charge-ordering phase transition in narrow-bandwidth compound, α'-(BEDT-TTF)2IBr2. Physica B 2010, 405, S232–S236.

178. Kowalska, A.A.; Yamamoto, K.; Yakushi, K. Ferroelectric polarization in α′-(ET)2IBr2 studied by second-harmonic generation microscopy. J. Phys. Conf. Ser. 2008, 132, 012006:1–012006:5.

179. Yamashita, A.; Watanabe, M.; Kobayashi, K.; Kumai, R.; Yamamoto, K.; Yakushi, K.; Noda, Y. SR-based study of successive phase transition of α′-(BEDT-TTF)2IBr2 II. Meeting Abstr. Phys. Soc. Jpn. 2012, 67, 907.

180. Sugano, T.; Saito, G.; Kinoshita, M. Conduction-electron-spin resonance in organic conductors: A and b phases of di[bis(ethylenedithiolo)tetrathiafulvalene]triiodide [(BEDT-TTF)2I3]. Phys. Rev. 1986, 34, 117–125.

181. The optical conductivities from 50–6000 cm−1 of α′-IBr2, α-I3 and α-NH4Hg were obtained in our research group. The low-frequency region from 0–600 cm−1 of β″-(BEDT-TTF)(TCNQ), θm-RbZn, θo-RbZn, and θ-CsZn were extrapolated to the dc conductivity value by straight line. The optical conductivities of θ-RbZn and θ-I3 were taken from the papers reported by Wang [95] and Takenaka [112], respectively.

182. Van der Marel, D. Optical Signatures of Electron Correlations in the Cuprates. In Strong Interactions in Low Dimensions; Baeriswyl, D., Degiorgi, L., Eds.; Kluwer Academic Publishers: Dordrecht, The Netherlands, 2004; pp. 237–276.

5
CHAPTER

GALLIUM NITRIDE-BASED SEMICONDUCTOR OPTICAL AMPLIFIERS

Rintaro Koda[1], Hideki Watanabe[1] and Shunsuke Kono[1]

[1] *Semiconductor Device Development Division, Sony Corporation, Japan*

ABSTRACT

GaN-based material can potentially cover a wide spectral emission range, and laser diodes emitting in the UV, violet, blue, green, and red wavelengths have already been demonstrated and/or commercialized. GaN-based semiconductor optical amplifiers (SOAs) have the ability to boost the output power of laser diodes and thus are candidates for a broad variety of potential uses. Applications that utilize short wavelength, ultrafast pulses, including microprocessing, orthoptics, and next-generation optical storage can most benefit from GaN-based SOAs since current ultrafast pulse sources rely on large, expensive solid-state lasers. GaN-based SOAs can generate high-energy, high peak power optical pulses when used in conjunction with mode-locked laser diodes. In this chapter, the basic characteristics of these devices are discussed, concentrating on pulse amplification. Early experimental work, as well as the latest results, is presented, and improvements in the SOA design allowing the generation of higher optical pulse energy are discussed.

Keywords: Gallium nitride, GaN, Semiconductor optical amplifier, SOA, Optical pulse generation

1. INTRODUCTION

Gallium nitride (GaN)-based optoelectronic devices have attracted significant research interest over the last two decades. Some of these applications, such as light-emitting diodes (LEDs) and ultraviolet (UV),[1] - blue-violet, blue, and green laser diodes (LDs), have already been commercialized and have had an impact on everyday life as solid-state lighting, high-capacity optical storage (such as Blu-ray discs), and display units. The GaN-based material AlGaInN has the potential to cover a particularly wide spectral range, from the deep UV to the infrared (IR). LED emissions at wavelengths as short as 210 nm have already been reported [1]. In the case of LDs, lasing actions over the range of 330–350 nm have been obtained with electrical pumping using GaN or AlGaN quantum wells [2, 3]. In addition, Frost et al. reported lasing at longer wavelengths in the red region of the spectrum (at 630 nm) from a InGaN/GaN quantum dot laser [4]. The fact that many researchers have focused on this topic indicates the significant potential of GaN-based optoelectronic devices. However, one important device is missing: the semiconductor optical amplifier (SOA).

An SOA is an important component of optical communication systems, and numerous reports regarding this type of device have been published since its invention in the 1960s. One of the main advantages of the SOA is its ability to be monolithically integrated with other devices. An SOA can be used as a pre-amplifier or a booster amplifier as well as in other functional devices, such as a wavelength converter or optical switch that utilizes its nonlinear nature [5, 6]. An SOA can also be applied as a gain medium to produce short optical pulses in conjunction with a saturable absorber [7]. In high peak power applications, such as two photon bioimaging, an SOA can be used to boost the short optical pulses generated by a mode-locked laser diode (MLLD) [8, 9].

High peak power and ultrashort optical pulses have a number of unique applications, many of which utilize the multiphoton processes that are discussed in detail in Section 4. The majority of pulse sources, however, consist of large, bulky solid-state lasers such as a mode-locked titanium sapphire (Ti:Sap) laser. Moreover, to generate optical pulses at near-UV wavelengths, most of the available pulse sources require wavelength conversion, such that a second- or third-order harmonic of an IR pulse is required. This means that more sophisticated laser systems are necessary to obtain UV pulses. Fiber lasers are an alternative to solid-state lasers and have the advantage of a smaller footprint. However, the requirements for wavelength conversion are the same since these devices can only generate photons in the IR wavelength. Such wavelength conversion systems have a number of disadvantages: a larger footprint, the need for precise alignment, less wall-plug efficiency, and increased complexity of the overall system, usually leading to higher costs and longer stabilizing times. For all these reasons, the direct generation of UV pulses is highly desirable, and thus, a GaN-based SOA could be an important device with regard to obtaining short wavelength optical pulses.

As noted, GaN-based devices potentially can generate a very wide range of photons, from the deep UV to the red wavelength of the spectrum. Optical pulse generation in this type of device was first demonstrated in 1997 by Professor Nakamura, the 2014 Noble laureate, using a self-pulsating LD (SPLD) [10]. Since then, many groups have explored GaN-based lasers with the aim of producing optical pulses. Earlier studies concentrated on self-pulsation for applications in optical storage systems since this approach can reduce the optical feedback noise [1--14]. Several pulse generation techniques exist, including gain-switching [1–17], self-pulsating [18-20], mode-locking [7, 17, 21], and superradiance [22]. Kono et al. demonstrated the first gain-switching (GS) operation of a GaN-based LD, producing a peak power of 12 W and 10 W with a pulse duration of 10 ps at 405 nm [15, 17]. Kuramoto et al. later improved the peak power to 55 W by optimizing the electron blocking layer in the LD [16]. SPLDs were also explored for applications in high peak power optical sources [18, 20, 23], and results included a peak power of 2.4 W with a bisectional LD [18], 10 W and 0.6 W with a triple section straight waveguide LD [19, 20], and 20 W and 10 W with a triple section bow tie structure [23, 24]. Mode locking is another means of generating optical pulses and was first demonstrated by Gee and Bowers using a GaN-based SOA as the gain medium [7]. Saito also demonstrated an external cavity passive mode-locked laser diode (MLLD), incorporating a bisectional waveguide structure with a 3-ps pulse duration and a 0.2-W peak power [25]. Oki et al. improved the peak power to 3 W in the optimized active region [26] and later to 20 W by employing a flared waveguide structure [21]. The reduction of the pulse duration is important for high peak power generation, and Kono et al. have demonstrated 200 fs pulse generation using a dispersion compensated external cavity and spectral filtering [27]. To further improve the peak power and pulse energy, the use of an SOA is essential. The SOA can be used to boost the output power generated by an LD in a master oscillator power amplifier (MOPA) configuration. Our group has previously developed GaN-based SOAs meant to improve the pulse energy as well as the peak power. In 2010, a peak power over 100 W with a corresponding pulse energy of approximately 330 pJ was demonstrated, employing an MLLD in conjunction with an SOA [28], and these values were later improved to approximately 300 W and 590 pJ [29]. Recently, an SOA incorporating a widely flared waveguide structure was developed, and the peak power and the pulse energy were further improved to 2.2 nJ and 630 W [30]. The pulse characteristics of published GaN-based devices are summarized in Figure 1 and Table 1 (note that earlier studies of SPLDs were excluded from these summaries since pulse durations were not available). It can be clearly seen that the use of an SOA significantly improves the peak power as well as the pulse energy.

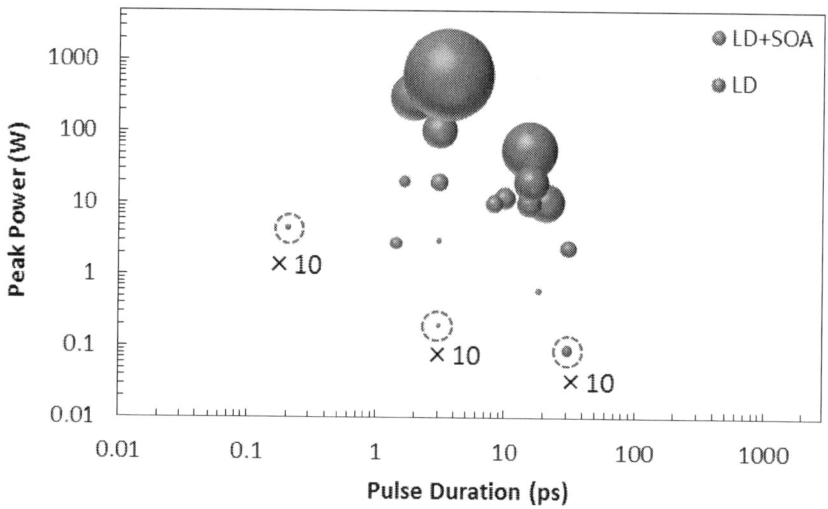

Figure 1. Peak power vs. pulse duration for GaN-based devices. The size of each circle indicates the optical pulse energy. The use of an SOA effectively increases both the peak power and pulse energy. Note that in the case of those data points indicated by dashed circles, the pulse energies are multiplied by 10 for illustrative purposes.

2. CHARACTERIZATION OF CONTINUOUS WAVE LIGHT AMPLIFICATION

In this section, the basic characteristics of GaN-based SOAs are reviewed.

Table 1. Summary of the GaN-based optical pulse sources developed by various research groups.

Device type	Pulse Energy (pJ)	Peak Power (W)	Average power (mW)	Pulse Repetition (GHz)	Pulse Duration (ps)	Wavelength (nm)	Reference
Gain-switching LD	116	12	0.01	0.0001	9.7	405	[15]
	820	55	.82	.001	15	401	[16]
Self-pulsating LD	72	2.4	65	0.9	30	407	[18]
	11[1]	0.6	-	4.5	18	414	[19]
	150	10	150	1	15	402	[20]
	80[2]	10	80[1]	1.7	8	-	[24]
	310	20	260	0.84	15.5	406	[23]
Mode-locked LD	2.8	0.09	2	0.72	30	408.5	[7]
	0.6	0.2	0.6	1	3	404.5	[25]
	8.9	3	8.9	3	1	404	[26]
	32	20	32	1	1.6	403.5	[21]
	0.9	4.5	0.9	1	0.2	401	[27]
Mode-locked LD + SOA	78	20	78	1	3	404	[31]
	329	103	329	1	3	404.3	[28]
	586	308	586	1	1.9	405.4	[29]
	2200	630	710	0.812	3.5	405.2	[30]

[1]: Estimated using the peak power, pulse repetition and pulse duration values provided in the paper.

[2]: Estimated by using values provided in the paper

2.1. Basic structures

In general, there are two types of SOA: the Fabry–Perot type and the travelling-wave type. A Fabry–Perot SOA has nonzero facet reflectivity so that signal light is amplified as the result of many passes through the amplifier. In contrast, a travelling-wave SOA has negligible facet reflectivity such that the signal passes through only once (single-pass). A travelling-wave SOA is simpler to use, and so these devices are discussed exclusively in the present chapter. GaN-based SOAs that we would discuss throughout this chapter consist of a double quantum well structure with $Ga_{0.92}In_{0.08}N$ quantum wells and $Ga_{0.98}In_{0.02}N$ barriers grown on an n-GaN substrate by metal organic chemical vapor deposition (MOCVD). The layer structure is shown in Figure 2. In a travelling-wave SOA, it is important to minimize facet reflectivity, and this is illustrated in Figure 3. Both devices shown in this figure are ridge waveguide structures with a length of 1 mm and both incorporate facets with anti-reflection (AR) coatings. Waveguides are straight (Figure 3(a)) and angled by 5° (Figure 3(b)). As the result of nonzero reflectivity, the straight waveguide exhibits significant resonance and a multilongitudinal mode, both of which are undesirable with regard to SOA operation. The incorporation of the angled waveguide effectively reduces back reflection from the facets and thus suppresses the resonance effect.

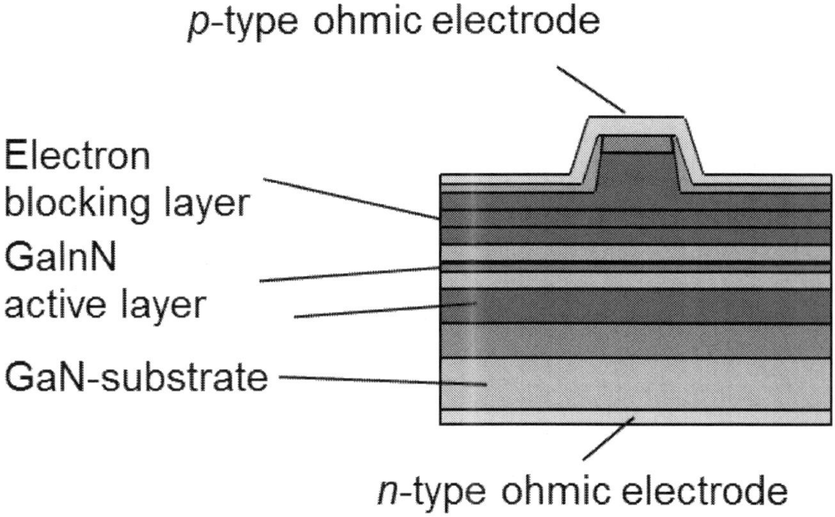

Figure 2. Schematics of the layer structure of the SOAs discussed in the present chapter.

The output power of an SOA is limited by its saturation power, P_{sat}, which is given by

$$P_{sat} = \frac{h\nu dw}{\Gamma a \tau} \tag{1}$$

where h is the Plank constant, ν is the signal optical frequency, d and w are the active region thickness and width, respectively, Γ is the confinement factor, a is the differential gain, and τ is the carrier lifetime. Equation (1)

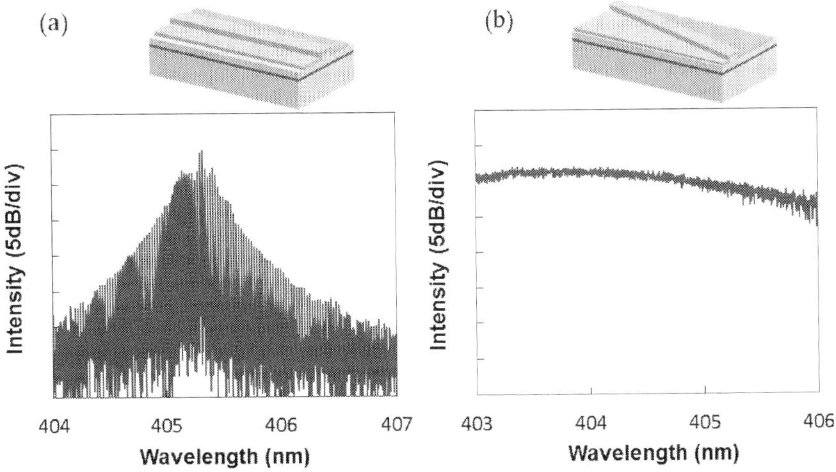

Figure 3.Optical spectra obtained from 1 mm long SOAs with either (a) straight or (b) angled waveguides. Although both structures had AR-coated faces, the straight waveguide exhibits strong resonance due to residual reflections from an imperfect AR coating.

indicates that a larger active region cross section, a lower confinement factor, a lower differential gain, and a reduced carrier lifetime are desirable for high power applications. However, the output power is independent of device length. Equation (1) can be interpreted qualitatively, that is, the amount of energy that can be stored in the device is important for higher output from the SOA.

The basic amplification characteristics of an SOA were assessed using device A shown in Figure 4, incorporating a flared waveguide. The input and the output waveguide widths were 1.4 and 5 μm, respectively, and both facets were AR coated. As noted earlier, the waveguide was angled at 5° to suppress the resonance effect.

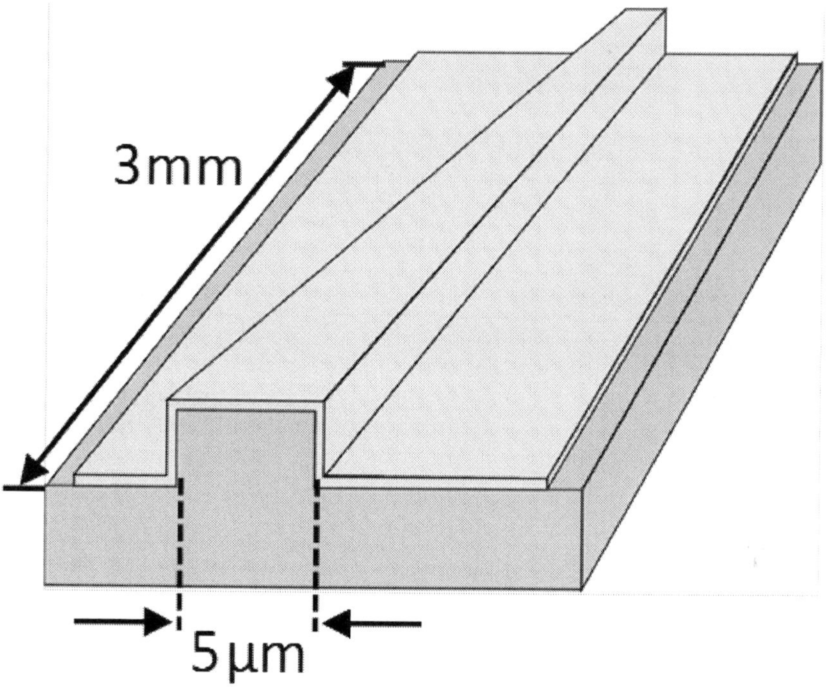

Figure 4. Schematics of device A.

2.2. Continuous wave amplification characteristics

In the amplification of continuous-wave (CW) light by the SOA, device A was assessed using the experimental setup shown in Figure 5. The resulting net gain (sometimes also known as the fiber-to-fiber gain) and output power at an input wavelength of 404 nm are presented as functions of input power at operating currents of 3, 4, and 5 kA/cm^2 in Figure 6. The net gain values were estimated using an optical band-pass filter (BPF) to reject the out-of-band components of the amplified spontaneous emission (ASE). It should be noted that the net gain values in this figure include the input and output coupling losses. It is evident that a gain of over 20 dB was obtained from a small signal input of less than 1 mW, a result that is comparable to the performance of SOAs fabricated with other materials [5, 6].

Wavelength dependencies were characterized by tuning the input wavelength to 403, 405, or 407 nm, and the corresponding spectra are presented in Figure 7, in which very different spectra are observed at different input wavelengths. At the shorter wavelength, the amplified spectrum shows that ASE is present, whereas less ASE is observed from amplification at longer wavelengths of 405 and 407 nm. Figure 8(a) plots the output power and ASE spectra, while Figure 8(b) presents the net gain spectra obtained at different input powers. It is interesting to note that the gain peak shifts to longer wavelength as the input power is increased. At lower input power, the wavelength of the gain peak and the maximum output power are nearly equal to the ASE peak, while higher input power results in a red shift of these values. This phenomenon can possibly be explained by considering that gain saturation due to carrier depletion leads to intra-band relaxation of the carriers. This results in more carriers available for longer wavelength amplification. In addition, GaN-based systems are known to have higher effective masses of electrons and holes [32, 33], and this may also play a role in this mechanism.

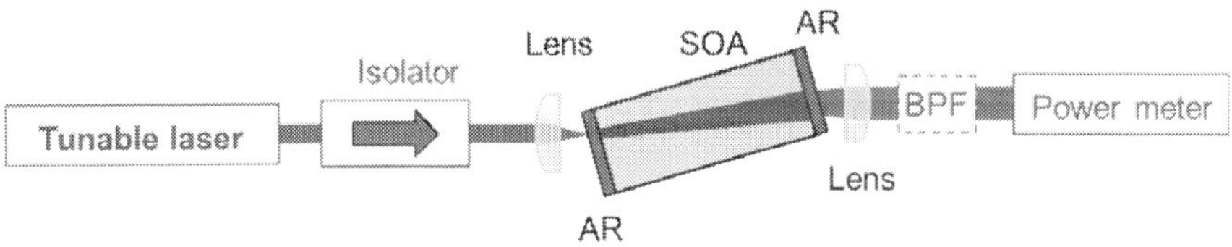

Figure 5.Experimental setup for CW amplification.

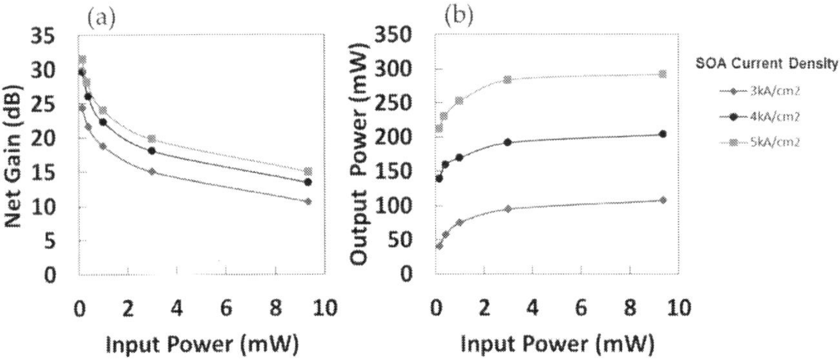

Figure 6.Net gain (left) and output power (right) obtained from the SOA (device A) as functions of CW input power for current densities of 3, 4, and 5 kA/cm^2 at an input wavelength of 404 nm.

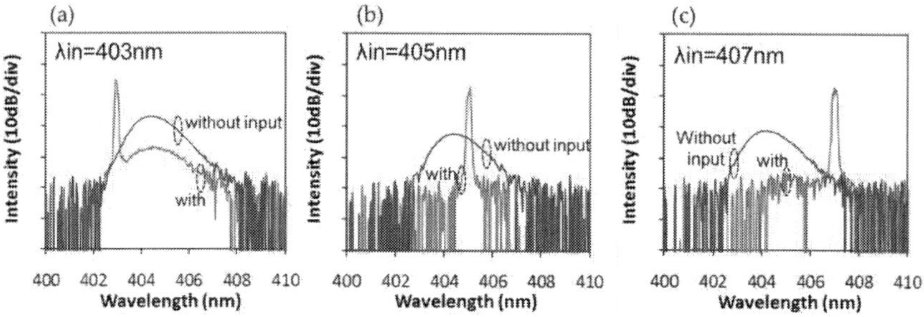

Figure 7.Optical spectra of the output of an angled SOA obtained with input wavelengths of (a) 403, (b) 405, and (c) 407 nm. Current density and input power were 4 kA/cm² and 11 mW, respectively. Spectra with (red) and without (blue) input are shown for each wavelength.

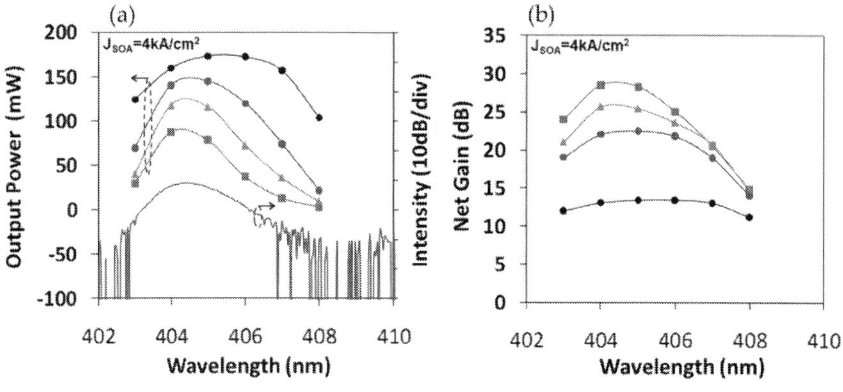

Figure 8.(a) Output power and ASE spectra and (b) net gain spectra obtained from device A at input powers of 11 mW (black), 1.2 mW (blue), 0.4 mW (red), and 0.2 mW (violet).

3. PULSE AMPLIFICATION

3.1. Limiting factor

Pulse amplification was initially studied using device A [31], employing a GaN-based external cavity MLLD developed by our group [26]. In this work, a GaN-based LD having the same GaInN active region as the SOA was grown by MOCVD. The resulting LD chip was composed of a forward bias gain section and a reverse bias saturable absorption (SA) section. The facets had a high reflection coating on the SA side and an AR coating on the other side.

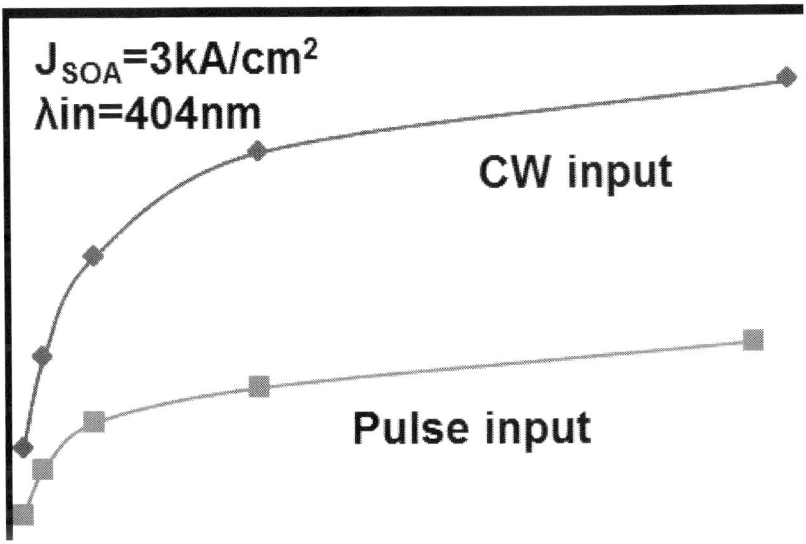

Figure 9.Average output power of the SOA (device A) as a function of input power for CW (diamonds) and pulse (squares) inputs at an SOA current density of 3 kA/cm^2 and an input wavelength of 404 nm. Significant reduction of gain is observed for pulse amplification.

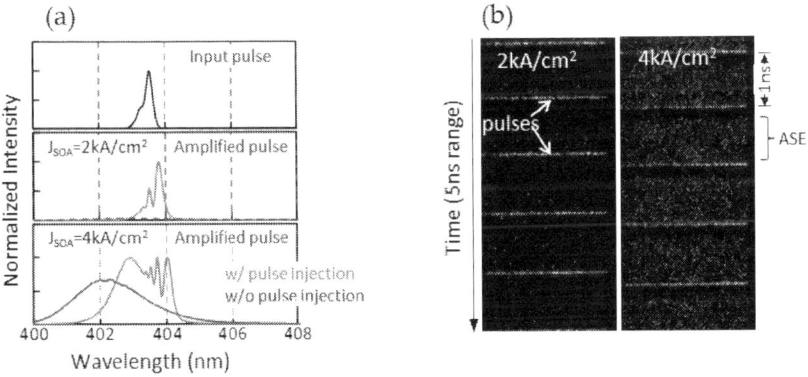

Figure 10.(a) Optical spectra obtained with input and amplified output and (b) streak camera images at SOA current densities of 2 and 4 kA/cm^2. At the higher current density, intense ASE is observed between pulses.

More details regarding this MLLD are available in the literature [26]. Figure 9 shows the output power of the device as a function of the average input power for both CW and pulse inputs at a 3-kA/cm^2 SOA bias current density. From this figure, it is evident that significantly lower average power was obtained when applying pulse amplification. To investigate the limiting factors for pulse amplification, we acquired optical spectra as well as temporal characteristics using an optical spectrum analyzer (OSA) and a streak camera. Figure 10(a) presents the optical spectra and Figure 10(b) shows streak camera images obtained from amplified pulses at 2 and 4 kA/cm^2 SOA bias current densities. The optical spectra of the amplified pulses exhibit a peak wavelength shift toward longer wavelengths as well as an oscillatory structure, representing self-phase modulation (SPM) [34]. High peak power optical pulses induce carrier depletion, which in turn varies the refractive index. At the lower current density of 2 kA/cm^2, the streak camera image shows clean optical pulses at a repeat rate of 1 GHz. In contrast, at the higher current density (4 kA/cm^2), ASE appears between pulses. The optical spectra also changes at the higher current density, such that a broad fifth peak appears near 403 nm. Careful examination of this short wavelength peak using the streak camera indicated that it is a combination of ASE and pulse elements. The carrier recovery time was determined to be less than 300 ps, and thus significantly faster than the pulse repetition of 1 ns. The duration of the amplified pulse was 3 ps and the peak power was estimated to be only 20 W.

3.2. Improving peak power

Investigation of device A indicated that intense ASE appeared as the operating current density was increased. Intense ASE leads to gain depletion, preventing efficient amplification of optical pulses. In this process, carrier depletion occurring via a stimulated emission process is induced by the intense ASE, which in turn results in fewer carriers available for pulse amplification. This generates the limited peak power of only 20 W. The fact that intense ASE is a limiting factor for an SOA has also been noted in previous studies of other systems, both theoretically and experimentally [35-37]. To overcome this limitation, intense ASE has to be reduced. However, a trade-off exists since ASE is also needed for optical gain. We have thus pursued the development of SOAs by optimizing device structures so as to obtain low ASE while maintaining sufficient gain. Peak powers were improved to approximately 100 W and later to 300 W by optimizing device structure such as the length and optical confinement factors [28, 29]. The corresponding optical pulse energies were approximately 330 and 590 pJ. Recently, further improvements gave a pulse energy of 2200 pJ (corresponding to a peak power of 630 W) [30]. This is the highest ever pulse energy reported for a GaN-based all-semiconductor pulse source. Details of this latest device are presented in the following sections. The SOAs described above are illustrated in Figure 11 as devices B, C, and D. It is known that optical pulse amplification is governed by the saturation energy, E_{sat}, of the SOA [34], and that it can be expressed as follows.

$$E_{sat} = \frac{h\nu dw}{\Gamma a} \tag{2}$$

The parameters *d*, *w*, and Γ can be readily tuned by adjusting the epitaxial layer structure and waveguide design. In the case of devices B, C, and D, the values of dw/Γ are 8×10^3, 9×10^3, and 6×10^4 μm², respectively. Each of these devices was grown by MOCVD, and all have a ridge waveguide structure. Both B and C employ waveguides that are linearly tapered from 1.4 μm at the input facet to 15 μm at the output facet. In the case of D, the waveguide consists of two sections: the straight preamplifier and the flared main amplifier sections. The lengths are 2, 2.5, and 3 mm for B, C, and D, respectively, and more detailed descriptions have been previously published [2--30]. Figure 12 summarizes the net gain and ASE power values as functions of the SOA current density. For each sample, the net gain is seen to increase with current density but to saturate as the ASE becomes too high. It should be noted that this discussion is not intended to offer a detailed comparison between the devices since the operating conditions under which these data were collected, such as temperature, input pulse width, and power, were not all the same. Rather, the intention is to show the effectiveness of varying the parameter dw/Γ, which governs the optical pulse energy. Figure 13 plots the optical pulse energy as a function of this term.

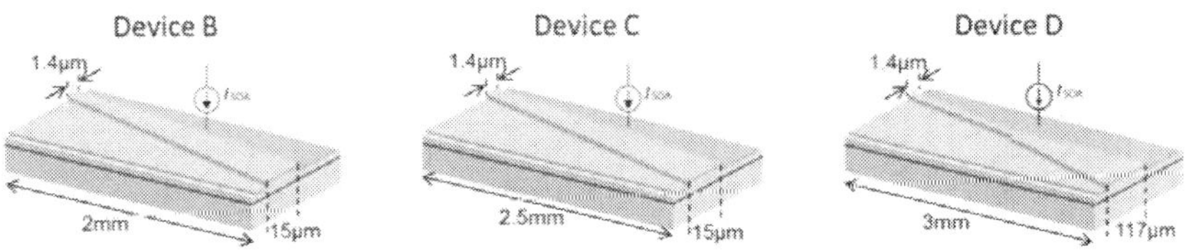

Figure 11.Schematics of three additional SOA designs. The peak power and pulse energy values were approximately 100, 300, and 630 W and 330, 590, and 2200 pJ for devices B, C, and D, respectively.

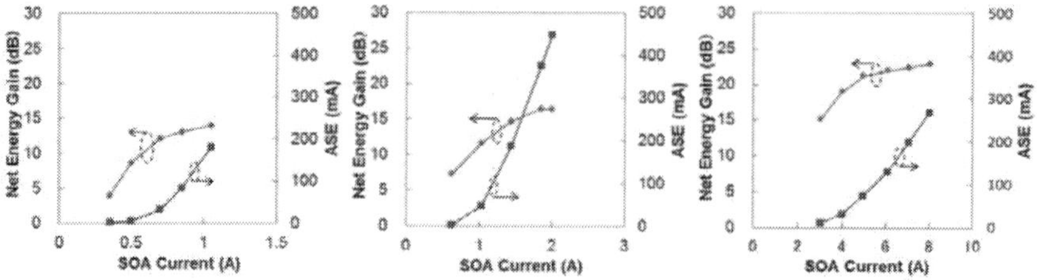

Figure 12. Net gain and ASE power values as functions of SOA current for devices B, C, and D.

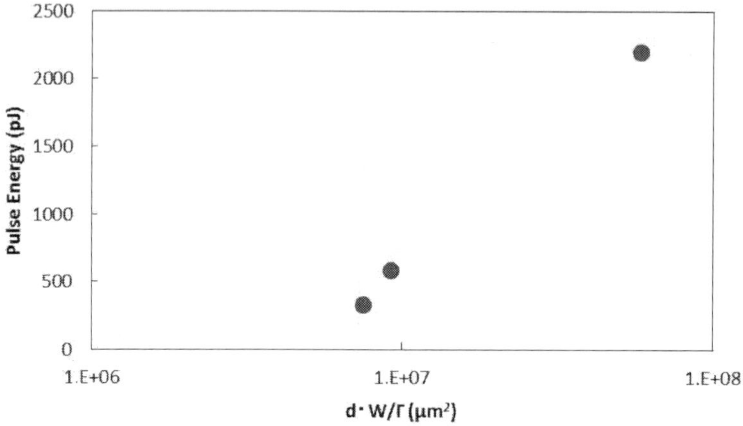

Figure 13. Pulse energy as a function of dw/Γ.

It is helpful to discuss in detail the result obtained with device D. This device employed a passively mode-locked GaN laser as the master oscillator, in the commonly used Littrow configuration. The gain section of the LD was forward biased by 90 mA, and the SA section was reverse biased at –9 V. The pulse repetition rate was 812 MHz, and the input power was 7.2 mW. The wavelength was tuned to 405 nm. Temporal characterization of the input pulse was performed using an in-house intensity autocorrelation measurement apparatus that utilizes the surface second harmonics of a β-BaB_2O_4 (BBO) crystal [38]. Autocorrelation measurements indicated an FWHM value of 4.2 ps, corresponding to a 3.0-ps pulse duration assuming a Gaussian waveform. The peak power was estimated to be 3 W.

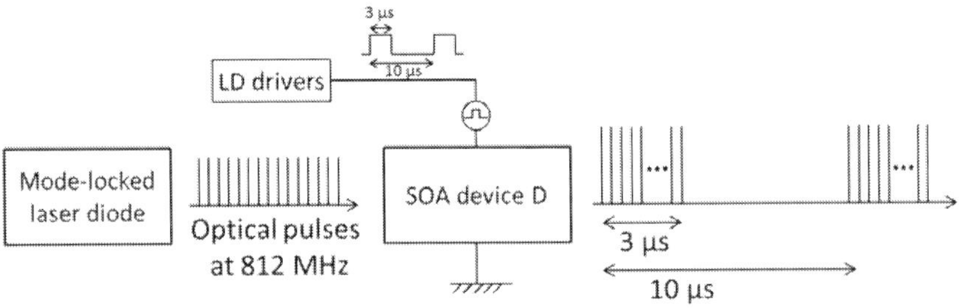

Figure 14. Schematics of the master oscillator power amplifier using SOA device D.

SOA D was driven by LD drivers at a 100-kHz repetition rate with 3 μs width rectangular current pulse, as illustrated in Figure 14, which presents the autocorrelation trace and optical spectrum acquired at a 8-kA/cm^2 bias current density. Here, the FWHM was 5.0 ps, and the pulse duration was estimated to be 3.5 ps, assuming a Gaussian waveform. Even though a high net gain of greater than 20 dB was obtained, the pulse duration was not significantly broadened. The peak wavelength of the amplified pulse was 405.2 nm, indicating a 0.2-nm red shift, as shown in Figure 15(b). The spectral bandwidth also increased from 0.08 to 0.26 nm after amplification. These spectral changes were due to SPM. It is important to point out that the amount of red shift exhibited in this case was less than that observed in previous trials; as an example, a shift of 0.4 nm was obtained with device C [29]. Figure 16 shows the average output power and estimated pulse energy values as functions of the SOA bias current density. The average output power values were 0.71 and 0.27 W with and without pulse injection. At 8 kA/cm^2, the pulse energy was estimated to be 2.2 nJ, and the corresponding peak power was estimated to be 630 W. These are the highest pulse energy and peak power ever generated by GaN-based MOPA.

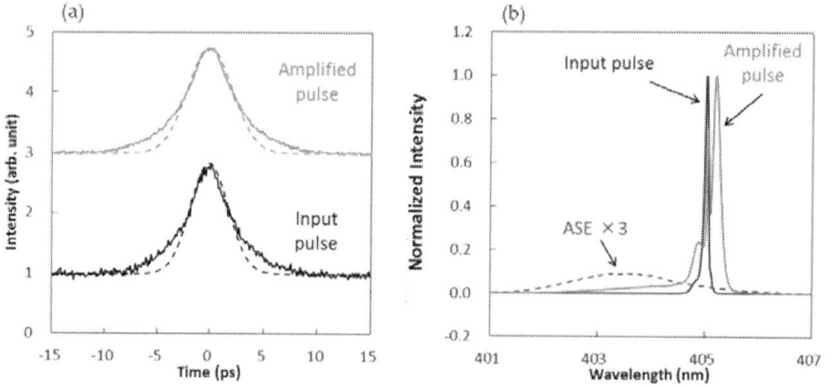

Figure 15. (a) Autocorrelation traces and (b) optical spectra of input and amplified pulses. The dashed lines show Gaussian function fitted to the autocorrelation traces in panel a. The ASE spectrum (with its intensity multiplied by three for illustrative purposes) is also shown in panel b as a dashed line.

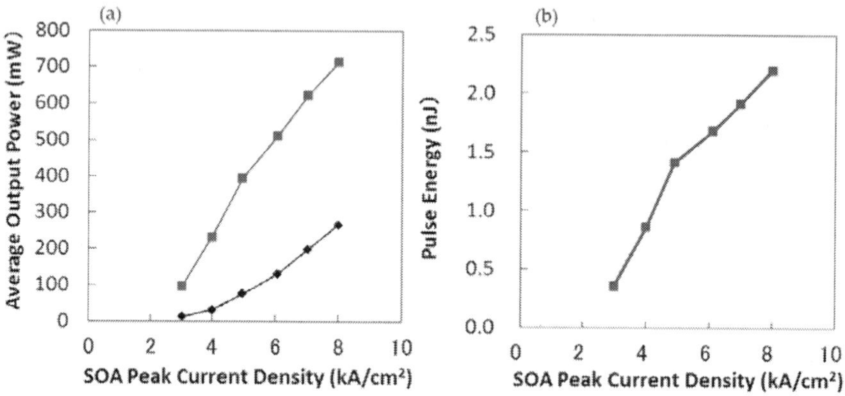

Figure 16. (a) Average output and (b) pulse energy values as functions of the SOA bias current density.

4. POTENTIAL APPLICATIONS

There are many potential applications for GaN-based SOAs, among which those utilizing optical pulses are the most prominent. In this section, these applications are discussed.

4.1. Micromachining applications

Recently, there has been a rapid growth in laser material processing, as the cost of lasers has decreased and fiber laser technology has advanced. In microscale machining, short optical pulses are desirable since picosecond/femtosecond pulses can effectively reduce the heat damage resulting from the laser light. In fact, it has been demonstrated that nanosecond pulses produce dull edges [39], as shown by Chichkov et al., who demonstrated this by comparing the pulse width dependence of laser ablation of steel.

Short wavelength optical pulses can also be beneficial with regard to precise processing since the spot size of the focused beam is proportional to NA/λ, where NA is the numerical aperture of the lens and λ is the wavelength of the laser. Kauf et al. published a systematic study of microprocessing using mode-locked picosecond and q-switched nanosecond UV lasers having the same average and peak power values, although with differing pulse energies [40]. These trials demonstrated that picosecond UV pulses offer an advantage compared to nanosecond pulses since they allow higher cutting speeds when working with dielectric materials such as polyimides. The significantly higher pulse repetition of the mode-locked laser operating at 80 MHz resulted in a higher cutting speed compared to that obtained from the q-switched laser working at 80 Hz.

4.2. Orthoptics

Laser refractive surgery, such as laser in situ keratomileusis (LASIK), corrects the refractive properties of the cornea. Two different types of pulse lasers are typically used in LASIK: an IR femtosecond laser for cutting corneal flaps and a UV nanosecond excimer laser for ablating corneal tissue. Ablation of the corneal tissue with UV pulses (typically at 193 nm) is able to change the shape of the cornea's surface to allow for vision correction. Recently, an alternative method for vision correction was demonstrated by Xu and Knox et al. Instead of physically removing a part of cornea, this new noninvasive technique modifies the refractive index of the cornea using femtosecond pulses at 400 nm, which is the most effective wavelength. The 400-nm pulses are provided by the second-order harmonics of a mode-locked Ti:Sap laser with an average power of 60 ± 1 mW and corresponding pulse energies of approximately 0.8 nJ at the cornea. A GaN-based SOA employing MLLD would seem to be a promising candidate for this application since the required average power and pulse energies can easily be generated by the type of SOA discussed in the present chapter and the associated semiconductor-based system offers energy efficiency and compactness.

4.3. Optical storage

The storage capacities of optical disks have improved from 0.7 gigabytes (GB) in the case of compact discs (CDs) to 4.7 GB for digital versatile discs (DVDs) and, more recently, to 25 GB with Blu-ray discs. These improvements were realized in conjunction with advances in laser diode technologies in terms of achieving shorter lasing wavelengths since the spot size of the laser beam is proportional to λ/NA. For next-generation optical data storage, three-dimensional (3D) optical data storage might be the best candidate since, as the name implies, this technique allows information to be stored three dimensionally, generating a significant increase in storage capacity [41]. Several groups have demonstrated such systems [42, 43], and 1 terabyte data storage has been demonstrated [44]. To store data three dimensionally, multiphoton processes, such as two photon absorption, must be induced using high peak power optical pulses. A GaN-based SOA incorporating MLLD would be highly suitable as a practical pulse source, and in fact, such applications have already been demonstrated [45, 46].

5. CONCLUSION

GaN-based optoelectronic devices are already widely used as optical storage lasers and solid-state lighting, and because of their ability to potentially cover a wide spectral range from the deep UV to the IR regions, many research studies concerning these devices are ongoing. However, very few reports on GaN-based SOAs exist. In this chapter, the basic characteristics of GaN-based SOAs were reviewed, focusing on pulse amplification. The particular SOA devices described herein generate a pulse energy greater than 2 nJ, a value that is comparable to the values obtained from widely used pulse sources based on solid-state lasers. There are many potential applications for these devices, such as microprocessing, orthoptics, and high-density optical storage, using UV picosecond and femtosecond pulses. These applications would all benefit from employing GaN-based SOAs.

6. ACKNOWLEDGEMENTS

The authors wish to gratefully acknowledge Professor Yokoyama of Tohoku University for helpful discussions and to note that earlier studies of the GaN-SOAs (devices A, B, and C) were performed in collaboration with Professor Yokoyama's research group. Thanks are also due to the author's research colleagues: T. Oki, Y. Takiguchi, M. Kuramoto, T. Miyajima, H. Nakajima, M. Shiozaki, Y. Hanzawa, N. Sugawara, and M. Ikeda. Finally, the assistance of Sony Semiconductor in fabricating the experimental devices is gratefully acknowledged.

REFERENCES

1. Taniyasu Y, Kasu M, Makimoto T: An aluminium nitride light-emitting diode with a wavelength of 210 nanometres. Nature. 2006;441(7091):32--8.
2. Yamashita Y, Kuwabara M, Torii K, Yoshida H: A 340-nm-band ultraviolet laser diode composed of GaN well layers. Optics Express. 2013;21:3133.
3. Yoshida H, Yamashita Y, Kuwabara M, Kan H: Demonstration of an ultraviolet 336 nm AlGaN multiple-quantum-well laser diode. Applied Physics Letters. 2008;93(24):241106.
4. Frost T, Banerjee A, Sun K, Chuang SL, Bhattacharya P: InGaN/GaN quantum dot red (λ = 630 nm) laser. IEEE Journal of Quantum Electronics. 2013;49:923.
5. Connelly MJ. Semiconductor Optical Amplifiers: Springer US; 2007.
6. Dutta NK, Wang Q. Semiconductor Optical Amplifiers: World Scientific Pub.; 2006.
7. Gee S, Bowers JE: Ultraviolet picosecond optical pulse generation from a mode-locked InGaN laser diode. Applied Physics Letters. 2001;79:1951.
8. Ding Y, Aviles-Espinosa R, Cataluna MA, Nikitichev D, Ruiz M, Tran M, et al.: High peak-power picosecond pulse generation at 1.26 μm using a quantum-dot-based external-cavity mode-locked laser and tapered optical amplifier. Optics Express. 2012;20(13):14308–20.
9. Kuramoto M, Kitajima N, Guo H, Furushima Y, Ikeda M, Yokoyama H: Two-photon fluorescence bioimaging with an all-semiconductor laser picosecond pulse source. Optics Letters. 2007;32(18):2726–8.

10. Nakamura S, Masayuki Senoh, Nagahama S-i, Iwasa N, Yamada T, Matsushita T, et al.: InGaN/GaN/AlGaN-based laser diodes with modulation-doped strained-layer superlattices. Japanese Journal of Applied Physics. 1997;36:L1568.
11. Obata T, Kitajima N, Ohta M, Ichinokura H, Masaru K: Low noise characteristics of AlGaInN-based self-pulsating laser diodes. Physica Status Solidi (A). 2008;205:1096.
12. Ohno T, Ito S, Kawakami T, Taneya M: Self-pulsation in InGaN laser diodes with saturable absorber layers. Applied Physics Letters. 2003;83:1098.
13. Tronciu VZ, Yamada M, Abram RA: Analysis of the dynamics of a blue-violet inxGa1-xN laser with a saturable absorber. American Physical Society. 2004;70(22):026604-1-6.
14. Tronciu VZ, Yamada M, Ohno T, Ito S, Kawakami T, Taneya M: Self-pulsation in an InGaN laser-theory and experiment. IEEE Journal of Quantum Electronics. 2003;39(12):1509–14.
15. Kono S, Oki T, Miyajima T, Ikeda M, Yokoyama H: 12W peak-power 10ps duration optical pulse generation by gain switching of a single-transverse-mode GaInN blue laser diode. Applied Physics Letters. 2008;93(13):131113.
16. Kuramoto M, Oki T, Sugahara T, Kono S, Ikeda M, Yokoyama H: Enormously high-peak-power optical pulse generation from a single-transverse-mode GaInN blue-violet laser diode. Applied Physics Letters. 2010;96(5):051102.
17. Oki T, Kono S, Kuramoto M, Ikeda M, Yokoyama H: Generation of over 10-W peak-power picosecond pulses by a gain-switched AlGaInN-based self-pulsating laser diode. Applied Physics Express. 2009;2:032101.
18. Miyajima T, Watanabe H, Ikeda M, Yokoyama H: Picosecond optical pulse generation from self-pulsating bisectional GaN-based blue-violet laser diodes. Applied Physics Letters. 2009;94(16):161103.
19. Scheibenzuber WG, Hornuss C, Schwarz UT, Sulmoni L, Dorsaz J, Carlin J-F, et al.: Self-pulsation at zero absorber bias in GaN-based multisection laser diodes. Applied Physics Express. 2011;4(6):062702.
20. Watanabe H, Miyajima T, Kuramoto M, Ikeda M, Yokoyama H: 10-W peak-power picosecond optical pulse generation from a triple section blue-violet self-pulsating laser diode. Applied Physics Express. 2010;3(5):052701.

21. Oki T, Koda R, Kono S, Miyajima T, Watanabe H, Kuramoto M, et al.: Direct generation of 20 W peak power picosecond optical pulses from an external-cavity mode-locked GaInN laser diode incorporating a flared waveguide. Applied Physics Letters. 2011;99(11):111105.
22. Olle VF, Vasil'ev PP, Wonfor A, Penty RV, White IH: Ultrashort superradiant pulse generation from a GaN/InGaN heterostructure. Opt Express. 2012;20(7):7035–9.
23. Watanabe H, Kuramoto M, Kono S, Ikeda M, Yokoyama H: Blue-violet bow-tie self-pulsating laser diode with a peak power of 20 W and a pulse energy of 310 pJ. Applied Physics Express. 2010;3(12):122103.
24. Holc K, Weig T, Köhler K, Wagner J, Schwarz UT: Impact of band structure and absorber dynamics on self-Q-switching in GaN-based multisection laser diodes at high reverse bias. Applied Physics Express. 2013;6(8):084101.
25. Saito K, Watanabe H, Miyajima T, Ikeda M, Yokoyama H: Mode locking of an external-cavity bisection GaInN blue-violet laser diode producing 3 ps duration optical pulses. Applied Physics Letters. 2010;96(3):031112.
26. Oki T, Saito K, Watanabe H, Miyajima T, Kuramoto M, Ikeda M, et al.: Passive and hybrid mode-locking of an external-cavity GaInN laser diode incorporating a strong saturable absorber. Applied Physics Express. 2010;3(3):032104.
27. Kono S, Watanabe H, Koda R, Miyajima T, Kuramoto M: 200-fs pulse generation from a GaInN semiconductor laser diode passively mode-locked in a dispersion-compensated external cavity. Applied Physics Letters. 2012;101(8):081121.
28. Koda R, Oki T, Miyajima T, Watanabe H, Kuramoto M, Ikeda M, et al.: 100 W peak-power 1 GHz repetition picoseconds optical pulse generation using blue-violet GaInN diode laser mode-locked oscillator and optical amplifier. Applied Physics Letters. 2010;97(2):021101.
29. Koda R, Oki T, Kono S, Miyajima T, Watanabe H, Kuramoto M, et al.: 300 W peak power picosecond optical pulse generation by blue-violet GaInN mode-locked laser diode and semiconductor optical amplifier. Applied Physics Express. 2012;5(2):022702.
30. Koda R, Takiguchi Y, Kono S, Watanabe H, Hanzawa Y, Nakajima H, et al.: Generation of a 2.2 nJ picosecond optical pulse with blue-violet wavelength using a GaInN master oscillator power amplifier. Submitted to Applied Physics Letters. 2015.

31. Koda R, Oki T, Miyajima T, Watanabe H, Kuramoto M, Ikeda M, et al., editors. Generation and amplification of 400 nm band picosecond optical pulses by GaInN laser diodes. International Conference on Ultrafast Phenomena; 2010 2010/07/18; Snowmass Village, Colorado: Optical Society of America.
32. Coldren LA, Corzine SW. Diode Lasers and Photonic Integrated Circuits. Wiley; 1995.
33. Piprek J. Nitride Semiconductor Devices: Principles and Simulation: Wiley; 2007.
34. Agrawal GP, Olsson NA: Self-phase modulation and spectral broadening of optical pulses in semiconductor laser amplifiers. IEEE Journal of Quantum Electronics. 1989;25(11):2297–306.
35. Baveja PP, Maywar DN, Kaplan AM, Agrawal GP: Self-phase modulation in semiconductor optical amplifiers: impact of amplified spontaneous emission. IEEE Journal of Quantum Electronics. 2010;46(9):1396–403.
36. Brosson P: Analytical model of a semiconductor optical amplifier. Journal of Lightwave Technology. 1994;12(1):49–54.
37. Lee CH, Delfyett PJ: Limits on amplification of picosecond pulses by using semiconductor laser traveling-wave amplifiers. IEEE Journal of Quantum Electronics. 1991;27(5):1110–4.
38. Kono S, Oki T, Kuramoto M, Ikeda M, Yokoyama H: Intensity autocorrelation measurement of 400 nm picosecond optical pulses from a GaInN mode-locked semiconductor laser diode using surface second harmonic generation of β-BaB2O4 crystal. Applied Physics Express. 2010;3(12):122701.
39. Chichkov BN: Femtosecond, picosecond and nanosecond laser ablation of solids. Applied Physics A. 1996;63(2):109–15.
40. Kauf M, Patel R, Bovatsek J, Gries W: High power UV q-switched and mode-locked laser comparisons for industrial processing applications. Proceedings of SPIE. 2008;6871:687123-10.
41. Gu M, Li X: The road to multi-dimensional bit-by-bit optical data storage. Optics and Photonics News. 2010;21(7):28–33.
42. Parthenopoulos DA, Rentzepis PM: Three-dimensional optical storage memory. Science. 1989;245(4920):843–5.
43. Strickler JH, Webb WW: Three-dimensional optical data storage in refractive media by two-photon point excitation. Optics Letters. 1991;16(22):1780–2.

44. Walker E, Dvornikov A, Coblentz K, Rentzepis P: Terabyte recorded in two-photon 3D disk. Applied Optics. 2008;47(22):4133–9.

45. Tashiro S, Takemoto Y, Yamatsu H, Miura T, Fujita G, Iwamura T, et al.: Volumetric optical recording using a 400 nm all-semiconductor picosecond laser. Applied Physics Express. 2010;3(10):102501.

46. Tashiro S, Yamatsu H, Takemoto Y, Fujita G, Miura T, Iwamura T, et al.: Volumetric optical recording with void marks using an all-semiconductor picosecond laser. Japanese Journal of Applied Physics. 2011;50(9):09MF3.

6
CHAPTER

DEPENDENCE OF THE CONTACT RESISTANCE ON THE DESIGN OF STRANDED CONDUCTORS

Youcef Zeroukhi [1,4], Ewa Napieralska-Juszczak [1,2,*], Guillaume Vega [4], Krzysztof Komeza [3], Fabrice Morganti [1,2] and Slawomir Wiak [3]

[1] *Lille Nord de France, F-59000 Lille, France*
[2] *LSEE, UA, Technoparc Futura, F-62400 Bethune, France*
[3] *Institute of Mechatronics and Information Systems Technical University of Lodz, Stefanowskiego 18/22, Lodz, Poland*
[4] *Nexans, 62300 Lens, Bd de Marais, France*

ABSTRACT

During the manufacturing process multi-strand conductors are subject to compressive force and rotation moments. The current distribution in the multi-strand conductors is not uniform and is controlled by the transverse resistivity. This is mainly determined by the contact resistance at the strand crossovers and inter-strand contact resistance. The surface layer properties, and in particular the crystalline structure and degree of oxidation, are key parameters in determining the transverse resistivity. The experimental set-ups made it possible to find the dependence of contact resistivity as a function of continuous working stresses and cable design. A study based on measurements and numerical simulation is made to identify the contact resistivity functions.

Keywords: stranded conductors; contact resistance; cable design; cable simulation; contact resistance measurement

1. INTRODUCTION

In the context of high metal prices—particularly of copper—the main material used in electrical conductors, a reduction in the use of raw materials is one of the challenges facing the cable industry today. Thus the usage of copper or aluminium (or any other conducting material) in the manufacturing of cables must be analysed to achieve the desired electrical conductance under the imposed conditions.

This paper presents partial results of our work concerning the development of a finite element model for three-dimensional modelling and design optimization of multi-strand conductors. One of the points hindering this study has been the lack of understanding in the area of contact between wires. Indeed, these areas of contact between the wires of the cable, are an essential problem for a model incorporating robust mechanical and electrical coupling. The characteristics of the contact electrical parameters between the conductor strands versus mechanical stress and the relative position of individual wires have a significant influence on the distribution of current lines in the cable [1–3]. This paper is focused on the characterization of the electrical contact resistance by experimental measurements and electromechanical simulation based on the finite element method. The authors' interest mainly focuses on the study of electrical Cu/Cu contact between two conductors under the influence of a compressive force. The attention is particularly concentrated on the study of the influence of the contact force and of the crossing angle between the wires, and the shape and the size of the contact area.

In the first phase, the experimental measurement device was prepared. First the bench for applying compression and measurement of the contact force was built. It also has a system for fixing the angle of intersection of the conductors. Measurements have been conducted for DC loads. The electrical contact resistance, for its part, is measured using a 4-pointmethod (Kelvin method) with amicro-ohmmeter. The measurement protocol and the experimental test conditions are specified.

In a second step, the results of several dozen tests are presented. The analysis of these results is performed in order to understand the influence and quantify the impact of each factor studied on the electrical behaviour of the contact interface, including its resistance. After each measurement the sample was examined under a microscope to first investigate the shape and the size of the imprint, and second to investigate the internal structure of the conductor.

The copper wires from which the measured samples have been made are produced using a process called cold deformation. The stranding and compacting process is a complex operation that combines two simultaneous displacements: translation and rotation. From these combined displacements, three distinct forces will act on the wires: traction, torsion and compression. To compare the measurement and simulation results, this process was modelled using the Abaqus program [4] which allows simulating the stranding and compacting operation to calculate the mechanical deformations, the residual stresses, the inter-strands contact pressure and the resultant plastic strain. This program detects the contact zone as well as calculates the stress distribution.

Thanks to the measured characteristics of the resistivity as a function of stress it is possible to takes into consideration the non-homogenous resistivity of the contact zone. The non-homogenous resistivity will be implemented and taken into account in the finite elements model by defining an electromechanical coupling strategy. This will allow us to perform a more detailed analysis of the electrical phenomena and their impact on the total electrical resistance of the conductive core.

Finally, the identification of the electrical contact resistance and resistivity of material according to acting stress were applied for simulation of the 1+6 cable. The simulation results for different samples are analysed and compared with experimental measurements to validate the numerical model.

2. EXPERIMENTAL SECTION

2.1. Method of Measuring of the Contact Resistance

There are many unknowns relevant to optimization of the electrical contact resistance and experimental techniques to measure the resistance and its dependence on various parameters, such as temperature, pressure and the contact angle [5–9]. The two main questions requiring an answer concern the value of the contact resistance between the strands of a multi-strand conductor at a given force, created by the stranding process or due to subsequent application of compression, and ways to reduce contact resistance by considering the shape, size and distribution of contact interface sand current paths.

It is necessary to use the theory of mechanical contact in the analysis of the electrical contact. For this purpose a dedicated experimental device has been developed ("experimental bench"). This device is used to evaluate the electrical contact resistance between two conductors under the action of a force of pressure on the contact area of the cables. The contact surface between two solid conductors contains a number of mechanical contact blocks due to the inevitable surface roughness and lay length. The number and size of the electrical contact depends on many parameters like the force, the macroscopic form of the surfaces, the microscopic topology of the surfaces (roughness), the mechanical properties of materials (hardness, Young's modulus) and the geometry. We need to measure the electrical resistance of a contact established between two metallic conductors arranged in a precise angle beam, subjected to a force of progressive compression. The electrical contact resistance is measured according to the "4-points method" (Kelvin method).The overall diagram of the experimental measurement is shown in Figure 1.

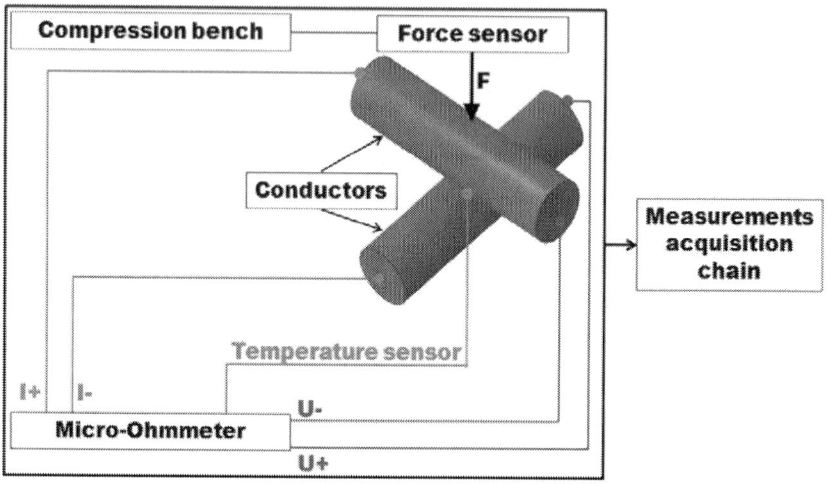

Figure 1. Diagram of the experimental measurement.

Two elements are linked together: a mechanical part that works as a compression machine, and a second part for electrical measurements using the Kelvin method to measure the resultant contact resistance. The bench has a rotary plate to allow the measurement of the contact resistance at a variable angle between the crossing wires (Figure 2b). It has the possibility to set different crossing angles corresponding to the desired lay length of

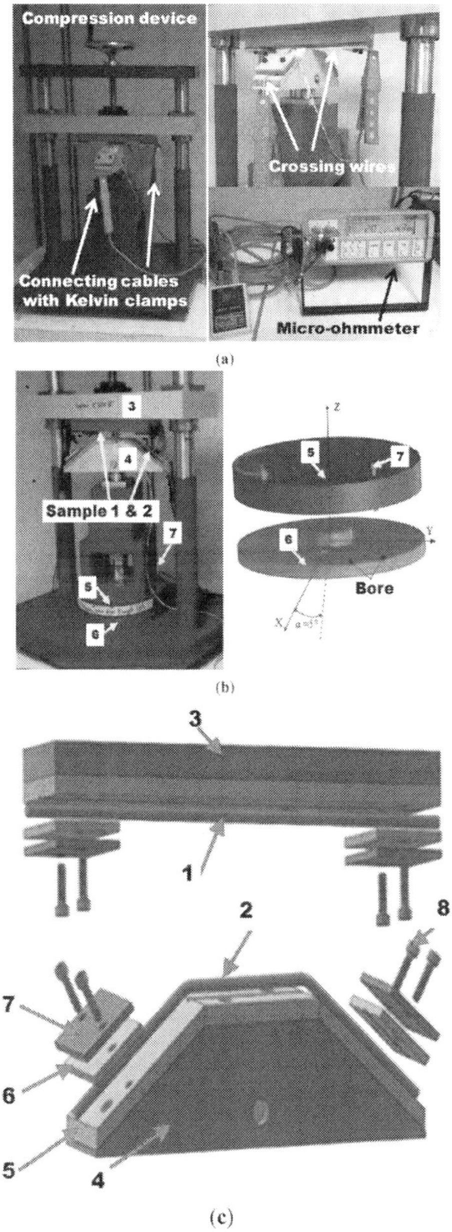

Figure 2. (**a**) The measuring bench; (**b**) Two disks allowing the adjustment of the crossing angle; (**c**) The system for fixing the wire.

the stranded conductor. The contact resistance is measured using the 4-points method: two points for current injection and two points for potential measurement. For this purpose a sensitive OM22 micro-ohmmeter is used. This apparatus offers the possibility for temperature compensation and automatic data collection.

The compression bench is shown in Figure 2; it uses a clamping force and its development was inspired by some publications [2,5,10]. Compression of the wires is achieved by turning the crank wheel at the top of the device as shown in Figure 2b; measurement of the compression force is possible for two intersecting strands. The measurement configuration (the force range of 0.5 N to 500 N) makes it possible to take several measurements on the same pair of samples. For this study we have 200 samples extracted from a copper coil manufactured by Nexans, from which we cut samples 40 cm in length in order to fit the test bench. The system for fixing the wire is shown in Figure 2c. The conductor (6) is placed on an insulating support (2) and held in position by insulating pressing blocks (3) and (4). The whole assembly is fixed to the frame (1) using two screws (5). The conductor needs to have a shape as shown in Figure 2c.

Once the mechanical contact is made between the two copper wires at a certain compression force, a pulsed current is injected on two opposite ends of the two wires. The current will flow from one wire to the other through the contact surface because of the difference of potential set between the two wires. The increase of the contact pressure will change the characteristics of the contact zone. Several sources of the measurement system error such as the external environment can affect the accuracy of low resistance measurements. To reduce these errors, the resistance measurements are made at room temperature using the Kelvin method [11]. The current circuit and the potential circuit are separated to eliminate the resistance of the connecting cables. The tests are performed with temperature compensation using sensitive thermocouple. A series of measurements according to the applied force and crossing angle are made. The contact force range is set from 0.5 N to 500 N with a contact angle variation from 0° to 90° under a pulsed current of 10 A with a time period of 0.5 s. The use of a pulsed current is to reduce the Joule heating in the contact interface. The contact force is measured using a tensile compression S-shaped type SM5424 strain gauge sensor, connected to a SM18 USB2 data acquisition type converter. Some of the parameters related to the conditions of measurements may be a source of measurement error. One of such cases arises when the angle between the conductors is 0°, as it is easy to overlook a slight misalignment of one axis against the other, resulting in the contact area to be reduced and thus the contact resistance to be overestimated.

2.2. Presentation and Analysis of Results

Several series of measurements have been conducted. The first series are mainly devoted to the identification of the impact force on the electrical resistance for different arrangements of wires. The average diameter of the wires is 2.06 mm. The second series are conducted to link the different parameters to each other to identify and understand their influence on the electrical contact resistance.

2.2.1. Influence of the Compression Force

This section mainly deals with the study of the impact of the compression force on the electrical contact resistance. The measurements are performed continuously in a 9 s interval between each measurement. The number of test specimens is 200. The number of samples was important to perform statistically relevant measures for each crossing angle. The first 10 tests were conducted on a perpendicular arrangement of the conductors (crossing angle 90°). At the end of these tests, the variation of the contact resistance as a function of the force was determined (see Figure 3).

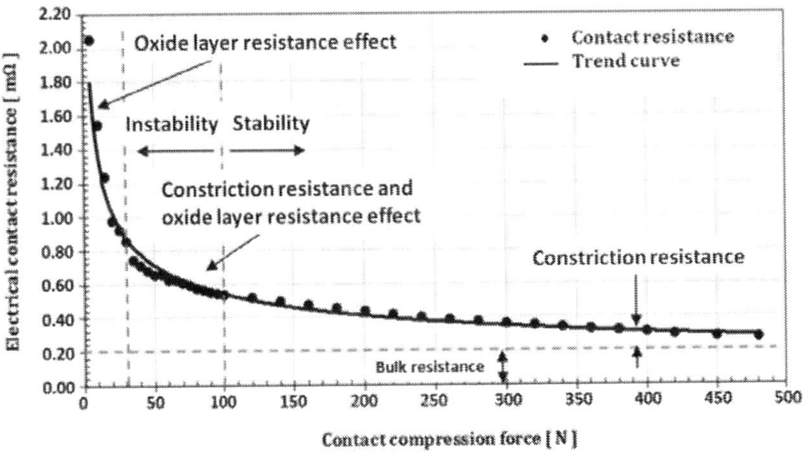

Figure 3. Electrical contact resistance as a function of the compression force (crossing angle of 90°) for copper (measuring current of 10 A, rated voltage drop of 20 mV, the diameter of wire 2.06 mm, crossing angle 90°).

This curve represents the average contact resistance obtained from the 10 tests. After each measurement the conductor was examined under a microscope. The imprint size was verified and its surface area was calculated. For a force of less than 30 N the imprint was invisible. This proves that the deformation was elastic. For forces greater of 30 N the imprint is visible, which means that we are dealing with plastic deformation. It may be observed that for low contact compression force (<30 N) we are dealing with elastic deformation and the current flow at the contact interface is controlled by the oxide layer, therefore the resultant contact area is extremely small. While, from 30 N to 100 N there is plastic deformation and a constriction resistance will appear in few contact points. In this zone, it may be observed that the oxide layer resistance and constriction resistance will coexist [12,13]. Finally, the stability of contact resistance can be observed up a compression force 100 N. (Table 1)

Table 1. Measurement of the electrical contact resistance (mΩ) as a function of the applied force (N) and the dimension of the contact area (mm^2) measured for crossing angles 90°, 60° and 45°, for the wire of 1.80 mm diameter.

Crossing Angle	Measured Average Force [N]	Contact Area [mm^2]	Contact Resistance [mΩ]
90°	50.58	-	3.878
	101.50	0.221	2.328
	202.14	0.478	1.500
	300.75	0.724	0.984
	401.37	1.021	0.785
	481.12	1.287	0.684
60°	50.58	//	2.854
	101.50	1.027	1.703
	202.14	1.979	0.945
	300.75	2.859	0.688
	401.37	4.134	0.479
	481.12	4.971	0.413
45°	50.58	//	2.242
	101.50	0.865	1.587
	202.14	2.009	0.814
	300.75	3.318	0.561
	401.37	4.366	0.309
	481.12	5.770	0.276

2.2.2. Influence of the Force Depending on the Angle of Intersection of Conductors

The study of the influence of the crossing angle is made at 10 angles. The studies were conducted to investigate the reproducibility of the curves and are presented in the following only for the curves for 10°, 20°, 40°, 50°, 60°, 80° and 90°. These curves are shown in Figure 4.

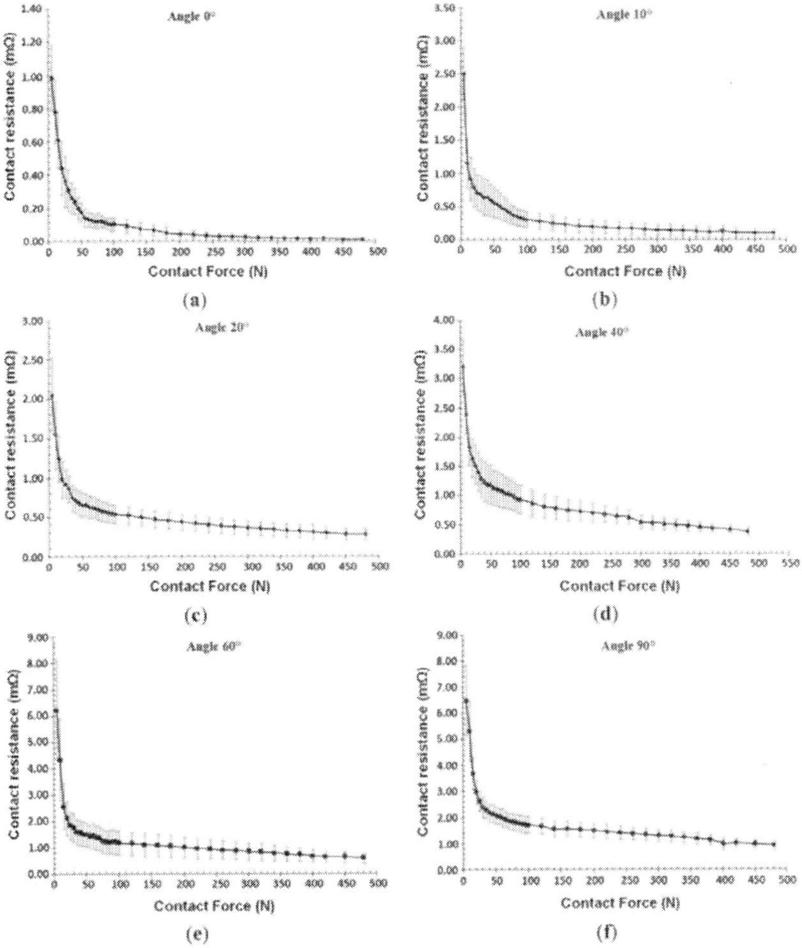

Figure 4. Resistance of the contact area according to the angle between the crossed wires, diameter of wire 1.8 mm.

For reasons of clarity, the results for some angles have not been presented. It can be seen that the dependence of the electrical resistance behaviour on the applied force is the same as identified above, and, for all specified angles. Figure 4 shows the effect of the crossing angle on the evolution of the electrical resistance. As the angle of intersection of the conductors in contact tends to a parallel arrangement of the conductors, something that occurs at the 0° angle, the more the resistance decreases and becomes more stable.

Summarising the results it can be observed that there are certain tendencies common to different measurements. In the initial part of the characteristic curves, for forces between 5 N and 30 N, there is an abrupt reduction of the contact resistance, followed by an 'inverted knee point' for forces in the range between 30 N and 100 N; this may be explained by a phenomenon of plastic deformation of the contact region while the rest of the material enjoys elastoplastic deformation. The last section of the curve clearly describes a fairly 'stable' behaviour up to the maximum force. Each of the above sections may also be related to the manufacturing process of multi-strands conductors where different forces are used when stranding the conductor. Generally, higher force the lower the contact resistance. At Figure 4 on each curve the dispersion of the measurement and the average values are shown.

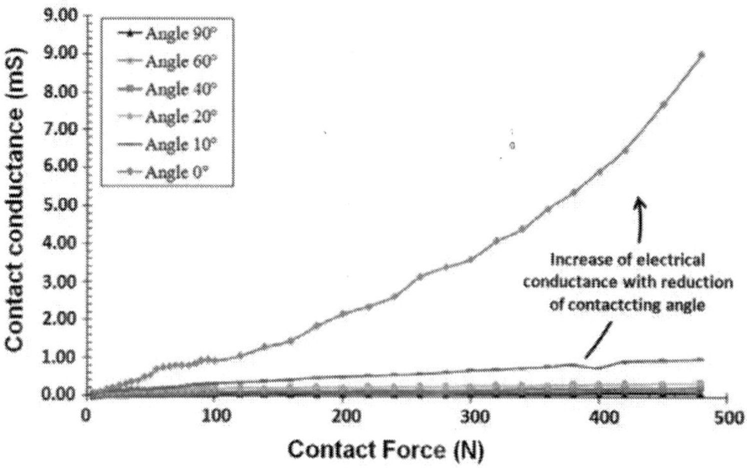

Figure 5. Variation of the electric conductance as a function of contact force to two extreme arrangements of conductors (angle 0° and 90°).

Figure 5 compares the electrical conductance of contact (1/RC) for the two extreme angles 0° and 90° depending on the applied force applied. As can be seen, electrical conductance is significantly higher for the cross angle 0°. The reason for this is an increase of the contact area with the reduction in angle of intersection of the conductors.

2.3. Influence of Variation of the Contact Area

Another effect also depending on the applied load is observed on changes in the contact resistance during testing. This effect relates to the contact area formed by the crossing angle between the two wires. It should be noted that not only is the distribution of the contact resistance non-uniform, but the depth of the penetration of one conductor into another is not constant. In order to incorporate this into our model, the mechanical analysis needs to be coupled with the electrical description as when two conductors penetrate each other their physical and mechanical properties changed. Accurate dimensions of the imprint were established under the microscope (vertical and lateral resolutions equal to 200 μm) and are shown in Figures 6 and 7. Figure 6 shows the parallel disposition of the conductors. The area of the contact surface under the maximum force at the angle of 0° may be found analytically by approximating the imprint between two parallel cylinders by a rectangle of known length equal to the length of the sample used in the measurements. Applying a large force of 480 N, results in smoothing out the unevenness of the contact surface and thus increasing the contact surface area up to the full area of the relevant rectangle, as shown in Figure 6. The rectangular area is then calculated as: $S = 100 \times [(0.386 + 0.258 + 0.279 + 0.274)/4] = 2.35\ mm^2$.

Another example of imprints measurements for a crossing angle of 45° are shown for different contact stresses in Figure 7. The imprints are always elliptical at all angles, except for 0° (parallel conductors). When the angle increases the 'width' of the imprint is reduced while the 'length' remains constant.

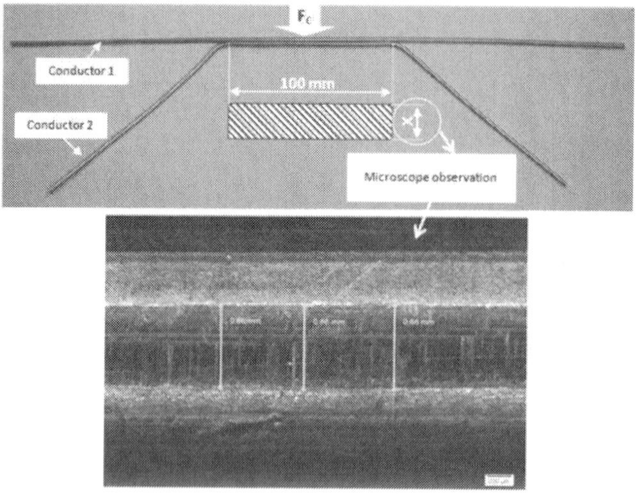

Figure 6. Measurement of the imprint at a crossing angle 0° for a contacting force of 480 N. and a stress of 7.27 N/mm².

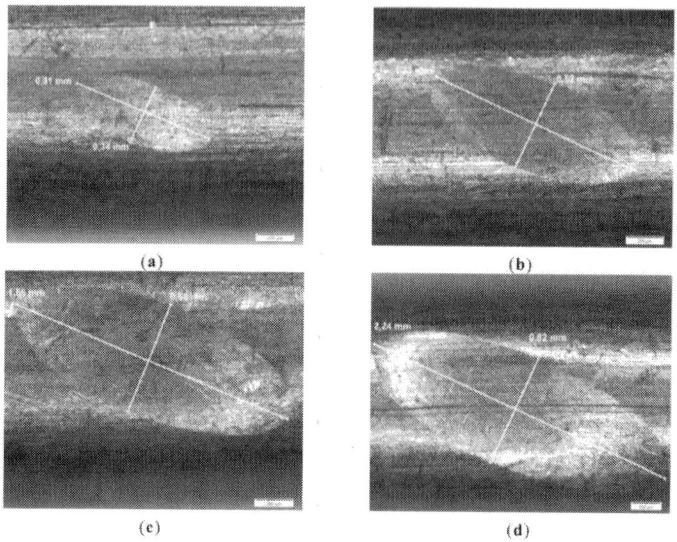

Figure 7. Residual imprints on the copper wire for a contacting angle of 45° under ascending contact forces. (**a**) F_C = 100 N, averaged stress =115 N/mm²; (**b**) F_C = 200 N, averaged stress = 99.5 N/mm²; (**c**) F_C = 300 N, averaged stress = 90.4 N/mm²; (**d**) F_C = 480 N, averaged stress =83.2 N/mm².

The uniform distribution of the stress on the contact area was established at this stage. The non-uniform stress repartition was taken into account during electro-mechanical simulations. From these experimental tests, we observe an increase of the dimensions of contact zone between the wires when the crossing angle is decreased (Figure 8). The contact resistance will then decrease to become stable for parallel wires as shown in Figure 9. In the case of stranded conductors, of the type supplied by Nexans and studied here, the stranding lay length determines the angle at which conductors intersect. Figure 8 and 9 show the resulting contact resistance and contact area as a function of the angle of intersection. Significant influence of this angle on the value of the resistance may be observed. The contact area shrinks as the angle increases; for example, the difference between the contact area at the angles of 0° and 90° is 1.491 mm^2.

A change in the contact area affects the current distribution. As the angle of intersection is related to the lay length with which the conductor is stranded, it is clear that we can influence the resultant resistance of the conductor by changing the lay length ratio.

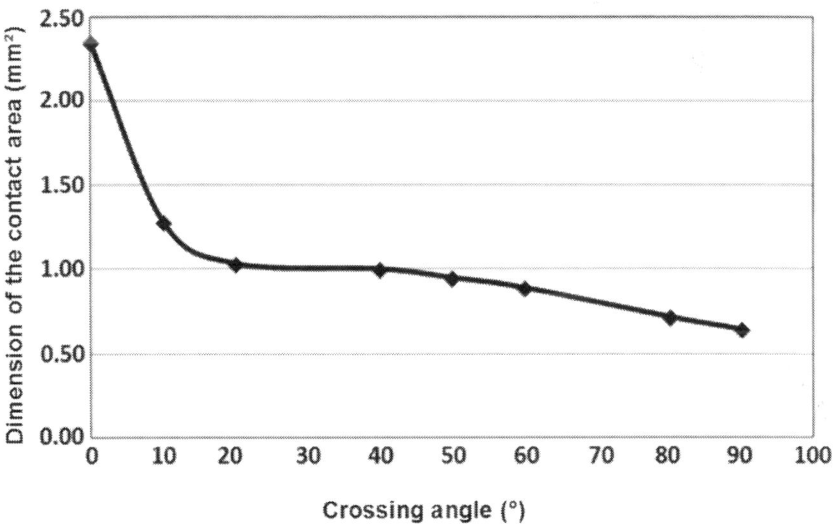

Figure 8. Dimensions of the contact zone as a function of crossing angle (compressive force 480 N).

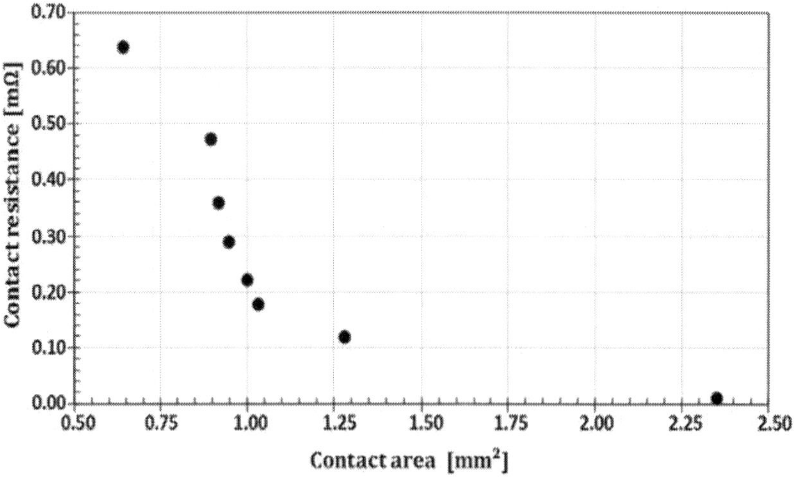

Figure 9. Contact resistances as a function of the areas' dimensions.

3. NUMERICAL MODEL

3.1. The Design of the Conductors

The stranding and compacting process is an operation that combines two simultaneous displacements: translation and rotation. The wires are subjected to three forces: traction, torsion and compression, as shown in Figure 10. The traction force is due to the linear speed provided by the winch, the torsion is generated from the yokes and the compression force from the area reduction at the lay point. The mechanical simulations have been performed to determine the actual deformed geometry of the conductor, the residual stresses, the plastic strain and material hardening due to area reduction of the strands. The results were linked to experimental measurements to quantify the changes on material electrical conductivity. These simulations were developed using the Abaqus program's explicit dynamic module. This module offers the possibility to analyse complex physical problems with large nonlinearities, complex geometry shapes and interface contact conditions. The total resistance of the stranded conductor depends on a number of parameters. The main parameters are the conductivity of the material and the overall design of the conductor (the diameter of the wires, the number of wires, the number of layers, the lay lengths of each layer, the direction of the layers, the compaction of conductors) and the contacts between the wires. Different stages of the production of the cable are sources of factors that may affect the electrical resistance. Indeed, the deformations due to the stranding process, the contact pressures between the wires, the reduction of area due to pulling forces, etc. strongly impact the electrical behaviour of the material. To facilitate identification of contact surfaces and definition of relevant equivalent volumes by Abaqus a procedure that can reproduce the twisting process was prepared,. This procedure is the following: first the geometry with a parallel beam of wires is prepared, secondly tensional forces are applied. The next step is a compaction and extrusion of the conductor. This process is shown in Figure 10.

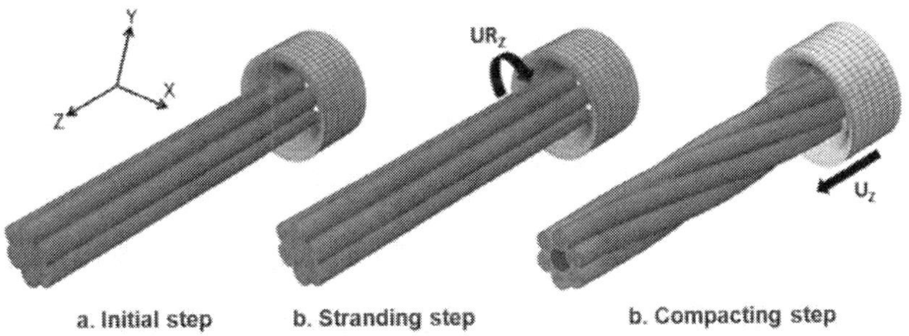

Figure 10. Abaqus stranding and compacting process boundary conditions.

The electrical phenomenon in the contact area between strands is complex; this interface is controlled by the contact pressure and the material's ability to undergo plastic deformation. It was determined that real surfaces are not flat, but have many bumps [10,14–17]. The current distribution in the cable depends mainly on the strength of each strand and the quality of the electrical contact between the strands. When the peripheral strands are subjected to tensile stress inherent in the manufacture of the conductor, the electrical resistance tends to increase slightly; the current lines will focus around the strands having the smallest resistance. Even small levels of mechanical stress, induce differences between the strands, that result in imbalances between the wires connected in parallel. Small imbalances can result in significant differences between local current densities. It is also to be considered that during crushing the contact zone changes its properties.

3.2. FEM Modelling of the Stranding Process with the Abaqus Explicit Modulus

The plastic deformations during the compacting process introduce local dislocations. Through extensive simulations and supporting measurements, conductor resistance data has been established for various conductor design parameters. The main objectives of the mechanical simulations consist of determining the actual deformed geometry of the conductor, the residual stresses, the plastic strain and the hardening range due to the area reduction. In fact, the mechanical results linked to the electrical experimental measurements will allow us to quantify the changes of the material electrical characteristics as the specific electrical resistivity. In this study we use a penalty contact algorithm with finite sliding. The advantage of a finite sliding contact is that all the rotations and translations are authorized in the contact interfaces. The simulations are calculated under quasi-static loadings. The generated mechanical stresses and plastic deformations are shows in Figure 11.

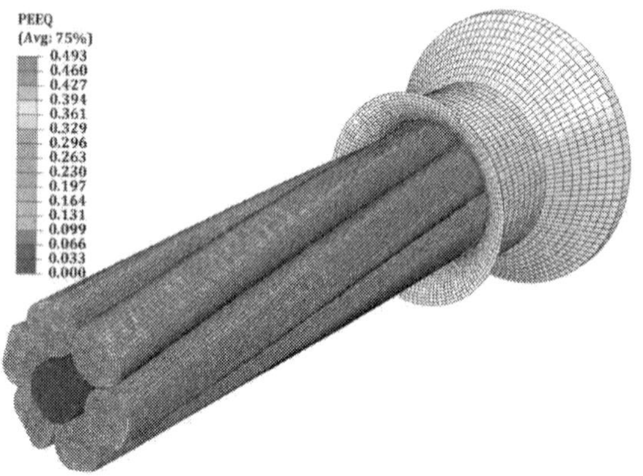

(a)

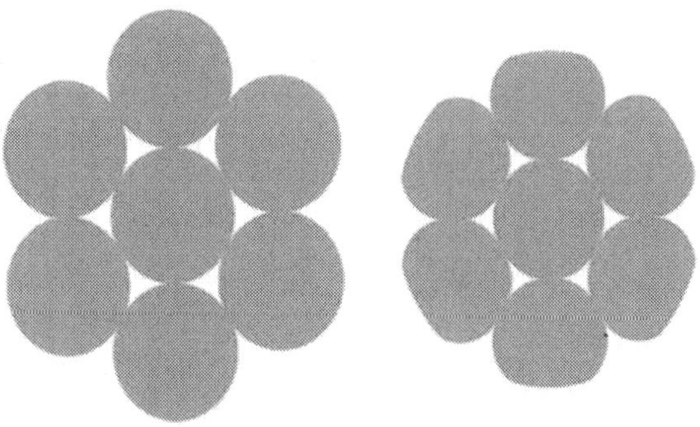

(b)

Figure 11. Simulation of the compacting process: (**a**) Equivalent plastic deformation; (**b**) conductor cross-section before and after the stranding and compacting process.

The relationship between the mechanical stress and copper electrical resistivity has been established. The curves shown in Figure 12 have been determined by a cold drawing process. The equivalent conductor hardening caused by compaction determined by the mechanical simulations were linked to the measured electrical resistivity of copper. After compaction, the equivalent resistivity of copper has been increased by 0.4% (1.71038E−08 $\Omega \cdot m^2/m$) from the initial copper resistivity (1.70382E−08 $\Omega \cdot m^2/m$).

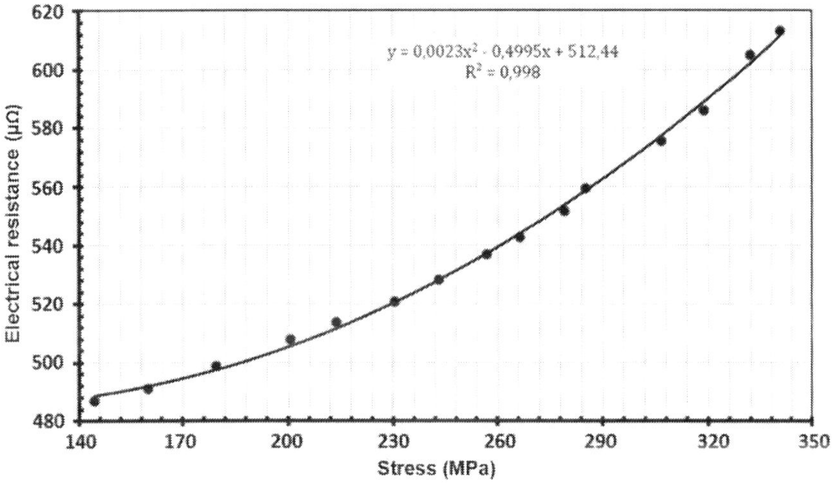

Figure 12. Electrical resistance in function of mechanical stress.

3.3. Electrical Model

To implement out the electrical model we use the Abaqus direct current flow module. The deformed geometry has been re-meshed. A difference of electrical potential is given between the two ends of the conductor. At the input surface we defined the ground U = 0 mV, and at the output surface a voltage equal to U = 1 mV. The geometry has been meshed with tetrahedral elements of 0.2 mm. The length considered for the DC model is 50 mm. The temperature is fixed at 20 °C. The behaviour law of the contact resistance as a function of contact pressure and crossing angle was implemented in Abaqus as an interpolated function. Based on the distribution of stresses in the conductor, the distribution of the conductivity and current density were calculated. Figure 13 shows the results of distribution of stresses at the cable cross-section and the resulting spatial distribution of electrical conductivity. The copper conductivity of the compacted wires has been decreased by 2%, but even so a small change can significantly affect the distribution of current density and resultant resistance of the cable.

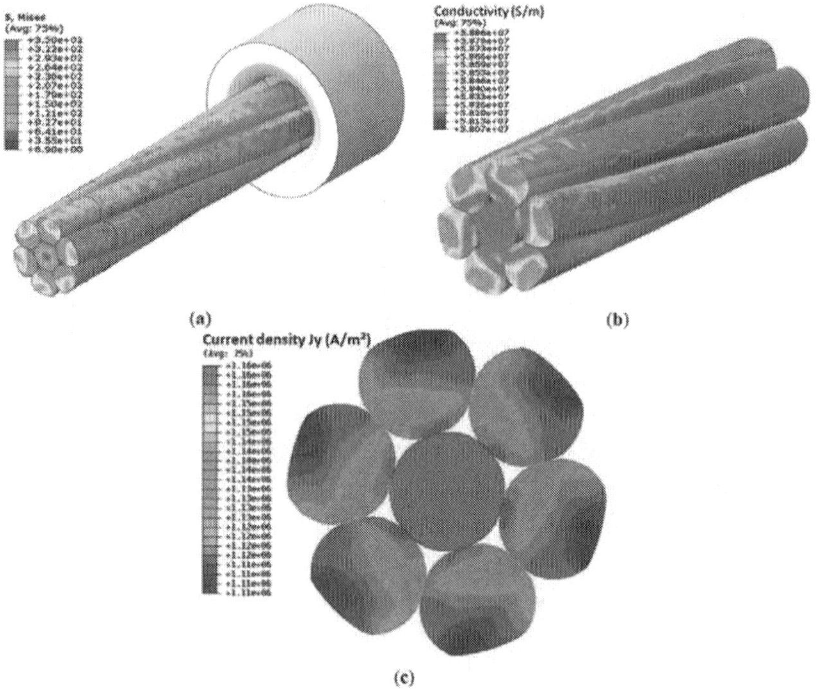

Figure 13. Distribution of residual stress (**a**); the resulting relative conductivity distribution (referred to the basic conductivity) (**b**) and resulting current density (**c**).

The electromechanical model was developed to understand and predict the influence of the stranding and compacting process on mechanical and electrical characteristics of a concentric conductor (1 + 6). Several mechanical and electrical simulations have been performed in order to quantify the influence of the area affected by mechanical deformation and hardening on the conductors' electromechanical behaviour. To validate this model, we compared the results obtained by experimental measurement and finite element simulations. Two main parameters are chosen to validate the numerical model: the weight and the electrical resistance of the conductor per kilometre. The differences between the two models has been identified, as shown in Table 2.

Table 2. Main results of actual and FEM models electrical resistance and conductor weight prediction.

Design Parameters	Actual Manufactured Model	FEM Model
Conductor weight [Kg/km]	218.03	218.00
Electrical resistance [Ω/km]	0.707	0.7086
Resistance average gap [%]	0.23%	

A comparison of measurement and simulation of an ideal transition zone indicates the existence of transition layers of different conductivity than the conductivity of copper. It has been shown that the voltage and current distribution depend strongly on the force used during the conductor manufacturing. Another problem is that during the crushing, the properties of the contact zone change. We have shown how important the contact zone is and how it changes when compacting forces are applied. The current density distribution is also heavily influenced by the structure of the contact regions between the wires and the resistance of the transient layer.

4. CONCLUSIONS

In this paper a new method has been presented to predict mechanical and electrical behaviour of a concentric conductor (1 + 6) according to the stranding and compacting process parameters. The original objective of this work was to consider the mechanical deformations of the conductor and their impact on material electrical conductivity. A relationship between copper electrical conductivity and stresses has been characterized by experimental measurements.

The condition of the conductor surface is very important and affects the reliability and repeatability of the measurements. Hundreds of samples were tested and results averaged to make them meaningful, so that relationships could be established showing the dependence of the contact resistance between stranded conductors on the acting compression force.

The experimental results show that the range of the electrical resistance is dependent of three major parameters: the compression force to which the conductor is subjected, the size of the two contact surfaces, which are directly related to the applied force, and the intersection angle between the strands of the cable. The resulting characteristics of contact resistance and conductivity of the single wire versus strain are used for modelling a real 1+6 cable wire system.

ACKNOWLEDGMENTS

The authors thank the Nexans company for making it possible to carry out measurements and for the preparation of a large variety of samples.

AUTHOR CONTRIBUTIONS

The work was done as part of the doctoral thesis of Y. Zeroukhi. The directors of the thesis are E. Napieralska-Juszczak and K. Komęza. The percentage share of the work performed by the various authors is: Y. Zeroukhi 30% (thisco-author has conducted the measurements and the simulations), E. Napieralska-Juszczak 20% (thisco-author has contributed at all stages of the reported work, including formulation of the initial concept of the project, performing calculations and discussion of the results), G. Vegas 15% (thisco-author has contributed at all stages of the reported work, including formulation of the initial concept of the project, performing calculations measurement and discussion of the results), K. Komęza 15% (thisco-author has contributed tothe formulation of the initial concept of the project and simulation), S. Wiak 10% (this co-author has contributed tothe formulation of the initial concept of the project), F. Morganti 10% (this co-author has conducted part of the measurements).

REFERENCES

1. Braunovic, M.; Konchits, V.; Myshkin, N. Fundamentals of Electrical Contacts; Taylor &Francis Group, LLC: London, UK, 2006.
2. Holm, R. Electric Contacts: Theory an Applications, 4th ed.; Springer: Berlin, Germany, 1979.
3. Slade, P.G. Electrical Contacts: Principles and Applications; CRC Press: New York: NY, USA, 1999.
4. Abaqus; Dassault Systems Simulia Corp: Paris France, 2011.
5. Nakamura, K.; Yamanouchi, M.; Yamaguchi, M.; Obana, T.; Hashimoto, K.; Takao, T.; Nishimura, A.; Mito, T. Influence of Intersecting Angles of Strands on Contact Resistance in Cable-in-Conduit Conductors. IEEE Trans. Appl. Superconduct. **2003**, 13, 2392–2395.
6. Nakamura, K.; Yamanouchi, M.; Hashimoto, K.; Takao, T. Effects of Compressive Force between Strands on Contact Resistance in

Cable-in-Conduit Conductors. IEEE Trans. Appl. Superconduct. **2007**, 17, 2466–2469.

7. Butta, M.; Ripka, P.; Atalay, S.; Atalay, F.E.; Li, X.P. Fluxgate effect in twisted magnetic wire. J. Magn. Magn. Mater. **2008**, 320, e974–e978.
8. Kwak, N.S.; Lee, J.; Jang, Y.H. Genetic-Algorithm-Based Controlling of Micro contact Distributions to Minimize Electrical Contact Resistance. IEEE Trans. Compon. Packag. Technol. **2012**, 2, 1768–1776.
9. Nakamura, M. Computer simulation for the constriction resistance depending on the form of conducting spots. IEEE Trans. Compon. Packag. Technol. **1995**, 18, 382–384.
10. Duvier, P.Y.; Mandrillon, V.; Polizzi, J.P.; Inal, K. Etude de la resistance electrique d'un contact Au/Au en function de la force appliqué. Proceedings of 19 Congrès Français de Mécanique, Marseille, France, 24–28 August 2009.
11. Fechant, L. Le Contact Electrique—Phénomènes Physiques et Matériaux; Hermès Science Publications: Paris, France, 1996.
12. Zeroukhi, Y.; Napieralska-Juszczak, E.; Komeza, K.; Morganti, F.; Vega, G. Coupled Mechanical—Electrical identification of the contact resistance in the stranded electric power cable. J. Achiev. Mater. Manuf. Eng. **2012**, 55, 921–923.
13. Zeroukhi, Y.; Napieralska-Juszczak, E.; Komeza, K.; Morganti, F.; Vega, G. Identification of current flow depending on the mechanical deformation in a stranded cable. COMPEL **2013**, 32, 1437–1450.
14. Timsit, R.S. Electrical contact resistance: Properties of stationary interfaces. IEEE Trans. Compon. Packag. Technol. **1999**, 22, 85–98.
15. Braunovic, M. Effect of connection design on the contact resistance of high power overlapping bolted joints. IEEE Trans. Compon. Packag. Technol. **2002**, 25, 642–650.
16. Kogut, L.; Komvopoulos, K. Electrical contact resistance theory for conductive rough surfaces. J. Appl. Phys. **2003**, 94, 3153–3162.
17. Ciavarella, M.; Murolo, G.; Demelio, G. On the elastic contact of rough surfaces: Numerical experiments and comparisons with recenttheories. Wear **2006**, 261, 1102–1113.

7
CHAPTER

CONTROL OF THRESHOLD AND GAIN OF PARAMETRIC AMPLIFICATION IN MAGNETOACTIVE III-V PIEZOELECTRIC SEMICONDUCTORS

Bhajan Lal[1,2*], P. Aghamkar[1]

Department of Physics, Chaudhary Devi Lal University, Sirsa, India
[2]Department of Physics, Government Polytechnic for Women, Sirsa, India

ABSTRACT

The effect of doping concentrations and a transverse external magnetostatic field on operational characteristics of parametric amplification of backward Stokes signal has been studied, using hydrodynamic model of semiconductors, in the far infrared regime. The model suggests three achievable resonance conditions: 1) lattice frequency and plasma frequency 2) stokes frequency and electron-cyclotron frequency 3) stokes frequency and hybrid (plasma and electron-cyclotron) frequency and these conditions have been utilised, on one hand, to substantially reduce the value of threshold intensity for onset of the parametric amplification and on the other hand, for switching of parametric large positive and negative gain coefficient (i.e. amplification and absorption). For example a strong transverse magnetostatic field 10.0 T with free carrier concentration 1.5 × $10^{19}m^{-3}$ enhances the gain by a factor of 10^3 as in its absence. Results also suggest that a weakly piezoelectric III–V semiconductor duly illuminated by slightly off-resonant not-too-high-power pulsed lasers with pulse duration sufficiently larger than the acoustic phonon lifetime is one of promising hosts for parametric amplifier/frequency converter.

Keywords: Parametric Gain, Piezoelectricity, Magnetostatic Field, Doped III-V Semiconductor

1. INTRODUCTION

The area of parametric interactions and their devices such as optical parametric amplifiers and oscillators (OPA/ OPO) occupies a special place in nonlinear optics due to their potential applications. Optical parametric amplifiers and oscillators are being extensively used in fabrication of tunable coherent radiation source with considerably high gain and high conversion efficiency [1]. Parametric interactions are also employed successfully to study the photon amplifiers [2,3] and to generate high peak power subpicosecond optical pulse [4]. The choice of a nonlinear medium and operating wavelength are crucial aspects in design and fabrication of parametric amplifiers and oscillators. Amongst different types of nonlinear media, semiconductors offer considerable flexibility for fabrication of optoelectronic devices because: 1) the large number of free electrons/holes available as majority charge carriers in doped semiconductors manifests many more exciting nonlinear optical processes [5]; 2) carrier relaxation times can be altered through design of materials and device structures; 3) large optical nonlinearities in the vicinity of band gap resonant transitions [6,7]; 4) either change in absorption or refractive index can be utilized; 5) devices may operate either at normal incidence or in waveguides; 6) devices are integrable with other optoelectronic components, 7) substantially transparent for photon energies less than the band gap energies [8]. Here it is worth mentioning that resonant and non-resonant optical nonlinearity of semiconductors have been utilised for fabrication of nonlinear optical devices. The resonant nonlinearities are large in magnitude [9], while, the magnitude of non-resonant optical nonlinearity (SNON) is small but its response time is fast due to virtual transition mechanisms. In order to design and develop efficient parametric amplifiers/oscillators, different types of nonlinear materials have been used [10,11]. However, recent trend reveals that still there is a great need to explore alternatives, particularly those compatible with semiconductor technology, due to the rapid growth of planar devices such as wavelength division multiplexers and add-drop filters in advanced networks. A hybridly integrated semiconductor optical amplifier Mach–Zehnder interferometer (MZI) was demonstrated using bidirectional data injection and an additional external continuous wave signal for differentially biasing the interferometer arms, optimizing gain and phase conditions during the switching functionality [12]. A long period linear grating is imprinted on semiconductor waveguide and using second-order susceptibility its frequency converters of high frequency is experimentally demonstrated [13].

It is well known that the manipulation of threshold condition and gain are important issues to improve the efficiency and functionality of parametric devices (e.g. amplifiers, oscillators and frequency converters). In case of semiconductors, doping (free carriers), composition and micro-structuring have been mostly exploited for improving the performance of optoelectronic devices [14, 15]. Besides, optical nonlinearities of semiconductors can be easily modified by externally applied electric and magnetic fields and hence external field dependent property has been exploited to understand the mechanisms involved in nonlinear processes such as electro-optic and magneto-optic effects and fabrication of devises based on it [16,17]. Makowska et. al. [18] pointed out that optical second harmonic generation (SHG) can be induced in both single and polycrystalline InAs by acoustic wave and found that efficiency of SHG depends on acoustic power. However, in case of layered periodic structure of III-V semiconductor-plasma, a significant increase in efficiency of nonlinear interaction is obtained near the nonlinear resonances [19]. Voigt geometrical configuration has been used to enhance the terahertz emission and the photo-refraction in III–V semiconductors [16,17]. Of-late authors have made systematic attempt to explore the influence of free carrier concentration, externally applied magnetostatic field and excitation intensity on second order susceptibility of weakly piezoelectric n-type III–V semiconductors [20]. In the present article, the model [20] has been further extended to study operational characteristic of parametric amplification and explore the possibility of switching of optical nonlinearity by applied magnetostatic field. Using the coupled-mode theory, expression for operational characteristic parameters of parametric amplifiers (i.e. threshold condition, parametric gain) has been derived. Finally, exhaustive numerical analysis is performed with a set of data appropriate for weakly piezoelectric semiconductor-plasma (n-InSb) duly irradiated by a 10.6 μm CO_2 laser to establish the application of the present model. The results indicate that the threshold condition and parametric gain can be manipulated by varying doping concentration, magnetostatic field and excitation intensity.

2. THEORETICAL FORMULATIONS

In the present section, expression of threshold condition for the onset of backward coherent scattered stokes wave and its gain coefficient are obtained, in weakly piezoelectric doped III-V semiconductors, under off-resonant transition regime in the absence and the presence of external applied magnetostatic field.

In doing so we assume the pump wave (laser) energy ($\hbar\omega_p$) < band gap energy ($\hbar\omega_g$) of semiconductor. The pulse duration of laser is much larger than the material damping time such that the interaction is treated to be of a steady-state type. Further we consider that there are many photons in the pump wave and it can be described by plane wave. The scheme of three-wave mixing (pump wave, backward scattered stokes wave and acoustic signal wave) mechanism within matter in the presence of an external magnetostatic field is shown in Figure 1. We assume the pump electric field

$$E_p(x,t) = E_0 \exp\left[i\left(k_p x - \omega_p t\right)\right]$$

is polarised along z-axis and incident on ($1\bar{1}0$) surface of the crystal. Let us assume the electric field of pump radiation produces longitudinal acoustic wave (i.e. idler wave), $u(x,t) = u_0 \exp\left[i\left(k_a x - \omega_a t\right)\right]$ and in turn it scatters the pump wave. We further assume that the backward scattered stokes wave, say

$E_s(x,t) = E_s \exp\left[i\left(k_s x - \omega_s t\right)\right]$ is also polarised along z-axis. In order to study parametric amplification in Voigt geometry an external transverse magnetostatic field (B_0) is applied parallel to z-axis.

The momentum and energy transfer between these waves can be described by phase matching conditions: $\hbar\vec{k}_p = \hbar\vec{k}_s + \hbar\vec{k}_a$ and $\hbar\omega_p = \hbar\omega_s + \hbar\omega_a$, where the frequencies {ω_p, ω_s, ω_a} and wave vectors {k_p, k_s, k_a} correspond to the incident laser, the backward scattered Stokes wave and the acoustic wave, respectively. We also consider hydrodynamic model of the homogeneous one component (viz., n-type) semiconductor, satisfying the condition

$k_a l \ll 1$ (k_a and l being the acoustic wave number and the electron mean free path, respectively). This condition implies that the sound wavelength is much greater than the mean free path of electrons, so that the motion of the carriers under the influence of the external field is averaged out. In addition, it allows neglecting the non-uniformity of the high frequency electric field under the dipole approximation [21].

The zeroth-order electron momentum transfer equation under the influence of magnetostatic field can be de-

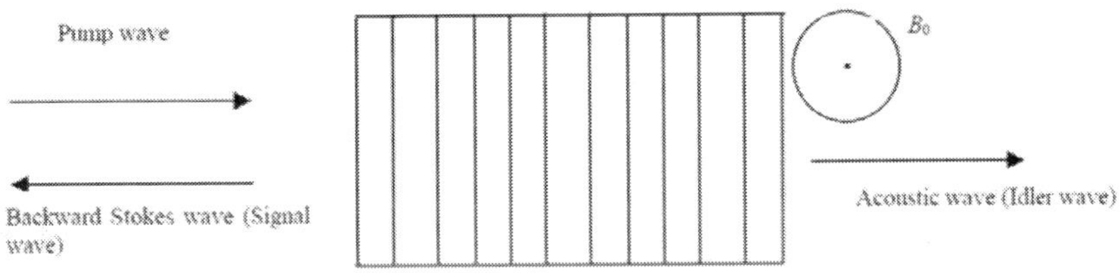

Figure 1. Schematic diagram for interaction of pump wave with doped III-V piezoelectric semiconducting crystal in the presence of magnetostatic field.

scribed as

$$\frac{\partial \vec{v}_0}{\partial t} + \gamma \vec{v}_0 = -e_1\left(\vec{E}_p + \vec{v}_0 \times \overrightarrow{B_0}\right) \tag{1}$$

where $e_1 = e/m$.

The first-order electron momentum transfer and continuity equation in the presence of external magnetostatic field are defined [20] by

$$\frac{\partial \vec{v}_1}{\partial t} + \gamma \vec{v}_1 + \vec{v}_0 \times \left(\nabla . \vec{v}_1\right) = -e_1\left(\vec{E}_1 + \vec{v}_1 \times \vec{B}_0\right) \tag{2}$$

$$\frac{\partial n_1}{\partial t} + n_0 \frac{\partial v_1}{\partial x} + n_1 \frac{\partial v_0}{\partial x} + v_0 \frac{\partial n_1}{\partial x} = 0 \tag{3}$$

where v_0 and v_1 are zerothand first-order oscillatory fluid velocities of the electron of effective mass m and charge −e. n_0 and n_1 are initial and perturbed electron density and γ is the electron collision frequency. It is a well-known fact that the effective mass of an electron in III–V semiconductors is smaller than the effective mass of a hole and therefore the drift velocity of a hole is less than the drift velocity of an electron and hence the effect of hole on nonlinearity can be neglected.

The pump wave produces stress in the medium and the linear relationship between stress and electric field is described by piezoelectricity. The origin of piezoelectricity lies in the first order force ($f_1 = \beta \frac{\partial E}{\partial x}$, where β is piezoelectric coefficient). In particular, in narrow bandgap weakly piezoelectric III-V semiconductors such as InSb and InAs, the high mobility of electrons (due to their low effective mass) allows one to work at drift velocities several times the sound velocity and thus keep the effects of inhomogeneities to a minimum [22].

Let us consider, if the deviation of lattice along x-direction is u(x, t), due to stress in the medium, in the presence of an external magnetostatic field then the strain can be described as $\partial u / \partial x$ and the resultant polarisation (P_{in}) can be given as

$$P_{in} = \pm \beta \left(\frac{\partial u}{\partial x} \right) \tag{4}$$

Taking into account the effect of piezoelectricity, the modified equation of motion of lattice vibrations may be defined as

$$\frac{\partial^2 u}{\partial t^2} - \frac{C}{\rho}\frac{\partial^2 u}{\partial x^2} + 2\zeta_a \frac{\partial u}{\partial t} - \frac{\beta}{\rho}\frac{\partial E_1}{\partial x} = 0 \tag{5}$$

where ρ, C and ζ_a being the mass density, material's elastic constant, acoustic phonon damping parameter and E_1 is being space charge electric field of the medium which can be obtained from Poisson's equation as

$$\frac{\partial E_1}{\partial x} = -\frac{n_1 e}{\varepsilon} - \frac{1}{C}\frac{\partial P_{in}}{\partial x}. \tag{6}$$

Now we are interested to derive expression for firstorder perturbed carrier-density of piezoelectric doped semiconductors due to three-wave mixing. Under rotating wave approximation and using Equations (1)-(6) a simplified expression for the density perturbation is obtained as

$$\frac{\partial^2 n_1}{\partial t^2} + \gamma_{eff}\frac{\partial n_1}{\partial t} + \Omega_p^2 n_1 A_1 \mp \frac{i\beta^2 k_a^3 \Omega_p^2 A_1 E_a}{\rho e \Delta_a^2}$$
$$\pm \frac{ik_p \Omega_p^2 E_p n_1 e_1}{\Delta_{p1}^2} \mp \frac{\beta^2 k_p \Omega_p^2 E_p k_a^3 E_a}{\rho m \Delta_a^2 \Delta_{p1}^2}$$
$$-e_1 A_0 E_p \frac{dn_1}{dx} - ie_1 n_1 A_0 k_p E_p = 0 \tag{7}$$

where $\Omega_p = \left(n_0 e^2/m\varepsilon\right)^{1/2}$ (electron-plasma frequency), $\omega_c = e_1 B_0$ (electron-cyclotron frequency), $\delta_p = \gamma - i\omega_p$,

$$\delta_1 = \gamma - i\omega_s + ik_s \mathrm{v}_0,\quad A_0 = 1 - \left(\omega_c^2 / \left(\delta_p^2 + \omega_c^2\right)\right),$$

$$A_1 = 1 - \left(\omega_c^2 / \left(\delta_1^2 + \omega_c^2\right)\right), \quad \gamma_{eff} = \gamma + ik_p \mathrm{v}_0,$$

$$\Delta_a^2 = -\omega_a^2 + \mathrm{v}_a^2 k_a^2 - 2i\zeta_a \omega_a$$

and $$\Delta_{p1}^2 = \frac{\left(\delta_p^2 + \omega_c^2\right)\left(\delta_1^2 + \omega_c^2\right)}{\delta_p \delta_1}.$$

In obtaining Equation (7) we have neglected the Doppler shift under the assumption that $\omega_p \gg k_a v_a$ (v_a being the velocity of an acoustic wave) and higher order terms like $n_0 \mathrm{v}_0 \frac{\partial^2 \mathrm{v}_1}{\partial x \partial t}$. Equation (7) states that density perturbation may oscillates at low (acoustic, ω_a) and high (stokes, ω_s) wave frequency components and that can be expressed as: $n_1 = n_{1s}(\omega_a) + n_{1f}(\omega_s)$. The higher-order terms with frequencies $\omega_{s,q}\left(= \omega_p \pm q\omega_a\right)$, for $q \geq 2$ will be off-resonant and are neglected in the study of second-order nonlinearities in doped semiconductors. Using phase matching conditions, low and high frequency density perturbation can be obtained from Equation (7) as

$$\frac{\partial^2 n_{1s}}{\partial t^2} + \gamma_{eff}\frac{\partial n_{1s}}{\partial t} + \Omega_p^2 n_{1s} A_1 \mp \frac{i\beta^2 \Omega_p^2 A_1 k_a^3 E_a}{\rho e \Delta_a^2} - ik_a e_1 A_0 n_{1f}^* E_p = 0 \tag{8}$$

and

$$\frac{\partial^2 n_{1f}}{\partial t^2} + \gamma_{eff}\frac{\partial n_{1f}}{\partial t} + \Omega_p^2 n_{1f} A_1 + ie_1 k_p \Omega_p^2 \Delta_{p1}^2 E_p n_{1s}^* \mp \frac{\beta^2 k_p \Omega_p^2 k_a^3 E_p E_a^*}{\rho m \Delta_a^2 \Delta_{p1}^2} - ie_1 A_0 k_s n_{1s}^* E_p = 0 \tag{9}$$

By solving coupled Equations (8) and (9), expression of n_{1s} (low frequency component) and n_{1f} (high frequency component) can be obtained as well as their values may be computed by using material parameters and electric field amplitudes in the absence and presence of transversely applied magnetostatic field. The expression of n_{1s} is obtained as

$$n_{1s}^{*} = -\frac{i\beta^2\Omega_p^2 k_a^3 E_a^*}{\rho\Omega_{pl}^2}\left[\pm\frac{A_1^*}{e\Delta_a^{2*}} \mp \frac{eA_0^* k_p k_a \left|E_p\right|^2}{m^2\Delta_a^2\Omega_{pd}^2\Delta_{p1}^2}\right] \tag{10}$$

where

$\Omega_{pl}^2 = \Omega_p^2 A_1 - \omega_a^2 - i\gamma_{eff}\omega_a$ and $\Omega_{pd}^2 = \Omega_p^2 A_1 - \omega_s^2 - i\gamma_{eff}\omega_s$.

By knowledge of the first-order carrier density perturbation, second-order polarisation at backward Stoke's frequency can be obtained by time integral of induced current density $J^{(2)}(\omega_s) = -n_{1s}^* e v_0$ as

$$P^{(2)}(\omega_s) = \varepsilon_0\chi^{(2)}E^2 = \int J^{(2)}(\pi)\, dt \tag{11}$$

Using Equations (10) and (11), effective second-order optical complex susceptibility is obtained as

$$\chi_{eff}^{(2)}(\omega_s) = \pm\left(\chi^{(2)}(\omega_s)\right)_{\beta} \mp \left(\chi^{(2)}(\omega_s)\right)_{\beta,I} \tag{12}$$

where

$$\left(\chi^{(2)}(\omega_s)\right)_{\beta} = \frac{i\beta^2\Omega_p^2 k_a^3 A_1^* e_1\delta_p}{\varepsilon_0\rho\Delta_a^{2*}\Omega_{pl}^2\left(\delta_p^2 + \omega_c^2\right)} \tag{12a}$$

$$\left(\chi^{(2)}(\omega_s)\right)_{\beta,I} = \frac{i2\beta^2\Omega_p^2 k_a^4 e_1^3 A_p^* k_p \delta_p I_p}{\varepsilon_0^2 \eta c \rho \Delta_a^2 \Omega_{pd}^2 \Omega_{pl}^2 \Delta_{p1}^2 \left(\delta_p^2 + \omega_c^2\right)} \quad (12b)$$

with $I_p = \frac{1}{2}\eta\varepsilon_0 c\left|E_p\right|^2$.

Using algebra, expression for real and imaginary parts of (12a) and (12b) can be obtained. It is well known fact that real and imaginary part of optical susceptibility describes refraction and gain/absorption phenomena in the crystal, respectively. By knowledge of propagation of light in refractive medium various optical devices have been designed and fabricated such as optical wave guide, filters, amplifiers, oscillators, coupler etc [11,17].

The present article is focused to study behavior of the threshold condition for parametric amplification and gain coefficient ($g(\omega_s)$) of backward coherent scattered stoke mode by varying free carrier density and externally applied magnetostatic field. In doing so, well known following expression [1] is used

$$g(\omega_s) = \frac{\omega_s}{\eta c}\left[\left(\chi^{(2)}(\omega_s)\right)_{imag}\right]\left|E_p\right| \quad (13)$$

3. RESULTS AND DISCUSSION

In order to explore the applicability of the model, a detailed numerical study of threshold condition and gain behavior of parametric amplification has been made. Efforts are also made to find out conditions for achieving large gain with low power laser. In doing so, we consider the irradiation of n-type doped III-V micro size semiconductor sample (such as InSb) by 10.6 μm CO_2 at liquid nitrogen temperature (77 K). Around this temperature, absorption coefficient of the sample is low around 10 μm and one may neglect contribution due to bandto-band transition mechanism. It is worth pointing that at 77 K temperature, the dominant mechanism for transfer of momentum and energy of the electron in scattered mode is due to the acoustic phonon scattering in semiconductors [23,24].

The following material parameters are taken as representative values to establish the theoretical formulation:

$\omega_p = 1.78 \times 10^{14}\ s^{-1}$, $\omega_a = 2 \times 10^{11}\ s^{-1}$, $\omega_s = 1.77 \times 10^{14}\ s^{-1}$, $m = 0.014m_0$ (m_0 being the rest mass of an electron), $\varepsilon_l = 15.8$, $\rho = 5.8 \times 10^3\ kgm^{-3}$, $\eta = 3.8$, $\beta =$

$0.054\ cm^{-2}$, $\gamma = 3 \times 10^{11}\ s^{-1}$, $v_a = 4 \times 10^3\ ms^{-1}$, $k_a = 5 \times 10^7\ m^{-1}$, $\zeta_a = 2 \times 10^{10}\ s^{-1}$, $c = 3 \times 10^8\ ms^{-1}$

3.1. Threshold Condition

The threshold pump intensity (I_{pth}) required for the onset of parametric amplification is obtained by setting imaginary part of Equation (12) equals to zero. I_{pth}, obtained

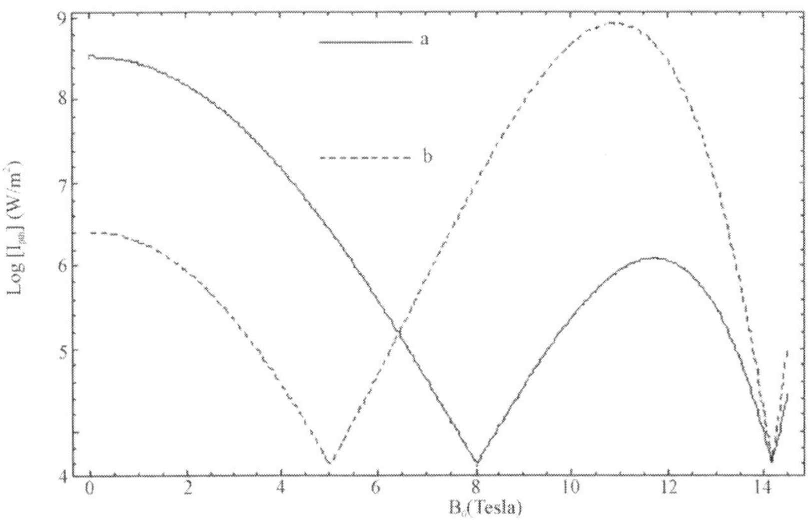

Figure 2.1. Variation of threshold pump intensity (Log [I_{pth}]), for Equation (12a), with magnetostatic field (B_0) at free carrier concentration (a) $n_0 = 1.5 \times 10^{19}\ m^{-3}$ and (b) $n_0 = 2.4 \times 10^{19}\ m^{-3}$.

from Equation (12a), versus with an external magnetostatic field for two different free carrier concentrations i.e. (a) $n_0 = 1.5 \times 10^{19}\ m^{-3}$ and (b) $n_0 = 2.4 \times 10^{19}\ m^{-3}$ is plotted in Figure 2.1. One may notice that in the absence of magnetostatic field, a large I_{pth} is required for onset of parametric amplification. For example, when $B_0 = 0.0$ T, we found $I_{pth} \sim 3.7 \times 10^8$ Wm^{-2} and 3.0×10^6 Wm^{-2} for $n_0 = 1.5 \times 10^{19}$ m^{-3} and 2.4×10^{19} m^{-3}, respectively.

When $B_0 > 0$, I_{pth} reduces considerably due to resonance conditions. Let us focus on the threshold condition for parametric amplification when piezoelectricity is solely responsible for parametric interaction. In this case, minimum threshold intensity can be achieved in two different range of externally applied magnetostatic field, i.e. 1) moderate and 2) strong magnetostatic fields. In former case, a minimum value of threshold intensity is found ~1.1×10^3 Wm^{-2} (for $B_0 = 5.0$ T, $n_0 = 2.4 \times 10^{19}$ m^{-3}). The minima of threshold intensity arises due to resonance condition; $\omega_s^2 = \left(\omega_c^2 + \gamma^2\right) / \left(1 - \left(\Omega_p^2 / 4\omega_a^2\right)\right)$. It is also noticed that the position of minima shifts toward lower magnetostatic field with increasing doping concentration. While in later case, I_{pth} can be achieved when

$\omega_p \approx \omega_s \approx \omega_c$. For example, when $B_0 = 14.1$ T, I_{pth}

~1.2×10^3 Wm^{-2}. Here it is worth pointing out that in later case a small change in free carrier concentration does not affect shifting position of minima. This result is well in agreement with the result of Palik and Furdyna [25].

In order to study the role of different physical mechanism leading to parametric interaction we have used Equation (12b) to obtain threshold condition for onset of parametric amplification due to coupling between the piezoelectric coefficient and pump electric field. The threshold pump intensity, determined from Equation (12b), versus external magnetostatic field for two different samples having free carrier concentration i.e. (a) $n_0 = 1.5 \times 10^{19}$ m^{-3} and (b) $n_0 = 2.4 \times 10^{19}$ m^{-3}, is plotted in Figure 2.2. One may observe that when $B_0 = 0.0$ T, I_{pth} required for onset of parametric amplification is 5.9×10^6 Wm^{-2}. The figure depicts that the minima of I_{pth} is independent of change of free carrier concentration. The model suggests that when $B_0 > 0$, a minimum I_{pth} can be achieved at three different values of magnetostatic field due to three different resonance conditions:

$$\omega_s^2 = \left(\Omega_p^2 \omega_c^2 + \gamma^2\right) / \gamma^2, \quad \omega_s^2 = 2\omega_c^2 + \gamma^2 + 2\zeta_a \omega_a$$

and $\omega_s = \omega_c$. Around the first resonance condition, the minimum I_{pth} is found to be 3.0×10^4 Wm^{-2}. This resonance condition also shifts the position of minima towards lower magnetostatic field with increase in free carrier concentration. However, the second and the third resonance conditions yield $I_{pth} \sim 1.1 \times 10^3$ Wm^{-2} and 1.9×10^6 Wm^{-2}, respectively. It is noticed that around the second and third resonance conditions the position of minima are nearly independent with the variation of doping concentrations. A comparison of the results obtained from Equation (12a) and (12b) reveal that the required minimum I_{pth} for the onset of parametric amplification due to parametric coupling of piezoelectric coefficient and pump electric field is nearly 100 times higher than the value obtained due to solely piezoelectricity. In the forthcoming analysis to study the behavior of parametric gain we consider $I_p = 1.5 \times 10^{11}$ Wm^{-2} as a representative excitation intensity which is well above the threshold intensity but below the damage threshold of the sample [26].

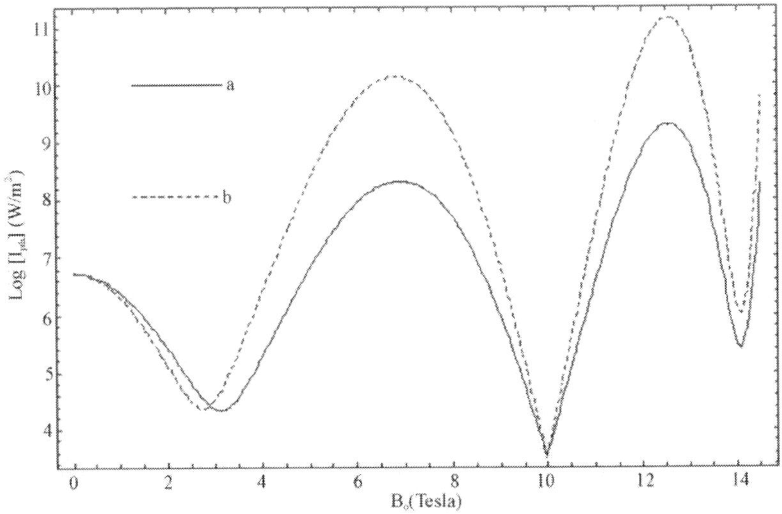

Figure 2.2. Variation of threshold pump intensity ($Log[I_{pth}]$), for Equation (12b), with magnetostatic field (B_0) at free carrier concentration (a) $n_0 = 1.5 \times 10^{19}$ m^{-3} and (b) $n_0 = 2.4 \times 10^{19}$ m^{-3}.

3.2. Parametric Gain

The Equation (13) with (12) shows the dependence of parametric gain on controllable physical parameters namely free carrier concentration (through plasma frequency, Ω_p), external applied magnetostatic field (through cyclotron frequency, ω_c) and pump Intensity (I_p).

Figure 3.1 shows the behavior of parametric gain at stokes frequency ($g(\omega_s)_\beta$), obtained using Equation (13) with Equation (12a), by varying B_0 and n_0 simultaneously. The figure clearly shows substantial enhancement of $g(\omega_s)_\beta$ and change in its sign. This typical behavior arises due to resonance condition:

$$\omega_s^2 = \left(\omega_c^2 + \gamma^2\right) / \left(1 - \left(\Omega_p^2 / 4\omega_a^2\right)\right)$$

(resonance between scattered stokes wave frequency and modified electron-cyclotron wave frequency). At the resonance condition, magnetostatic field dependent drift velocity becomes many times larger than the acoustic wave velocity and due to which interaction between plasmon and acoustic phonon becomes strong. The resonance condition allows one to select typical value of B_0 and n_0 at which maximum gain can be achieved. A small change in typical value of B_0 and n_0 leads to decrease in the value of positive gain coefficient and further change gives rise to maximum negative gain (i.e. maximum absorption).

By selecting appropriate value of B_0 (say 9.0 T) and n_0 (say 1.5×10^{19} m^{-3}), we obtained $g(\omega_s)_\beta \sim 396$ m^{-1}. By varying B_0 and n_0 simultaneously, the change in sign of $g(\omega_s)_\beta$ can also be observed. Besides, it is noticed that when the sample is subjected to moderate magnetostatic field, i.e. $0.0 < B_0 < 10.5$ T, the value of $g(\omega_s)_\beta$ varies almost parabolically with rising n_0. The theoretical model also suggests possibilities of shifting in frequency of coherent backward stokes modes towards higher value of B_0 (i.e. blue shift) as plasma frequency (through n_0) approaches twice of acoustic wave frequency. After departure from the resonance condition, the parametric gain becomes negative (absorption) and saturates to very small value.

However, in the absence of external magnetostatic field ($B_0 = 0$) and $n_0 = 1.5 \times 10^{19}$ m^{-3}, the value of $g(\omega_s)_\beta$ is 7.4 m^{-1} which is nearly 50 times smaller than the value obtained around the resonance condition.

The salient feature of result is the switching of positive and negative gain between high value to low value by varying independently/simultaneously B_0 and n_0 and also achieving its large value in III-V weakly piezoelectric semiconductors. The result suggests the possibility of devise of parametric switches in far infrared region.

Figure 3.2 shows variation of the gain coefficient $g(\omega_s)_{\beta,I}$, using Equation (12b), with respect to B_0 and n_0. This figure also depicts the enhancement of $g(\omega_s)_{\beta,I}$ as well as change in its sign. In this case, enhancement and switching of the gain occurs due to resonance conditions: 1) $\omega_s^2 = (\Omega_p^2 \omega_c^2 + \gamma^2)/\gamma^2$ and 2) $\omega_s^2 = 2\omega_c^2 + \gamma^2 + 2\zeta_a \omega_a$. Interesting feature of this case is the interaction between plasmon oscillator and electron-cyclotron oscillator in the presence of excitation intensity. It is often advantageous, moving the frequency of coherent stokes wave to higher and more accessible spectral region in proportion to: a) plasma frequency (or n_0) for fixed electroncyclotron frequency (or B_0), b) electron-cyclotron frequency for fixed plasma frequency and c) combination of both frequencies. For example, in moderately doped semconconductor ($n_0 = 1.5 \times 10^{19}$ m^{-3}), low externally applied magnetostatic field (1.5 - 3.0 T) yields a large value of $g(\omega_s)_{\beta,I}$ and also alters its sign as well as shifts the stokes frequency towards higher frequency (Blue shift). With further increase in B_0, the second resonance condition ($\omega_s^2 = 2\omega_c^2 + \gamma^2 + 2\zeta_a \omega_a$) yields sharp enhancement and change of sign of $g(\omega_s)_{\beta,I}$ at $B_0 = 10.0$ T.

In addition to the B_0 and n_0 dependence, the parametric gain can also be enhanced by increasing the value of excitation intensity. Here it is worth mentioning that practically the excitation intensity can not be increased arbitrarily because it may optically damage the sample [26].

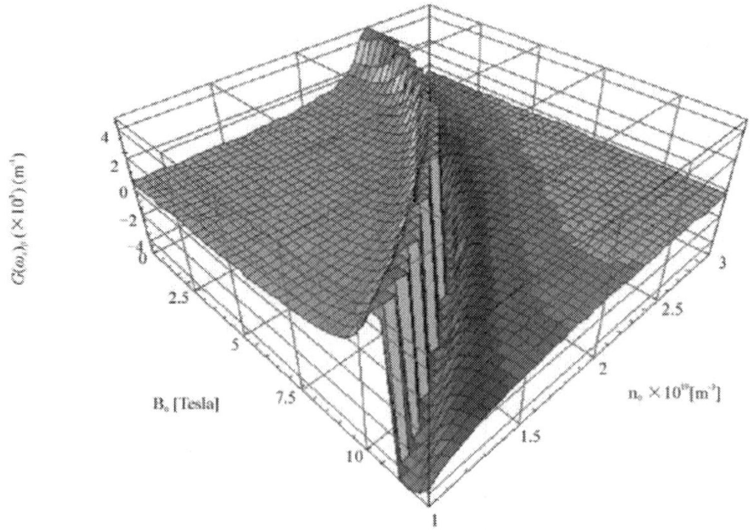

Figure 3.1. Variation of $g(\omega_s))_\beta$ with magnetostatic field (B_0) and free carrier concentration (n_0).

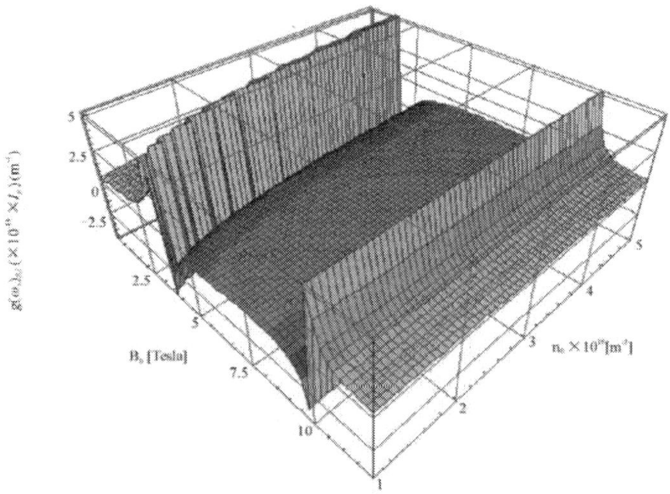

Figure 3.2. Variation of $g(\omega_s))_{\beta,I}$ with magnetostatic field (B_0) and free carrier concentration (n_0).

4. CONCLUSIONS

In the present study, using electromagnetic treatment the operational characteristic of the parametric amplifier has been studied in weakly piezoelectric narrow direct-gap III–V doped semiconductors like n-InSb subjected to a transverse magnetostatic field under off-resonant transition regime. The following important conclusions may be drawn:

1) The hydrodynamic model of semiconductor-plasma has been successfully applied to study the effect of doping level, transverse magnetostatic field and pump intensity on operational characteristic of parametric amplifiers (i.e. threshold condition and optical parametric gain/ absorption) in piezoelectric III–V doped semiconducting crystals duly illuminated by slightly off-resonant not-toohigh-power pulsed lasers with pulse duration sufficiently larger than the acoustic phonon lifetime.

2) Resonance conditions between 1) lattice frequency and plasma frequency 2) stokes frequency and electron-cyclotron frequency 3) stokes frequency and hybrid (plasma and electron-cyclotron) frequency may be utilized, on one hand, to reduce the threshold intensity for onset of parametric amplification. On the other hand, switching of parametric gain between the low and high value and hence open up the possibility of devising optical switch. A strong transverse magnetostatic field (10.0 T) enhances the gain by a factor of 10^3 as in its absence.

3) The technological potentiality of a magnetoactive III-V semiconductor-plasma as the host for parametric devices like parametric amplifiers and oscillators are established. In III-V crystals, parametric amplification and oscillation in the infrared regime appears quite promising under the resonance conditions and replaces the conventional idea of using high power pulsed lasers.

5. ACKNOWLEDGEMENTS

One of the authors (PA) acknowledges financial support from Department of Science and Technology (FIST), New Delhi, India. Dr. K.C. Bhardwaj (Vice-chancellor) is also acknowledged for his constant encouragement.

REFERENCES

1. R. L. Bayer, "Nonlinear Optics," Academic Press, London, 1975, pp. 47-159.
2. C. K. Hong and L. Mandel, "Theory of Parametric Frequency down Conversion of Light," Physical Review A, Vol. 31, No. 4, 1985, pp. 2409-2418. doi: 10.1103/PhysRevA.31.2409
3. Z. Y. Ou, L. J. Wang and L. Mandel, "Photon Amplification by Parametric Down Conversion," Journal Optical Society America, Vol. B7, 1990, pp. 211-214.
4. J. Zyss, I. Leodoun, J. Badan, J. L. Oudar, J. Etchepare, D. Hulin, A. Mingus and A. Antonnetii, "Amplification at Emission Paramétriques à l'Échelle Subpicoseconde dans un Crystal Organique: Application à la Spectroscopy InfraRouge," Revue de Physique Appliquée, Vol. 22, No. 10, 1987, pp. 1229-1238.
5. D. A. B. Miller, "Bistable Optical Devices: Physics and Operating Characteristics of Nonlinear Optics in Semiconductors," Laser Focus, Vol. 19, 1983, pp. 61-68.
6. R. K. Jain, "Degenerate Four-Wave Mixing in Semiconductors: Application to Phase Conjugation and to Picoseconds Resolved Studies of Transient Carrier Dynamics," Optics Engineering Bellingham, Vol. 21, 1982, pp. 199-218.
7. S. D. Smith, "Lasers: Nonlinear Optics and Optical Computers," Nature, Vol. 316, 1982, pp. 319-324.
8. P. A. Wolff, "Nonlinear Optics," Academic Press, London, 1977, pp. 169-212.
9. E. Garmire, "Resonant Optical Nonlinearities in Semiconductors," IEEE Journal Quantum Electronics, Vol. 6, No. 6, 2000, pp. 1094-1110.
10. R. W. Boyd, "Nonlinear Optics," Academic Press, San Diego, 1992, Ch. 9, pp. 409-450.
11. M. J. Connelly, "Semiconductor Optical Amplifiers," Springer-Verlag, Boston, 2002.
12. D. Apostolopoulos, K. Vyrsokinos, P. Zakynthinos, N. Pleros and H. Avramopoulos, "An SOA-MZI NRZ Wavelength Conversion Scheme with Enhanced 2R Regeneration Characteristics," IEEE Photonics Technology Letter, Vol. 21, No. 19, 2009, pp. 1363-1365.

13. A. Hayat, Y. Elor, E. Small and M. Orenstein, "Phasematching in Semiconductor Nonlinear Optics by Linear Long-Period Gratings," Applied Physics Letter, Vol. 92, No. 18, 2008, pp. 181110-112.
14. G. Lutz, "Semiconductor Radiation Detectors: Device Physics," Springer, Berlin, 1999.
15. Y. Fu and M. Willander, "Physics Models of Semiconductor Quantum Devices," Springer, Berlin, 1999.
16. M. B. Johnston, D. M. Whittaker, A. Dowd, A. G. Davies, E. H. Linfield, X. Li and D. A. Ritchie, "Generation of High-Power Terahertz Pulses in a Prism," Optics Letters, Vol. 27, No. 21, 2002, pp. 1935-1937.
17. G. Meinert, L. Banyai, P. Gartner and H. Haug, "Theory of THz Emission from Optically Excited Semiconductors in Crossed Electric and Magnetic Fields", Physical Review B, Vol. 62, 2000, pp. 5003-5009.
18. J. M. Makowska, K. J. Plucinski, A. Hruban, J. Ebothe, J. I. Fuks and I. V. Kityk, "Acoustically Induced Optical Second Harmonic Generation in InAs," Semiconductor Science & Technology, Vol. 19, No. 11, 2004, pp. 1285- 1290.
19. A. A. Bulgakov and O. V. Sharmkova, "Nonlinear Interaction of Waves in Semiconductor Plasma," Journal Physics D: Applied Physics, Vol. 40, No. 9, 2007, pp. 5896- 5901.
20. B. Lal, P. Aghamkar, S. Kumar and M. K. Kashyap, "Second-Order Optical Susceptibility in Doped III-V Piezoelectric Semiconductors in the Presence of a Magnetostatic Field," European Physics Journal D, Vol. 61, No. 3, 2011, pp. 717-724.
21. P. K. Kaw, "Parametric Excitation of Ultrasonic Waves In Piezoelectric Semiconductors," Journal Applied Physics, Vol. 44, No. 4, 1973, pp. 1497-1498.
22. D. L. Spears, "Brillouin Scattering Study of Propagating Acoustoelectric Domains in n-GaAs," Physical Review B, Vol. 2, No. 6, 1970, pp. 1931-1951. doi: 10.1103/PhysRevB.2.1931
23. D. L. Rode, "Semiconductors and Semimetals," Academic, New York, 1975, Ch. 1, pp. 5-17.
24. G. E. Stillman, C. M. Wolfe and J. O. Dimmock, "Hall Coefficient Factor for Polar Mode Scattering in n-Type GaAs," Journal Phys Chem Solids, Vol. 31, No. 6, 1970, pp. 1199-1204.

25. E. D. Palik and J. K. Furdyna, "Infrared and Microwave Magnetoplasma Effects in Semiconductors," Report Progress in Physics, Vol. 33, No. 3, 1970, pp. 1193-1322.
26. J. M. Mayer, F. J. Bartoli and M. R. Kruer, "Optical heating in Semiconductors," Physical Review B, Vol. 21, No. 4, 1980, pp. 1559-1568.

8

CHAPTER

SEMICONDUCTOR CDF_2:GA AND CDF_2:IN CRYSTALS AS MEDIA FOR REAL-TIME HOLOGRAPHY

Alexander I. Ryskin, Alexander S. Shcheulin and Alexander E. Angervaks *

National Research University of Information Technologies, Mechanics and Optics, 49, Kronverkskiy pr., St. Petersburg 197101, Russia

ABSTRACT

Monocrystalline cadmium fluoride is a dielectric solid that can be converted into a semiconductor by doping with donor impurities and subsequent heating in the reduction atmosphere. For two donor elements, Ga and In, the donor ("shallow") state is a metastable one separated from the ground ("deep") state by a barrier. Photoinduced deep-to-shallow state transition underlies the photochromism of CdF_2:Ga and CdF_2:In. Real-time phase holograms are recorded in these crystals capable of following up optical processes in a wide frequency range. The features of photochromic transformations in CdF_2:Ga and CdF_2:In crystals as well as holographic characteristics of these media are discussed. Exemplary applications of CdF_2-based holographic elements are given.

Keywords: cadmium fluoride; bistable centers; photoinduced metastable state; holographic grating decay; real-time holography; dynamic correction of wavefronts and optical images; holographic correlator

1. INTRODUCTION

A change in the optical properties of a crystal caused by a change of defect configuration in the lattice allows the crystal to be used as a holographic medium. In particular, such defects are DX-centers in III–V and II–VI semiconductors. This impurity center has a ground state with two valence electrons localized at the atom-like orbital of the impurity and excited hydrogenic (donor) state. Photoinduced change of center state is accompanied by a drastic center reconstruction. This reconstruction creates a barrier separating two states of the center and makes the excited state a metastable one. Features of DX-centers underlie photochromy of semiconductor crystals containing these impurities.

In 1994, the recording of holographic phase grating in an AlGaAs:Si layer created by the photoionization of the ground state of DX-center was demonstrated [1]. The resulting metastable conducting state persisted at sufficiently low temperature, T, for an immeasurably long time. Localized modification of free-carrier concentration produced the observed refractive index grating. Similar gratings were also observed in GaAlAs:Te and CdZnTe:Cl [2]. Also, absorption gratings in AlSb:Se were recorded using a bistable defect which did not change free-carrier levels [3]. In 1995, the phase grating was recorded in a wide-gap (7.6 eV) ionic semiconductor, CdF_2:In, for which variation of the refractive index was produced by photoinduced depopulation of the ground state of In center and the resulting population of its metastable state [4]. Similar grating was recorded in CdF_2:Ga [5]. As in compound semiconductors, persistent hologram recording is possible in CdF_2 at a sufficiently low temperature due to a barrier between the metastable and ground states of the impurity center. This barrier determines the time interval for which the grating persists. The temperature increase results in grating decay. Using temperature as a managing parameter ensures realization of the wide range of decay times, i.e., wide range of frequencies of optical processes that can be followed up with these holographic media. Besides, these media have a number of valuable properties, including high spatial resolution of holographic gratings, laser radiation tolerance (unlimited number of recording/readout cycles), and optical isotropy. Large, high-quality CdF_2 crystals, whose properties make them promising media for the creation of volume elements of real-time holography, are available.

In this review, we consider the mechanism of photoinduced transformation of bistable center state in semiconductor CdF_2 crystals and the use of crystals with these centers for recording holograms in the real time scale.

We argue that these centers are identical to DX-centers in conventional III–V and II–VI semiconductors and point out which features of Ga and In determine their bistable nature in CdF_2 crystals. The natures of bistable center transformations are discussed in Section 2. Section 3 deals with CdF_2:In and CdF_2:Ga crystals as holographic media. In Section 4, some examples of holographic element applications based on these crystals are given.

2. BISTABLE CENTERS IN CDF$_2$ CRYSTALS AND THEIR TRANSFORMATIONS

2.1. The Nature of Bistable Centers in CdF$_2$ Crystals

The cadmium fluoride crystal has a fluorite structure (the space symmetry $05h(Fm3m)$) that may be presented as a sequence of anion (fluorine) cubes, half of the central positions of which are occupied with cations (cadmium) and the other half are empty, forming interstices. CdF_2 is an ionic dielectric crystal that can be converted into a semiconducting state via doping with column-III elements of the periodic table (donors). These donors are introduced in the raw material for crystal growth. During subsequent annealing of as-grown crystals in a reducing atmosphere of Cd vapor or in hydrogen (an "additive coloration" of the crystal [6,7]), interstitial fluorine ions, F^-, which are charge compensators for the excess "+1" charge of the dopants, diffuse out of the volume of the crystal to its surface, where they recombine with the reducing agents (Cd^{2+} or H^+ ions). The charge neutrality of the crystal is maintained by an opposite current of electrons that are segregated off this agent and diffuse into the crystal volume. These electrons are localized in the conduction band or at hydrogenic donor orbitals (e_{hydr}) centered on the trivalent impurity ion, thus converting the crystal into a semiconductor state [8,9].

The additive coloration is carried out in the unsoldered quartz ampoule or in a vacuum-processed set-up at temperature T = 350–500 °C (the melting temperature of CdF_2 is 1050 °C).

The possibility of converting the ionic dielectric crystal into a semiconductor state is a unique feature of CdF_2 among predominantly ionic crystals. Due to the great electron affinity of CdF_2, the energy of donor s-levels occurs near the bottom of the conduction band and the levels are collectivized with it, forming the hydrogenic orbitals. The binding energy of these orbitals is ~0.1 eV for any donor [10,11]. A wide and intense infrared

(IR) absorption band corresponds to the donor photoionization process (Figure 1a; see [11] for calculation of the band's shape). The band maximum is located at 7–8 μm, depending on the specific impurity.

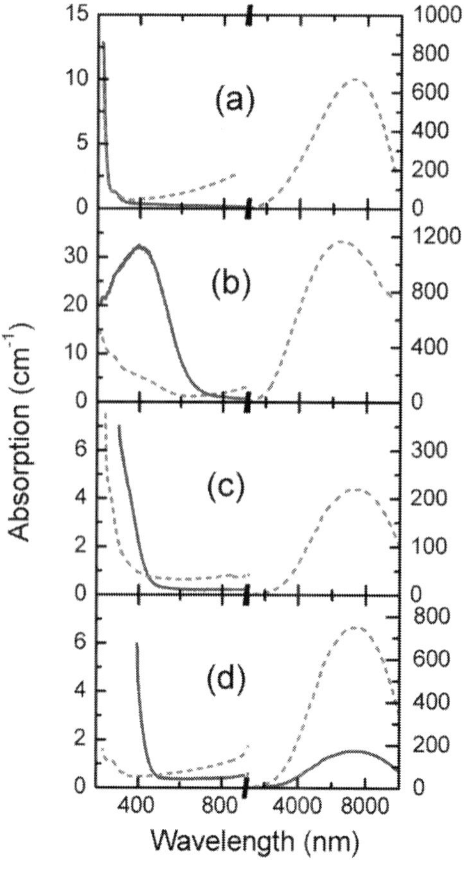

Figure 1. Absorption spectra of (**a**) CdF_2:Y ($N_Y = 2.2 \times 10^{18}$ cm^{-3}) at $T = 77$ K before (solid line) and after (dotted line) additive coloration, and spectra of (**b**) CdF_2:In ($N_{In} = 4 \times 10^{18}$ cm^{-3}); (**c**) CdF_2:Ga ($N_{Ga} = 7 \times 10^{17}$ cm^{-3}); (**d**) CdF_2:Ga,Y ($N_{Ga} = 1.7 \times 10^{18}$ cm^{-3}) additively colored crystals cooled in the dark down to T = 77 K (solid lines) and the same crystals after subsequent illumination in the UV-VIS absorption band at the same temperature (dotted lines).

The constant of interaction of conductivity electrons with longitudinal optical phonons α is equal to 3.3 [12]. It follows from this value that the mobile carriers in the conduction band and carriers bound at hydrogenic donor orbitals are a free and bound polaron, respectively.

During the additive coloration procedure, not all interstitial F− ions present in as-grown crystal are substituted for electrons. Similarly to conventional n-type semiconductors in which acceptors compensate donors, thus reducing the effective concentration of electrons, $\tilde{n}_\Sigma$, one may define for CdF_2 the compensation degree, K, by the equation

$$\tilde{n}_\Sigma = N_\Sigma(1-K) \tag{1}$$

where N_Σ is the total donor concentration at sufficiently low temperature (all electrons introduced at additive coloration are localized on donors), whereas $N_\Sigma K$ is the concentration of interstitial F^- ions that remain in the crystal after the additive coloration procedure. Probably, this compensation is of non-local nature.

Two donor impurities, Ga and In, form bistable centers in semiconductor CdF_2 crystals. Like DX-centers, they have the ground and the metastable excited states.

It was initially assumed that, unlike DX-centers, the two states of these impurities in predominantly ionic CdF_2 crystal correspond to electron localization either at the intrinsic atomic-like orbital (Me^{2+} valent state of the impurity, Me = Ga, In; "deep" state) or at the hydrogenic orbital (Me^{3+} + e_{hydr}; "shallow" donor state) [13,14,15,16]. Thus, transformation of the center state was assumed to proceed without change in its charge. Within this framework, Ga and In bistable centers were treated as examples of intrinsic self-trapping after Toyozawa (see, for instance, [17]).

Subsequent studies of optical and thermal transformations of Me centers in CdF_2 showed that, in fact, a change in a center's state is accompanied by a change in the charge. The bimolecular kinetics of thermal destruction of non-equilibrium shallow centers [18] and the quantum yield η = 2 of the photoinduced reaction of deep-to-shallow center conversion clearly indicate that two shallow centers participate in the formation of one deep center and vice versa [19]. This means that the ground state of the Me center corresponds to localization of two electrons at the intrinsic atomic-like orbital of the center, thus testifying to the formally single-valent nature of Ga and In ions in the deep state. Accordingly, the photoinduced deep-to-shallow center conversion can be described by Reaction (2):

$$Me^{1+} + Me^{3+} \xrightarrow{h\nu} 2\left(Me^{3+} + e_{hydr}\right) \tag{2}$$

As seen from Equation (2), an "empty" Me^{3+} ion is an indispensable element of the conversion process.

Direct evidence for the two-electron nature of the deep center was found in measurements of the magnetic moment, J, of CdF_2:In crystals. In these experiments, no magnetic moment was observed in the deep state of the In center. However, it appeared (J = ½) when the shallow state occurred, to be populated at photoexcitation of the deep state [20,21].

The above facts prove that Ga and In centers in ionic semiconductor CdF_2 are identical to DX-centers in III–V and II–VI semiconductors. As well as for these centers, the barrier separating the deep and shallow states of the Ga and In centers is due to the center reconstruction when its charge is changed. The microscopic nature of this reconstruction was identified by the first-principle calculation [22]. Formation of the deep state was found to be accompanied by displacement of an impurity from the site position (surrounded by the cube of F^- ions) into an adjacent empty cube of anions for a distance of about ¾ of the cube edge (Figure 2a). The presence of vacancy in the deep center structure (after displacement of an impurity) was supported by experiments on positron annihilation in CdF_2:Ga and CdF_2:In [23]. The binding energy of the deep state was found to be 0.70 eV for Ga and 0.25 eV for In [22].

A peculiarity of Ga and In, which belong to the boron sub-column of column III of the periodic table, is that for these ions the filled electron shell exists not only for the trivalent Me^{3+} state, which is typical for all column-III elements, but also for the univalent Me^{1+} state. This state, with two electrons localized with opposite spins on the ns-orbital of the impurity (n = 4, 5 for Ga and In, respectively), corresponds to the ground state of the center.

Prior to Ga, elements of this sub-column, Al and B, are too small to be incorporated in the site position of the CdF_2 lattice. T1, which is next after In in the sub-column, was never used as a dopant of this crystal. Elements of the scandium sub-column of column III, in particular rare-earth elements, form only shallow donor states in CdF_2. Single-valent states of these elements contain one electron in an ns-shell and are unstable.

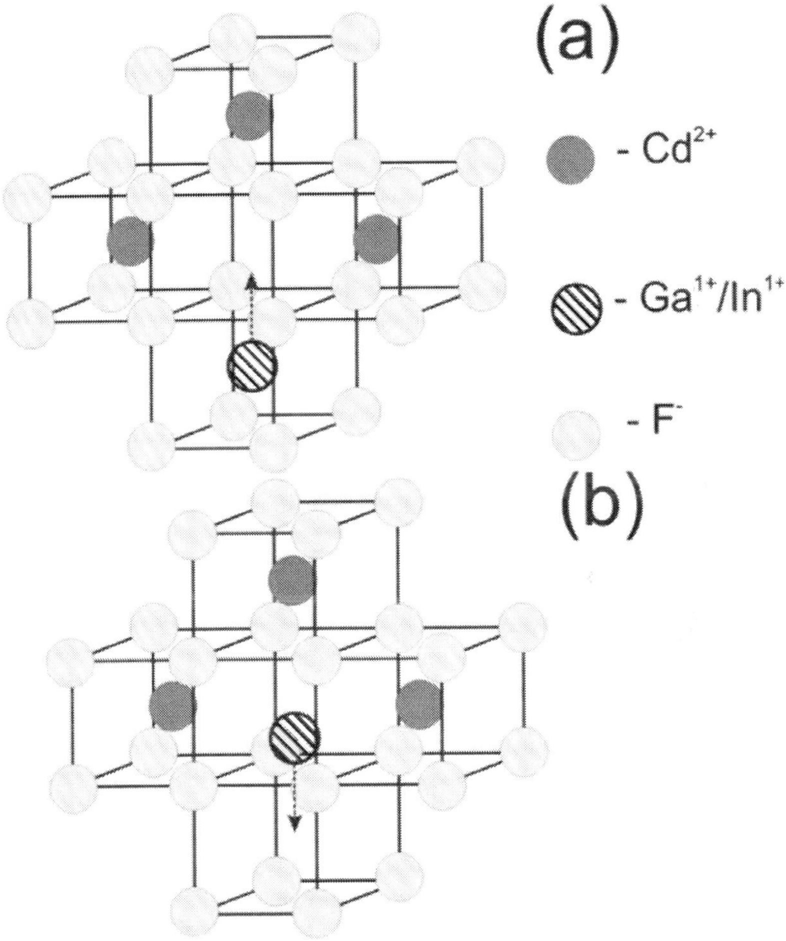

Figure 2. Displacement of a bistable impurity ion at (**a**) formation of a deep center; and (**b**) conversion of a photoionized deep center into a shallow center.

The large lattice relaxation accompanying the deep state formation creates a barrier that separates the shallow donor state from the deep state and makes the shallow state a metastable one (Figure 3). The main feature of this relaxation, i.e., the configuration coordinate of the center, is shown in

Figure 2a. The large lattice relaxation is typical for DX-centers and is a consequence of the linear electron-lattice coupling.

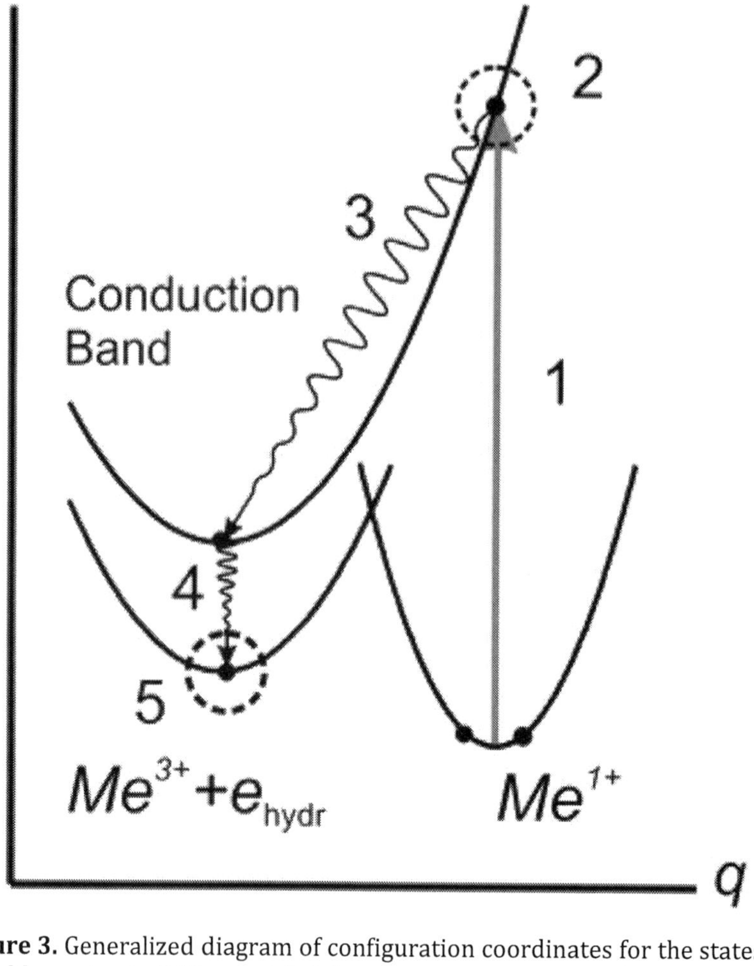

Figure 3. Generalized diagram of configuration coordinates for the states of bistable center in CdF_2 crystal. Schematic representation of processes 1–5 at deep-to-shallow center transformation are shown (see below in the text). Lattice relaxations that correspond to other configuration coordinates than q, are indicated by dotted circles. Non-radiation transitions with photon emission are shown by wavy lines.

Figure 4 shows configuration coordinate diagrams for both ions as calculated in [24] using deep center energies found in [22]. Note the large difference in the barrier height for Ga and In. The main reason for the large barrier in CdF_2:Ga is the lattice contraction around the relatively small Ga ion [25] that results in more elastic coupling of the impurity with neighboring ions as compared to In.

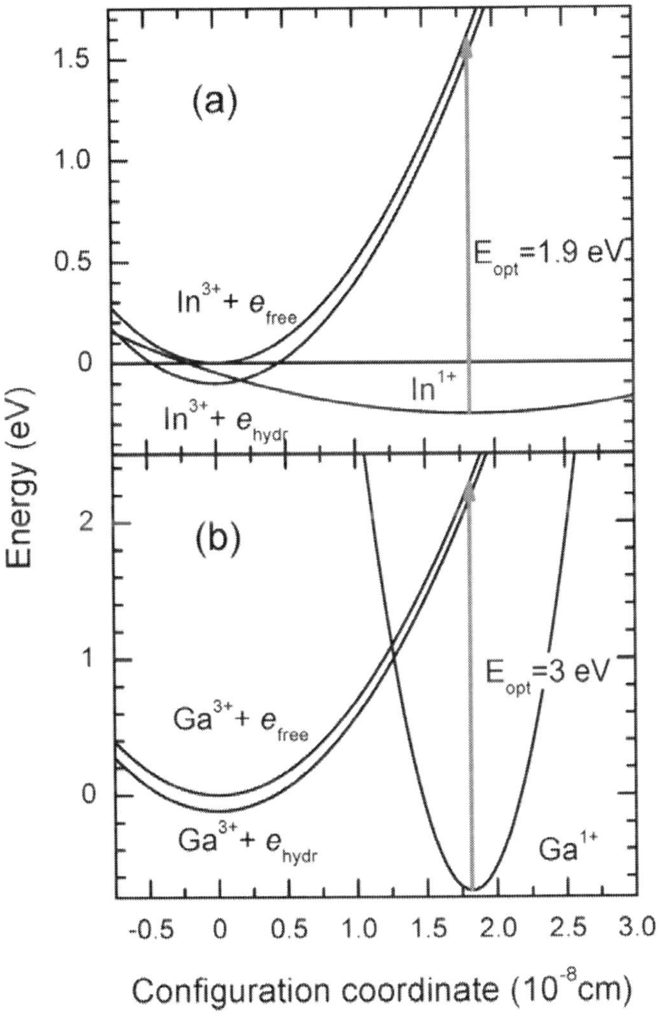

Figure 4. Diagram of configuration coordinates for (**a**) CdF_2:In; and (**b**) CdF_2:Ga. The configuration coordinate is the displacement of the impurity ion along the fourth-order axis as it is shown in Figure 2a. e_{free} is a free carrier in the conduction band.

In absorption spectra of CdF_2:In and CdF_2:Ga crystals, the band in the ultraviolet-visible (UV-VIS) spectral range exists, as well as the IR band (Figure 1(b,c); the shape of this band in CdF_2:Ga crystal is clearly seen in the differential spectrum of this crystal, see Figure 13 below). This wide band has a maximum of ~390 nm for CdF_2:Ga and ~460 nm for CdF_2:In; for both crystals it covers the whole of the visible spectrum range and extends up to 830–850 nm. This band is due to photoionization of deep centers. Its shape represents the convolution of the vibronic line-shaped function of optical transition from the deep centers to the conduction band with the density of electronic states in this band.

CdF_2:In and CdF_2:Ga crystals cooled in the dark are in a semi-isolating state, since electrons introduced in the course of additive coloration are predominantly located in couples at deep centers. Only the UV-VIS band exists in the absorption spectra of the crystals (Figure 1(b,c)). At excitation in this band, deep-to-shallow state conversion occurs in accordance with the reaction (2) accompanied by UV-VIS band disappearance and IR band occurrence (Figure 1) [26]. The photoexcited shallow state persists below ~220 K in CdF_2:Ga and below ~40 K in CdF_2:In [27,28]. At higher temperatures, the thermo-destruction of shallow centers and formation of deep centers occur, both types of center being in equilibrium with each other. Thus, deep-to-shallow and shallow-to-deep center conversion under the impact of temperature is a mutual process that is described by reaction (3):

$$Me^{1+} + Me^{3+} \xleftrightarrow{kT} 2\left(Me^{3+} + e_{hydr}\right) \quad (3)$$

Figure 5 and Figure 6 show the temperature dependencies of relative population of deep and shallow states for both crystals. In principle, electrons introduced in the crystal at its additive coloration are distributed between two states of the impurity and the conduction band. However, for both crystals, the free carrier concentration is very small even at T = 400 K [29]. Since the Ga deep state has a higher binding energy as compared with In, and binding energy of the shallow state is nearly equal for both ions, one may propose that the relative population of the shallow state of Ga increases with temperature much more slowly than that of In. Meanwhile, experiment reveals a reverse situation. The difference in the distribution of electrons over levels of two bistable centers was explained in [29] proceeding from statistics of this distribution with allowance for typical dopantcompensation degree and relative concentration of impurity centers capable of two-electron deep state formation. It was found that the compensation degree for both dopants is very high: K = 0.90–0.97 for CdF_2:In and K = 0.996 for CdF_2:Ga. This suggests that for both ions the concentration of Me^{3+} ions significantly exceeds the maximal concentration of (Me^{3+} + e_{hydr}) centers.

The most important difference between In and Ga dopants in CdF_2 crystal lies in the number of impurity ions that can form both shallow and deep centers. Whereas for In practically all dopant ions can form these centers, a very small portion of Ga ions present in the crystal (not exceeding one percent) can form the deep centers; meanwhile, most or all of these ions can form the shallow center. This feature of Ga shifts the electron distribution between deep and shallow centers towards the shallow centers and provides faster destruction of Ga deep centers with temperature compared to In. A possible source of this Ga feature is discussed in [29].

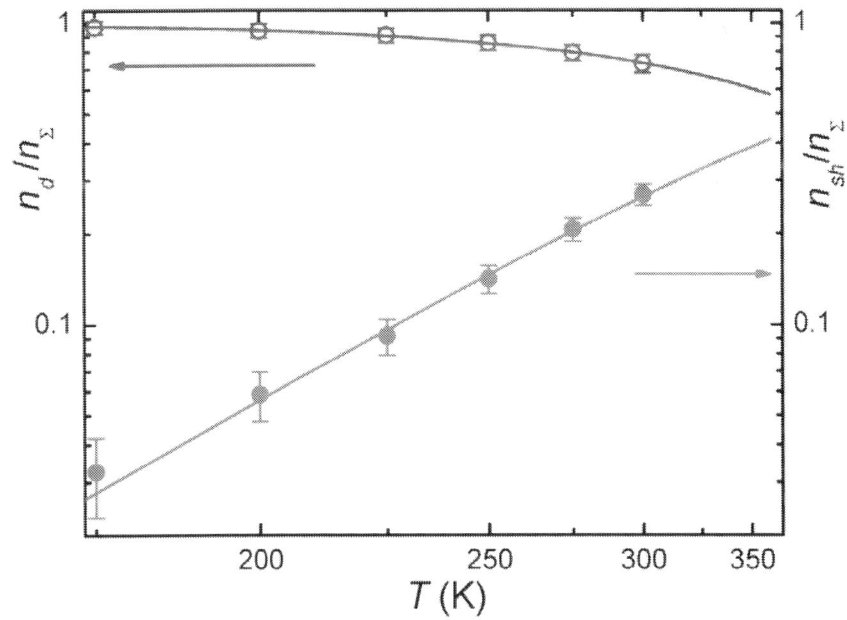

Figure 5. Temperature dependencies of relative concentrations of carriers at deep, $\tilde{n}_d/\tilde{n}_\Sigma$,and shallow, $\tilde{n}_{sh}/\tilde{n}_\Sigma$ centers for CdF_2:In. Open circles (deep centers) and dark circles (shallow centers) are experimental data; solid lines show theoretical dependency as calculated in [28].

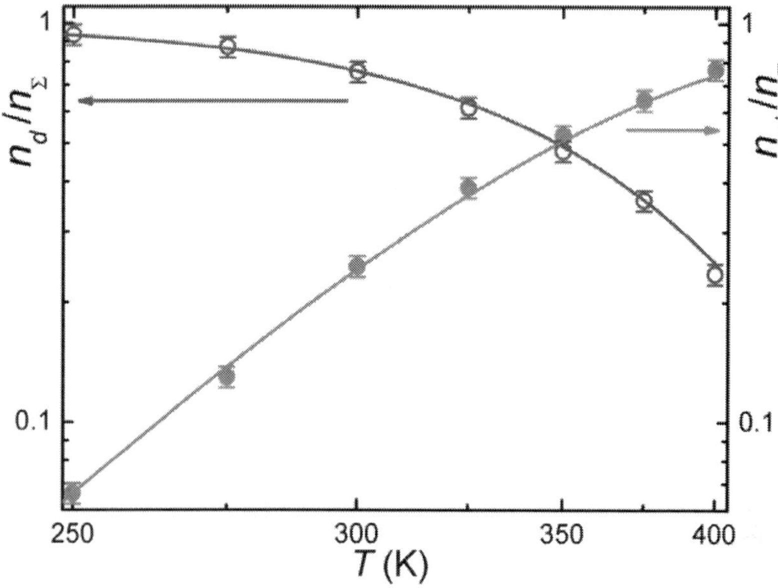

Figure 6. The same as in Figure 5, but for CdF_2:Ga crystal.

Doping CdF_2 with In, and especially with small-size Ga, increases light scattering in crystals. To reduce this effect, the Ga-doped crystals are co-doped with Y. Co-doping not only decreases the scattering but also diminishes the non-photochromic absorption of these crystals (Figure 1d) [30].

The concentration of centers that can form the deep as well as the shallow state in CdF_2:Ga crystals ("optically active centers", which are responsible for the photochromic effect) coincides practically with the concentration of electrons introduced into the crystal during additive coloration, whereas the total Ga concentration in the crystals in accordance with mass-spectrometric data is $\sim 10^{20}$ cm^{-3}. This concentration is determined by the growth conditions and is nearly the same in all Ga-doped crystals.

Co-doping CdF_2:Ga crystals with Y at the level of 10^{19} cm^{-3} (typical value) increases the concentration of "optically active" Ga ions to approximately twice as many. Co-doping with Y insignificantly changes the temperature dependency of distribution electrons between deep and shallow centers, as shown in Figure 6.

Indium concentration depends on the doping level; it can reach 10^{21} cm^{-3} [31].

The concentration of optically active centers in the specific sample depends on the regime of its additive coloration. As a rule, this concentration does not exceed 10^{18} cm^{-3} for CdF_2:Ga crystals and 10^{19} cm^{-3} for CdF_2:In crystals [32].

Shallow donor states in crystals under consideration form the narrow impurity band. For CdF_2:Ga crystals, the width of this band was estimated as ≤0.2 eV [29] (see also [33]).

Analysis of dependencies shown in Figure 5 and Figure 6 gives lower values of binding energy compared to [23]: this energy was found to be ~0.38 eV for CdF_2:Ga and ~0.17 eV for CdF_2:In [29]. These values based on experimental data are probably more trustworthy than values found via the first-principle calculations.

Bearing in mind the following discussion of holographic properties of CdF_2 crystals with bistable centers one should note such features of semiconductor CdF_2 crystals as presence of far-IR absorption (10–150 cm^{-1}) [34]. It was proposed that this absorption is due to the existence of so-called "ionized donor pairs" in the crystals, near-by disposed couples of donors having one electron (the analog of the molecular hydrogen ion) [35]. It is absent for CdF_2:Ga and CdF_2:In crystals cooled in the dark but appears on their illumination in the UV-VIS band.

2.2. Kinetics of Photo- and Thermal-Transformations of Bistable Centers

A time evolution of the photoinduced deep-to-shallow center conversion was studied by recording the absorption spectrum of the crystals at excitation by short UV pulses [36]. Under the impact of 150-femtosecond laser pulses with wavelength of 395 nm, a decrease of absorption in the band of deep centers and an increase of absorption in the band of shallow centers occurs, as recorded by test pulses. The time-resolved changes in absorption of CdF_2:In and CdF_2:Ga crystals at several wavelengths in the spectral range of 430–1100 nm are shown in Figure 7 and Figure 8, respectively.

The curve 3 in Figure 7 and Figure 8 are of special interest in terms of the holographic properties of these crystals. These curves confirm the existence of a so-called "isobestic point", the wavelength for which the increase in the shallow center absorption upon illumination in the UV-VIS band is practically equal to the decrease in the deep center absorption [37]. Actually, this balance is realized not only for a fixed wavelength but for a spectral range in which the absorption bands of both centers overlap (an "isobestic gap"). Within this gap, photoinduced deep-to-shallow state transformation practically does not change the absorption of the crystal.

Mutual compensation of absorption of two centers within this gap takes place not only under photoexcitation of the crystal but even at the change of the crystal temperature.

Curves 4 and 5 in Figure 7 show that the photoinduced transmission band of deep centers is completely formed in a time ~1 ps. This time significantly exceeds both the physical photoionization time of the centers (units of femtoseconds) and the time-resolution function of the spectrometer (~400 fs). This means that the deep center photoionization does not complete the formation of the transmission band of deep centers, which reflects the deep-to-shallow center conversion. The point is that the photoionization time is too short for a noticeable change in the space configuration of the ionized deep center. The one-electron state of such a center is unstable because its nuclear configuration corresponds to the deep (two-electron) state. This instability leads to rearrangement of the lattice, shown in Figure 3b. The rearrangement occurs in a time equal to reverse vibrational frequencies (10^{-13}–10^{-12} s). This process determines the dynamics of UV-VIS transmission band formation.

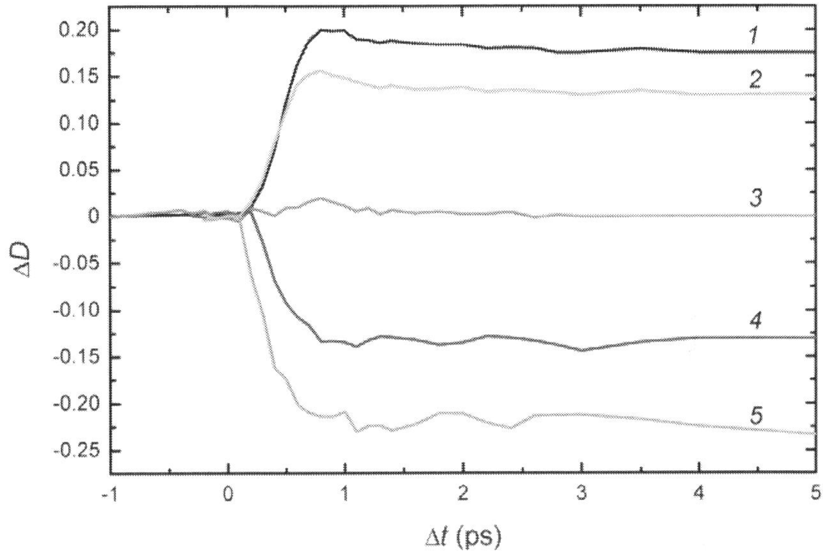

Figure 7. Kinetics of photoinduced time-resolved changes in the absorption spectrum of CdF_2:In (N_{In} = 2.9 × 10^{18} cm^{-3}) crystal under the impact of 150-femtosecond laser pulses with wavelength of 395 nm at room temperature. Probe wavelengths are curve *1*: 1100 nm; curve *2*: 1000 nm; curve *3*: 730 nm; curve *4*: 570 nm; and curve *5*: 530 nm. Curve 3 corresponds to minimal absorption in the gap between the UV-VIS and IR bands.

The nature and origin of transient processes in the photoinduced IR absorption (curves 1 and 2 in Figure 7 and Figure 8) are more complex. The IR response is formed by a number of processes, including the following: (1) photo-detachment of an electron from a deep center to the conduction band; (2) subsequent transformation of this electron into a free polaron; (3) polaron relaxation to the bottom of the conduction band; (4) capture of the polaron by an Me^{3+} ion; (5) transformation of this complex into the hydrogen-like shallow center (the bound polaron); (6) formation of a similar center as a result of the transformation of an ionized deep center; the corresponding lattice relaxation is shown in Figure 3b; and (7) establishment of temperature equilibrium in the system of bistable centers in the deep and shallow states and free polarons. Processes 1–5 are shown schematically in Figure 3.

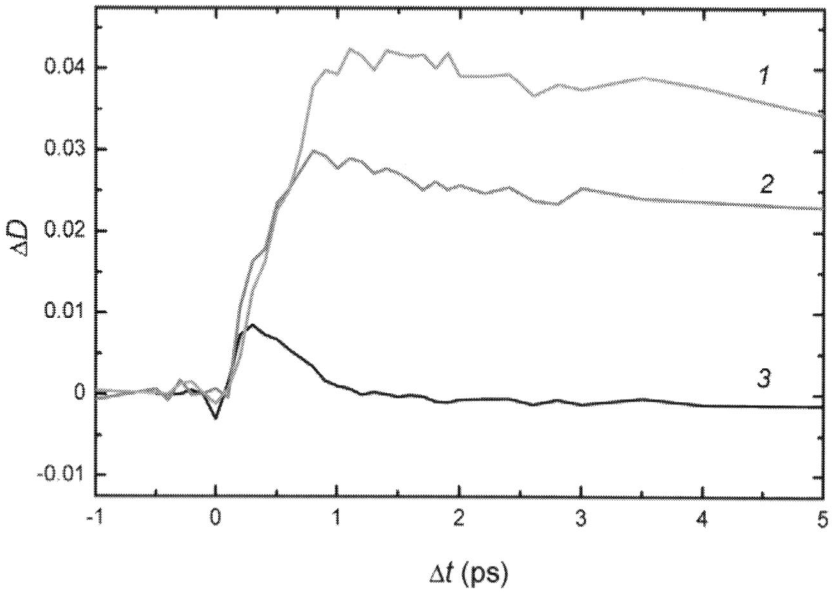

Figure 8. Kinetics of photoinduced time-resolved changes in the absorption spectrum of CdF_2:Ga (N_{Ga} = 4 × 10^{17} cm^{-3}) crystal under the impact of 150-femtosecond laser pulses with wavelength of 395 nm at room temperature. Probe wavelengths are curve 1: 1000 nm; curve 2: 850 nm and curve 3: 530 nm.

The important circumstance for interpreting the photoinduced IR absorption is that the free-carrier absorption cross-section in the near-IR spectral range exceeds the cross-section of shallow centers by a factor of three to four [38]. Hence, the initial stage of the rise of optical density with time for the IR band up to the maximum is mainly due to absorption by photoinduced free carriers (electrons and polarons). A kink (shoulder) on the increasing edge of IR response (curves 1 and 2 in Figure 8) probably divides stages corresponding to free-electron and free-polaron absorption. This feature allows estimation of the time of polaron formation in CdF_2 as 0.8–1.2 ps; this time is similar to that of lattice rearrangement at the conversion of the ionized deep center into the shallow one. Its presence in CdF_2:Ga and absence in CdF_2:In crystals is explained in [36].

The relative contribution of the shallow-center absorption increases as free polarons are captured by Me^{3+} ions. The time of this capturing is comparable with polaron formation and lattice rearrangement times due to a large concentration of Me^{3+} ions (the high compensation degree).

The temperature-dependent decay of photoinduced (non-equilibrium) shallow centers is a much slower process compared with the process of formation of these centers. According to reaction (3), it obeys bimolecular kinetics, for which shallow center concentration is low compared to the concentration of Me^{3+} ions. The above-mentioned difference between In and Ga dopants (see Section 2.1) results in different equations describing the shallow center decay for these two impurities (Equations (4) and (5), respectively [39]:

$$\frac{d\tilde{n}}{C(T)dt} = -\tilde{n}^2 + A(T)(1-\tilde{n}) \tag{4}$$

$$\frac{d\tilde{n}}{C(T)dt} = -\tilde{n}^2(1+\tilde{n}) + A(T)(1-\tilde{n}) \tag{5}$$

Here $\tilde{n} = \tilde{n}_{sh}/\tilde{n}_{\Sigma}$ is the relative concentration of shallow centers, C(T) is the temperature-dependent parameter related to the shallow center decay, and A(T) is the parameter that characterizes the equilibrium concentration of these centers. The solution of Equation (4) is of the hyperbolic-cotangent type and it converts into an exponent at the final stages of decay. For Equation (5), only a numerical solution is possible. As well as in Equation (4), its final stages are exponential ones. As temperature increases, the exponential stage embraces the growing part of the decay curve. At sufficiently high temperature, the decay is practically a purely exponential one.

The C(T) parameter in Equations (4) and (5) determines the rate of formation of deep centers from the shallow centers. The rate of the inverse process is determined by the parameter *B*(T) = *A*(T)C(T). The dimensionless *A*(T) parameter is determined by the relative concentration of shallow centers in thermal equilibrium conditions, $\tilde{n}^0(T) = \frac{\tilde{n}^0_{sh}}{\tilde{n}_{\Sigma}}$, where $\tilde{n}^0_{sh}$ is an equilibrium concentration of these centers:

$$A(T)=\frac{\left[\tilde{n}^{0}\right]^{2}}{1-\tilde{n}^{0}(T)} \tag{6}$$

A(T) dependency is calculated from the temperature dependencies of equilibrium concentrations (Figure 5 and Figure 6). Then, Equation (4) or Equation (5) is solved for a set of temperatures, and the solutions are compared with experimental decay curves. C(T) is found as a best-fit parameter. Then, B(T) dependence is calculated in accordance with the above-given relation. Temperature dependencies of all parameters lie on the Arrhenius plot.

As an example, the decay of photoinduced concentration of shallow centers at T = 254 K for CdF_2:Ga crystal and its approximation by Equation (5) are shown in Figure 9. The "high-temperature" (nearly exponential) decay for this crystal is shown in Figure 10.

Using C(T) and B(T) dependencies, the barrier height, E_{bar}, can be found (see [39] for details). For CdF:Ga, E_{bar} = 0.630 ± 0.010 eV; for CdF:In, E_{bar} = 0.125 ± 0.010 eV.

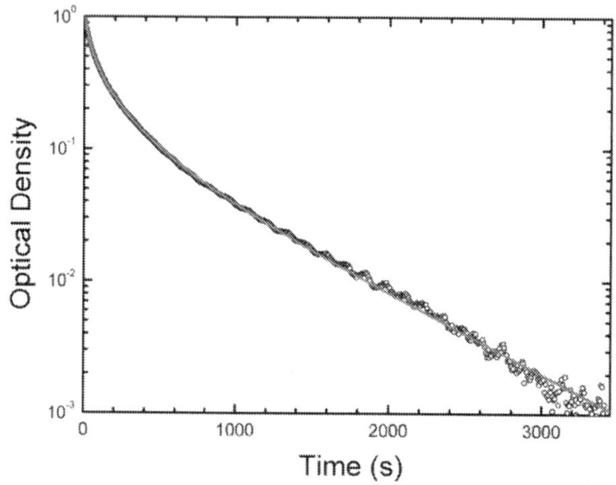

Figure 9. Restoration of the equilibrium concentration of shallow donor states after stoppage of excitation in the UV-VIS absorption band (at t = 0) in CdF_2:Ga (N_{Ga} = 4 × 10^{17} cm^{-3}) crystal at T = 254 K. Points are experimental data; solid line represents the solution of Equation (5) with best-fit parameters.

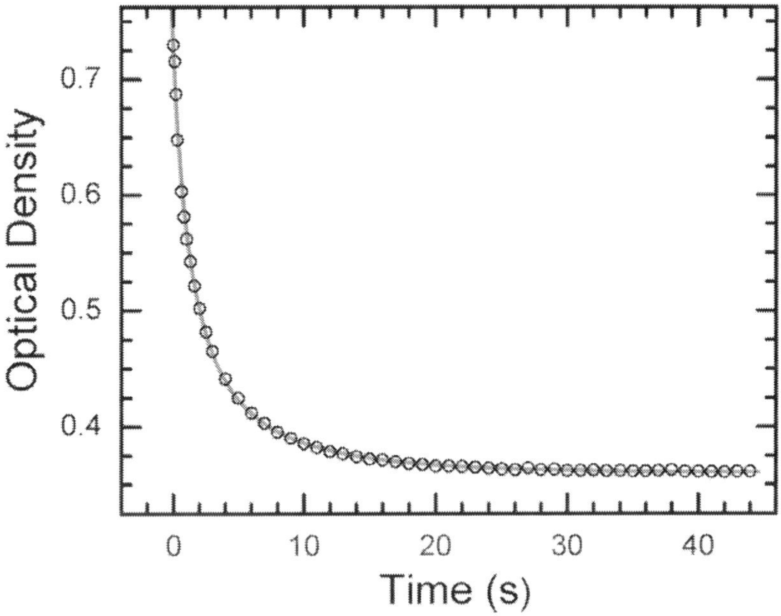

Figure 10. Restoration of the equilibrium concentration of shallow donor states after stoppage of excitation in the UV-VIS absorption band (at t = 0) in CdF_2:Ga ($N_{Ga} = 4 \times 10^{17}$ cm^{-3}) crystal at T = 294 K. Points are experimental data; solid line represents an exponent.

3. CDF$_2$ CRYSTALS WITH BISTABLE IMPURITY CENTERS AS A HOLOGRAPHIC MEDIA

3.1. Diffraction Efficiency and Hologram Decay

As follows from Section 2, the photochromy of CdF_2 crystals with bistable impurity centers is based on photoinduced conversion of the deep centers into the shallow centers. A hologram recording is possible in the spectral range of the UV-VIS absorption band. When exciting the crystal in this band, electrons tightly bound at the deep centers are substituted for weakly bound electrons located at shallow donor centers or in the conduction band that is in equilibrium with these centers. The spatial modulation of photoinduced shallow centers concentration results in modulation of optical constants of the crystal, i.e., formation of holographic grating. Because of the metastable nature of the shallow state, the processes of grating creation (due to deep-to-shallow center conversion, the "direct" process) and decay (due to thermal-induced destruction of non-equilibrium shallow centers, the "reverse" process) progress simultaneously [40].

The photoinduced change of absorption spectra of CdF_2 crystals with bistable centers is shown in Figure 1(b–d). The existence of the "isobestic gap", in which the long-wavelength tail of UV-VIS is superimposed on the short-wavelength tail of the IR band (see Section 2.2) so that deep-to-shallow center conversion does not change an absorption, allows the reading out of predominantly phase holograms. This gap embraces the range of 450–650 nm for CdF_2:Ga and CdF_2:Ga,Y and 680–730 nm for CdF_2:In. Figure 11 shows angular dependencies of the zero- and the first-order diffraction responses for two readout wavelengths in the CdF_2:Ga,Y crystal isobestic gap. The hologram was recorded by 532 nm radiation in a 10 × 10 × 10 mm^3 sample and readout by 532 nm and 660 nm radiation. As is seen from Figure 11, the minima in the zero-order response coincide with the maxima of the first-order response that proves the phase nature of the holograms. One can see significant signal oscillations in Figure 11. Their origin is explained below.

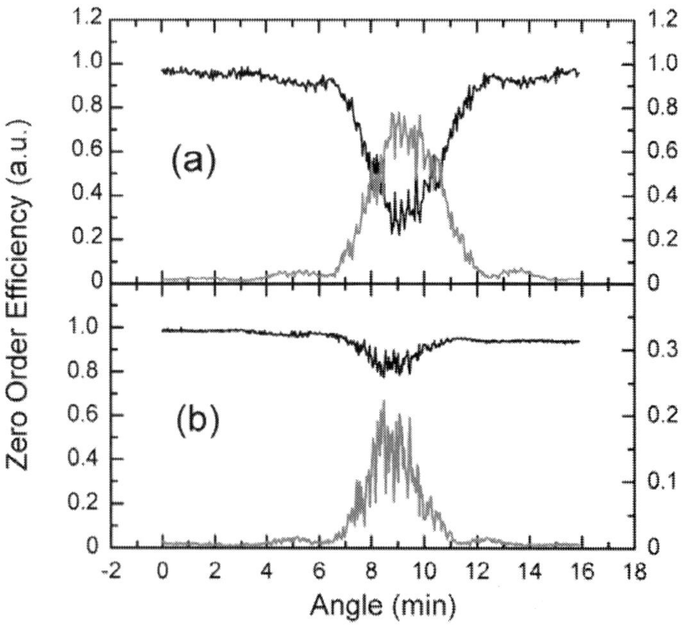

Figure 11. Angular dependencies of the zero-(black lines) and first-order (red lines) diffraction responses of CdF_2:Ga,Y crystal with hologram at readout by (**a**) 532 nm and (**b**) 660 nm radiation.

Figure 12 shows the spectral dependency of δn for CdF_2:Ga crystal. The measurements in the spectral range of 457–837 nm were executed by means of a Mach-Zehnder interferometer [41].

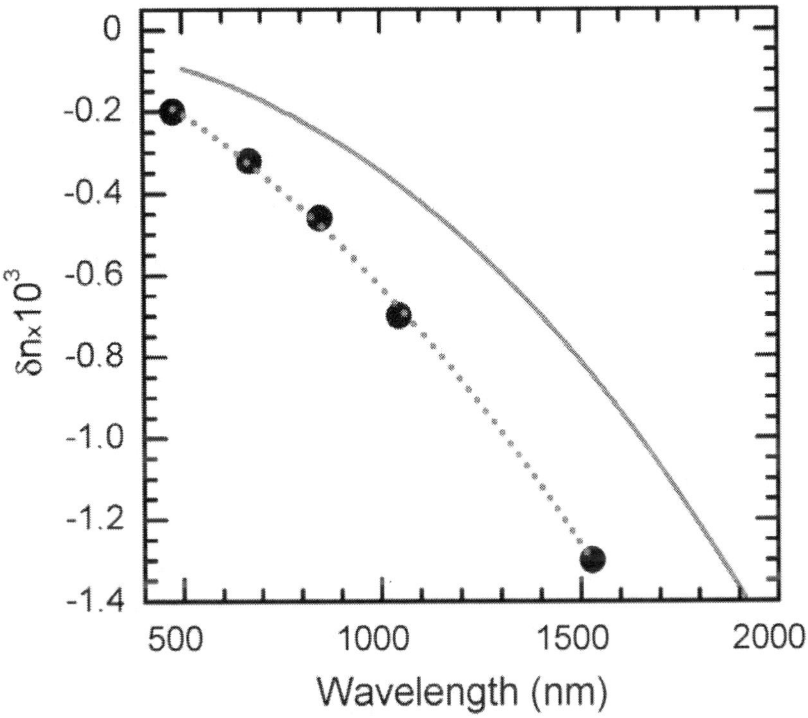

Figure 12. Spectral dependency of the photoinduced change of refractive index of CdF_2:Ga (N_{Ga} = 4 × 10^{17} cm^{-3}) crystal determined at T = 100 K (black circles) and its fit with the two-oscillator model (dotted line). The solid line shows this dependence calculated by the Kramers-Kronig transformation using the photoinduced modification of absorption spectrum of the crystal that is shown in Figure 13.

This dependency can be fit by a two-oscillator model that includes high- (ω_1) and low-frequency (ω_2) effective oscillators [4,41]:

$$\delta n(\omega) = \frac{2\pi e^2}{mn}\left(\frac{\Delta N_1 f_1}{\omega_1^2 - \omega^2} + \frac{\Delta N_2 f_2}{\omega_2^2 - \omega^2}\right) \tag{7}$$

where e and m are the electron charge and mass, respectively, ΔN_i is the photoinduced change of the i-th oscillator concentration, and f_i is the i-th oscillator strength (i = 1, 2). Let us suppose that for frequencies in the spectral range for which δn is determined, the following inequality is satisfied: $\omega_1 >> \omega >> \omega_2$. Then, after neglecting ω and ω_2 in the denominator of the first and second terms of Equation (7), respectively, and substituting ω for $2\pi c/\lambda$ (c is the light speed in vacuum) in the second term, Equation (7) can be presented in the form:

$$\delta n(\lambda) \cong a - b\lambda^2 \tag{8}$$

where $a = 2\pi e^2 \Delta N_1 f_1 / mn\omega_1^2$ and $b = e^2 \Delta N_2 f_2 c / 2\pi mn$ the best-fit parameters of the model. The negative sign of the experimentally determined δn testifies that the parameter a is negative and b is positive, which corresponds to a decrease of "high-frequency" centers ($\Delta N_1 < 0$) and an increase in "low-frequency" centers ($\Delta N_2 > 0$) in the process of holographic grating recording. It is evident from Equation (8) that the high-frequency oscillator is responsible for the constant (spectrally independent) shift in δn whereas the low-frequency oscillator determines the quadratic character of the δn(λ) dependency.

The experimental spectral dependency of δn is well described by Equation (8) (Figure 12). Figure 12 shows that the low-frequency oscillator brings the major contribution to δn (also see below).

Assuming the total completion of reaction (2) and using the Kramers-Kronig transformation

$$\delta n(\nu_1) = \int_0^\infty \frac{\delta\alpha(\nu)\nu d\nu}{\nu^2 - \nu_1^2} \tag{9}$$

where ν, ν_1 are wave numbers, one can calculate from the change of the absorption coefficient δα(ν) the spectral dependency of the photoinduced modification of the refractive index of the crystal δn(ν), which determines the diffraction efficiency of grating recorded in the crystal.

The modification of the absorption spectrum (differential spectrum) of CdF_2:Ga crystal in the photon-energy range of 0.09–5.5 eV at T = 77 K is shown in Figure 13. The dependency δn(λ) calculated from Equation (9) using this spectrum is illustrated by the solid line in Figure 12. The comparison of such dependency with the best-fit spectral dependency (8) shows that the photoinduced IR band is responsible for no more than 70% of refractive index modification. In other words, the contribution of this band to the low-frequency oscillator does not exceed 70%. One can propose that the other 30% is due to ionized donor pair absorption that arises at photoexcitation of deep centers (see Section 2.1).

Another approach to the calculation of the refractive index variation is based on the fact that for probing photon frequencies much higher than frequencies of optical transitions of the weakly bound electrons of hydrogenic centers, the contribution of donor electrons to the refractive index change can be represented by the same expression as for free-electron plasma [42]:

$$\delta n = -\frac{2\pi N_{hydr} e^2}{nm^* \omega^2} = -\frac{N_{hydr} e^2 \lambda^2}{2\pi c^2 nm^*} \tag{10}$$

with N_{hydr} = 7 × 10^{17} cm^{-3} [41], n = 1.575, and polaron mass m^* = 0.9, we find, for the readout wavelength of 476 nm, δn = −2 × 10^{-4}, which is in reasonable agreement with the experimental value. This conclusion states that the change in the shallow center concentration determines the strength of the phase grating that is read out in the isobestic gap.

In accordance with Kogelnik's theory [43], the diffraction efficiency, η, for the phase sinusoidal grating is determined by the formula

$$\eta = \sin^2\left(\frac{\pi n_1 d}{\lambda_0 \cos\theta}\right) \tag{11}$$

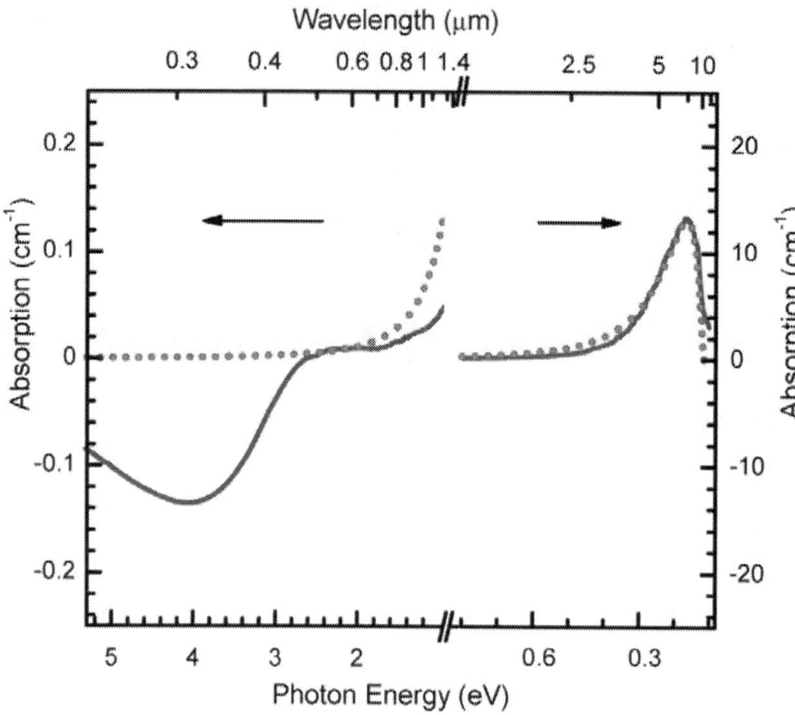

Figure 13. Photoinduced modification of the absorption spectrum of CdF_2:Ga (N_{Ga} = 4 × 10^{17} cm^{-3}) crystal cooled in the dark down to T = 77 K after illumination in the UV-VIS band at the same temperature (solid line). The dotted line shows an approximation of the IR band using the theory of [11].

where the refractive index is assumed to be spatially modulated in the form $n(x) = n_0 + n_1 \sin x$. In Equation (11), d is the grating thickness, λ_0 is the readout wavelength in a vacuum, and θ is the angle between the readout beam and the normal to the grating surface inside the crystal volume. For not too high values of η (practically, for $\eta < 0.7$), this quantity has quadratic dependency on δn or, according to (10), on the concentration of the photoinduced shallow centers.

In accordance with Equations (8) and (11), for $\eta < 0.7, \eta = (a - b\lambda^2)^2$. A quadratic dependency of η on the exposure at the initial stage of the dependency means that both a and b coefficients depend linearly on the exposure. This conclusion agrees with the nature of η dependency on the recording light intensity [44].

Discussing the dependency of η on exposure and on temperature (see below), one should bear in mind that the crystal with DX-center is a reversible medium. The maximum value of the diffraction efficiency at the given power density of radiation and given temperature corresponds to equilibrium of the direct and reverse processes of center transformation.

Figure 14 shows the dependency of the diffraction efficiency of the grating recorded in CdF_2:Ga crystal at T = 100 K on exposure; the recording and readout of the grating were executed at wavelength λ = 453 nm. At the recording temperature of 100 K, the photoinduced shallow centers are persistent (see Section 2.1). It means that the reverse process is "frozen". The dependency in Figure 14 exhibits typical saturation behavior associated with this type of holographic material. The diffraction efficiency increases with the square of exposure for weak exposures, saturates as the deep centers become depleted and the shape of the grating departs from pure sinusoidal, and diminishes at further exposure growth. The decrease of η after attaining the saturation value is due to the deep-to-shallow center conversion in the minima of the fringe pattern because of the light scattering in the sample.

At T > 220 K, the reverse process works and diffraction efficiency depends not on the exposition but on the power density of the recording radiation. To get the maximum diffraction efficiency at the given temperature, the power density should be high enough to realize deep-to-shallow center conversion for a time shorter than the decay time of the hologram. As a rule, such power density demands a pulse excitation.

The reversible nature of this media displays itself in oscillations of the diffraction response at deficient mechanical stability of the optical scheme during the hologram recording process (Figure 15) [40]. At bench vibrations, displacement of the fringe pattern occurs and the hologram recording begins in a new position, while the hologram in the previous position does not erase. As a result, the hologram recorded in a thick sample is a noticeably inhomogeneous one that leads to chaotic change of diffraction response during the recording process. This inhomogeneity also appears in the form of oscillations of the angular dependencies of diffraction response (Figure 11). According to the phase nature of holograms, the zero-order and first-order oscillations at any angular position have equal but oppositely-signed amplitudes.

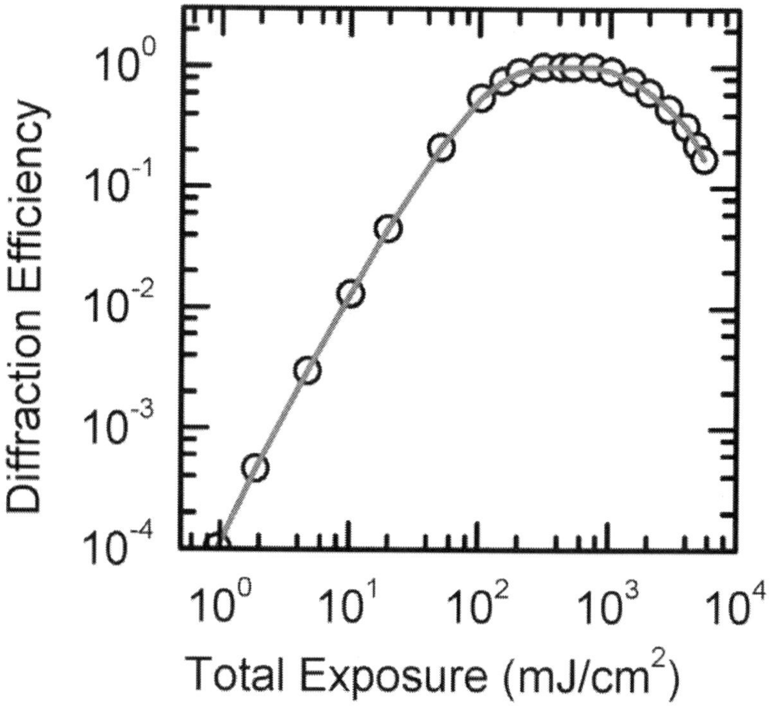

Figure 14. Diffraction efficiency of holographic grating recorded at T = 100 K in CdF_2:Ga (N_{Ga} = 4 × 10^{17} cm^{-3}) crystal vs. exposure. Recording and readout at λ = 453 nm.

The diffraction efficiency of grating recorded on bistable centers in CdF_2:In crystal reaches a maximum at temperature T_{max} that depends on the power density of the recording radiation. Both decrease and increase of T relative to T_{max} result in decrease of η (Figure 16). The "low-temperature" decrease is due to erasing the grating as a result of accumulation of shallow centers in minima of the fringe pattern induced by light scattering in the sample. The "high-temperature" decrease is determined by the increasing role of the reverse process with the temperature growth. A considerable part of this wing lies on the Arrhenius plot; in this part, the decay is nearly exponential (see Section 2.2).

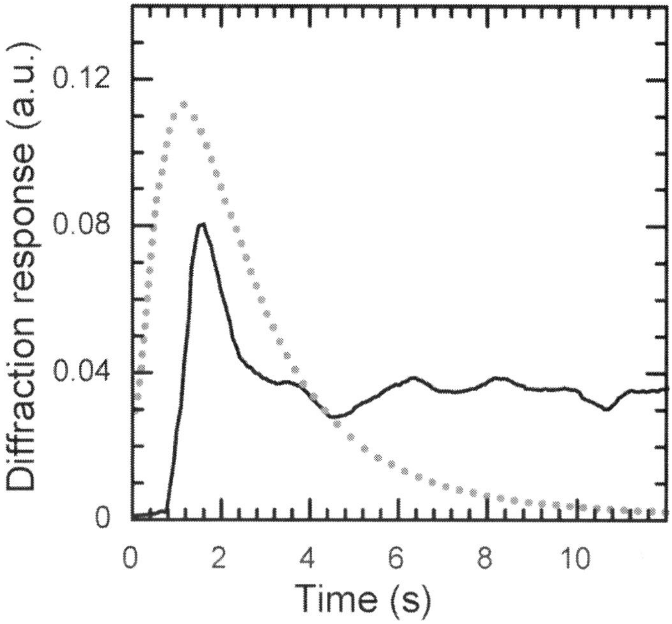

Figure 15. Kinetics of the hologram recording (solid line) and decay (dotted line) in CdF_2:Ga,Y (N_{Ga} = 1.1 × 10^{18} cm^{-3}) crystal at room temperature. Recording at λ = 532 nm, readout at λ = 632.8 nm. The maximal diffraction response at both switching and switching-off the recording radiation correspond to the phase incursion of λ/4. The one-second delay at switching the recording radiation is due to the response time of the gate cutting off the radiation.

It was stated above that the diffraction efficiency of a hologram recorded in CdF_2 crystals with bistable centers depends on the photoinduced concentration of shallow centers. Figure 17 shows IR absorption of CdF_2:Ga,Y crystals in the temperature range of 193–344 K in the dark and after illumination by argon laser (λ = 488 nm) up to saturation of the optical density [45]. Absorption was measured at the wavelength of 1.3 μm. The shape of the IR band weakly depends on temperature so the temperature dependency of IR absorption can be measured at the fixed wavelength. The "dark" absorption follows the temperature dependency of shallow center concentration (Figure 6); its increase reflects the thermal population of shallow levels. The light-induced absorption at T = 193 K corresponds to the total deep-to-shallow center conversion.

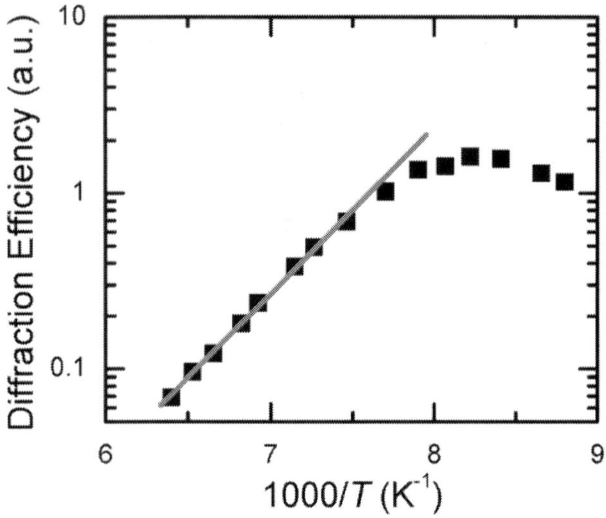

Figure 16. Temperature dependency of diffraction efficiency of the hologram recorded in CdF_2:In (N_{In} = 2 × 10^{18} cm^{-3}) crystal by argon laser radiation (λ = 514 nm).

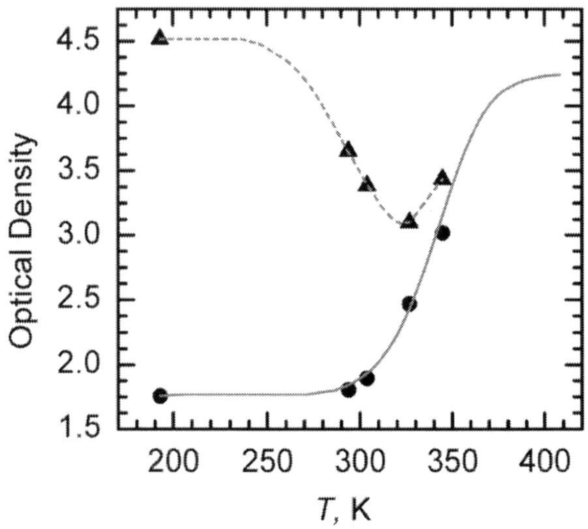

Figure 17. Temperature dependency of IR absorption at λ = 1.3 μm in CdF_2:Ga,Y (N_{Ga} = 5.7 × 10^{17} cm^{-3}) crystal in the dark (dark circles are experimental points; solid line is the result of calculation based on Figure 6 data) and under excitation of argon laser radiation (dark triangles are experimental points; dashed line shows a trend).

Figure 17 reveals that beginning with T ~ 250 K a total deep-to-shallow center conversion becomes impossible (at finite power density of exciting light of 1.5 W/cm^2) due to thermal destruction of deep centers; this is the origin of the decrease of the photoinduced absorption with increasing T in the range of 250–320 K. However, a further increase of T results in an increase of this absorption because of the growth of the equilibrium population of the shallow centers. The difference between the upper and lower curves shows the photoinduced absorption; the corresponding refractive index change determines the diffraction efficiency of the hologram.

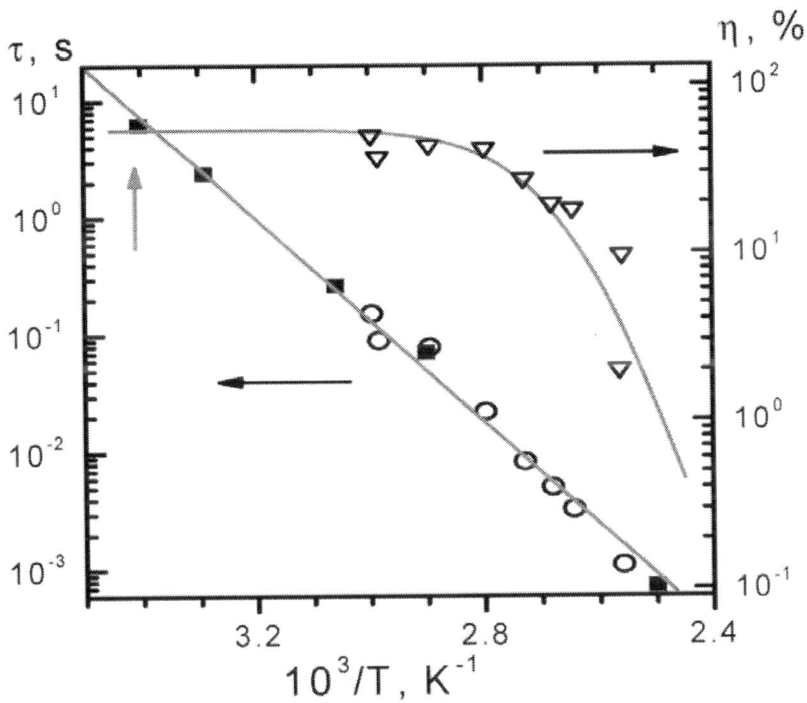

Figure 18. Temperature dependencies of diffraction efficiency (open triangles are experimental data; solid line is the result of calculation based on temperature dependence of the shallow center concentration shown in Figure 6) and the time of ten-fold decay of IR absorption (dark squares) and hundred-fold decay of initial diffraction efficiency (open circles) of the hologram recorded in CdF_2:Ga,Y (N_{Ga} = 5.7 × 10^{17} cm^{-3}) crystal by pulses of the second harmonics of Nd:YAG laser. The solid line is an approximation of decay data by an Arrhenius plot. Data that correspond to room temperature are indicated by the vertical arrow.

To characterize the crystal response for real-time holography, the temperature dependencies of diffraction efficiency and decay time of phase holograms were studied for the temperature ranges of 300–400 K (CdF_2:Ga,Y, Figure 18) and 77–300 K (CdF_2:In, Figure 19) [45,46,47]. The hologram recording was executed using single pulses of second harmonic of Nd:YAG laser at 532 nm of 20 ns duration with pulse energy equal to 400 mJ/cm^2 (CdF_2:Ga,Y), and single pulses of ruby laser at 693 nm of 50 ns duration with pulse energy equal to 300 mJ/cm^2 (CdF_2:In). The hologram readout for these crystals was executed using He-Ne laser. For both crystals, the hologram recording and readout was performed within the isobestic gap. The hologram decay down to 1/100 of the initial value of diffraction efficiency was determined using the oscillographic technique. For CdF_2:Ga,Y, the photoinduced IR absorption decay was recorded together with the hologram decay, taking into account that, in accordance with Equation (11), the hundred-fold decrease of the diffraction efficiency corresponds to a ten-fold decrease of the refractive index (the shallow center concentration).

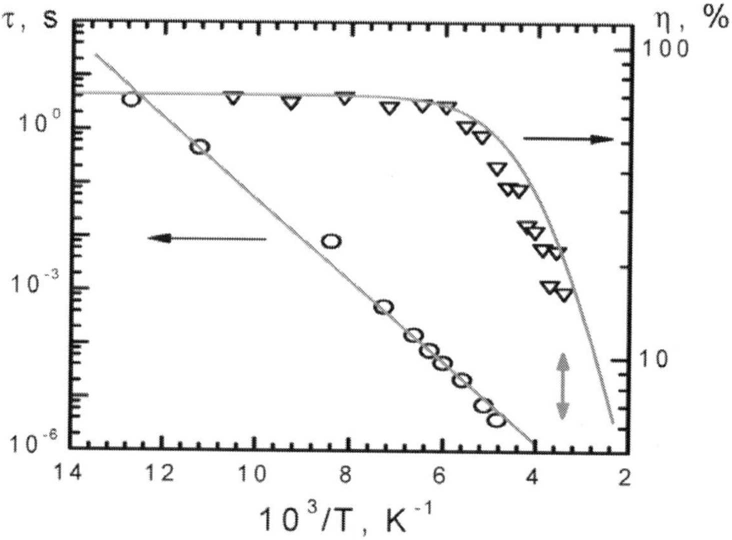

Figure 19. Temperature dependencies of diffraction efficiency (open triangles are experimental data; solid line is the result of calculation based on temperature dependence of the shallow center concentration shown in Figure 5) and the time of hundred-fold decay of initial diffraction efficiency (open circles) of the hologram recorded in CdF_2:In ($N_{In} = 4 \times 10^{18}$ cm^{-3}) crystal by pulses of the ruby laser. The solid line is an approximation of decay data by an Arrhenius plot. Data that correspond to room temperature are indicated by the vertical arrow.

The crystal thickness was 5 mm for CdF_2:Ga,Y and 2 mm for CdF_2:In. These thicknesses ensure, at sufficiently low temperature, the $\pi/2$ nonlinear phase shift, i.e., the maximal diffraction efficiency of holograms.

The low-temperature diffraction efficiencies of holograms are limited by the transmission of the crystal at the readout wavelength (54% for CdF_2:Ga,Y and 70% for CdF_2:In). The temperature dependency of diffraction efficiency is determined by the equilibrium population of shallow centers; the growth of this population with temperature decreases their photoinduced modification. It is seen from Figure 18 and Figure 19 that for both crystals (1) the pulse energy is enough for the total deep-to-shallow center conversion within the pulse and (2) the pulse duration is sufficiently short and does not influence the diffraction efficiency for the studied range of decay times of the hologram.

For both crystals, decay time lies on the Arrhenius plot, indicating that in the temperature range under consideration the hologram decay is practically exponential. Activation energies of decay time dependencies on the temperature are 0.84 eV for CdF_2:Ga,Y and 0.15 eV for CdF_2:In.

Outside the isobestic gap, holograms are of amplitude-phase nature. Allowing for the fact that δn is negative and $\delta\alpha$ is negative for the UV-VIS absorption band and positive for the IR band, one may conclude that the grating in the spectral range abutted on the short-wavelength border of the isobestic gap has π-out-of-phase character, whereas the grating in the spectral range abutted on the long-wavelength border of this gap has in-phase character.

Hologram readouts in the spectral range of the IR absorption band (~0.8–12 μm) could be of special interest. Due to quadratic dependency of the photoinduced refractive index on the readout wavelength, δn may vary with λ increase in this range from units of 10^{-4} up to units of 10^{-2}, i.e., it has a relatively large value. In spite of the large optical density of the IR band (see Figure 1), the transmission at the readout wavelength can be sufficiently large due to the Borrmann effect. In fact, if the equilibrium population of shallow centers is small the total deep-to-shallow center conversion creates high contrast of the diffraction fringe patterns, which favors this effect, i.e., ensures relatively high crystal transmission at the Bragg angle. Therefore, the forming of effective amplitude-phase holograms is possible in the IR spectral range. The diffraction efficiency of a hologram should be maximal at low temperature. It should diminish with increasing temperature due to both a decrease in photoinduced modulation of optical constants and deterioration of conditions for the Borrmann effect display.

One should note that, unlike UV-VIS, IR radiation does not provoke the photochromic effect, i.e., it does not cancel the hologram [48]. However, the decrease of diffraction efficiency of the hologram can be accounted for by heating of the crystal by readout radiation [44].

3.2. Holographic Media Resolution, Sensitivity and Operation Mode

The recording locality ensures high resolution of CdF_2 crystals with bistable centers. Figure 20 shows the diffraction efficiency of holograms recorded by ruby laser in CdF_2:In crystal, with convergence angles of recording beams in the ranges of 10–60° (transmitting holograms) and 150–160° (reflecting holograms). As follows from this figure, this angle does not practically influence the diffraction efficiency. For the maximal angle of 160°, the space frequency of the grating is 2800 mm^{-1} in air (4300 mm^{-1} inside the crystal). Reflecting holograms were recorded in this crystal by argon laser at angle convergence close to 180°. This shows that the space frequency of gratings in this crystal exceeds 5000 mm^{-1}.

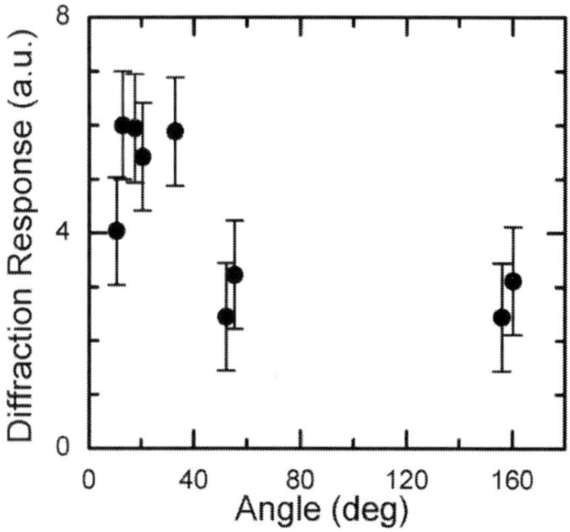

Figure 20. Dependency of diffraction efficiency of hologram in CdF_2:In (N_{In} = 4×10^{18} cm^{-3}) crystal on the convergence angle of recording beams.

The sensitivity of CdF:Ga,Y crystal to grating recording, S ≅ 4 cm/J, is much higher than that of the typical photorefractive crystal LiNbO3:Fe (S = 0.02–0.07 cm/J) and comparable with that of Bi12TiO20 (S ≅ 10 cm/J); however, it is still lower than the sensitivity of Polaroid photopolymer (S ≅ 20 cm/J). This comparison is, however, not completely correct because at room temperature CdF_2:Ga,Y crystal exhibits rather short storage time, whereas $LiNbO_3$:Fe and Polaroid photopolymer ensure the recording of permanent holographic gratings. A more suitable characteristic of this crystal would be a nonlinear susceptibility $\chi^{(3)}$, that can be estimated in CdF_2:Ga,Y as ~– 1.4×10^{-10} m^2/V^2 [44].

The high nonlinear susceptibility $\chi^{(3)'}$ of CdF_2 crystals with bistable centers and high spatial frequency of recorded gratings makes these crystals a suitable object for the observation of backward-wave four-wave mixing, which is the underlying process for optical phase conjugation and also for various kinds of frequency-degenerate coherent optical oscillators. Phase conjugation and coherent optical oscillations in CdF_2:Ga,Y crystal were reported in [49]. The interference of a relatively weak light beam coherent with two counter-propagate powerful (pump) beams formed the phase-conjugate mirror that, together with a conventional mirror, created the semi-linear cavity in which a coherent oscillator arose as a result of the interaction of signal and reversal weak waves. The temporal dynamics of oscillation intensity as a function of exposition time is shown in Figure 21. This dependency is characteristic for a coherent oscillator. The onset of oscillation proves unambiguously that the phase-conjugate reflectivity is higher than unity, i.e., amplified reflection is reached.

Holograms can be recorded over the whole visible range of the spectrum, since for both crystals the UV-VIS band embraces all of this range. Their readouts with differing or the same wavelength determine two principle ways of using holographic elements based on CdF_2 crystals with bistable centers. Both readout modes can be used in hologram recording by beams with the plane wavefront. Such recording allows formation of a static or dynamic (light-controlled) holographic mirror operating in both the visible and IR spectral ranges. The IR light-controlled holographic mirror based on CdF_2:Ga crystal was used in a widely distributed Bragg reflector laser to set the lasing wavelength in the range of 1.260–1.285 μm [50].

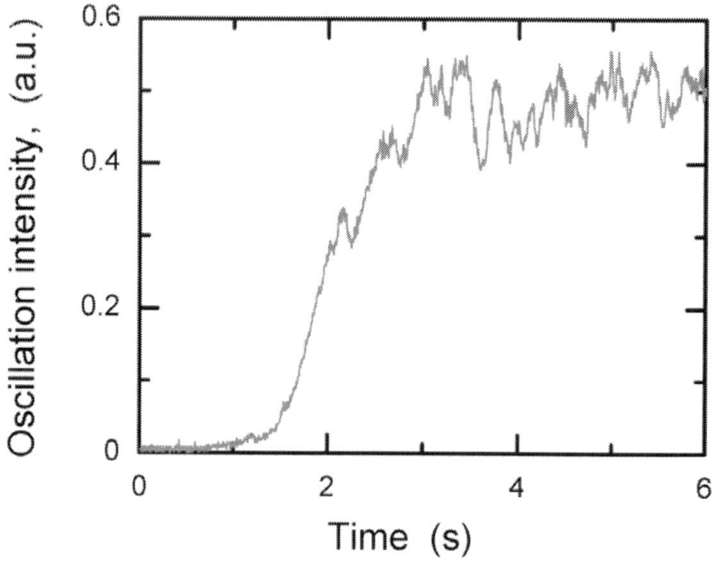

Figure 21. Temporal dynamics of coherent oscillation intensity in CdF_2:Ga,Y (N_{Ga} = 5.7 × 10^{17} cm^{-3}) crystal. Total intensity of counter-propagate pump beams ~ 200 mW/cm^2 at equal intensities of these beams. Intensity fluctuations in the saturation area are due to poor mechanical stability of the optical table.

Using the same wavelength for hologram recording and readout allows recording of dynamic information holograms operating in the visible range of the spectrum.

As demonstrated in [44], CdF_2 crystals with bistable impurity centers can be used for dynamic recording of information patterns. In this paper the dynamic hologram of the 1951 USAF resolution test chart was successfully recorded and read-out in CdF_2:Ga,Y crystal by continuous-wave Nd:YAG laser operating at λ = 532 nm at room temperature, which confirmed the high resolution of the medium.

In [51], holographic recording and readout of a binary information target ("chessboard-like" pattern) in CdF_2:In crystal was performed at room temperature by frequency-doubled Nd:YAG-laser pulses with duration of 20 ns. According to the kinetics of the hologram decay (see Section 2.2), the decay time was about 0.1 ms in this case.

4. HOLOGRAPHIC OPTICAL ELEMENTS BASED ON CDF_2:GA AND CDF_2:IN CRYSTALS

4.1. Wavefront Correction

The dynamic phase conjugation ability of photochromic CdF_2 crystals is demonstrated by an experiment on correction of model phase distortions with the use of a CdF_2:In based wavefront-conjugating mirror [52]. The optical scheme of the experiment is shown in Figure 22. A ruby laser, 1, is used as a source of pump and signal waves. To divide the output laser beam into signal and reference beams, a high-quality plane-parallel plate, 2, was used. The angle of incidence of the beam on the plate and its orientation in space were chosen experimentally to provide the required ratio of beam powers. The cross sections of the concurrent pump beam, E_1, and the signal beam, E_3, were aligned in the CdF_2:In crystal, 3, which was in the form of a disk of 12 mm diameter and 1 mm thickness. The beam pass distances between the front face of the plate, 2, and the crystal, 3, for the waves E_1 and E_3 were the same and equal to 1540 mm. As a result, the beams were coherent in the crystal. The wave E_2, formed due to reflection of the wave E_1 from the plane mirror, 6, was used as a readout beam.

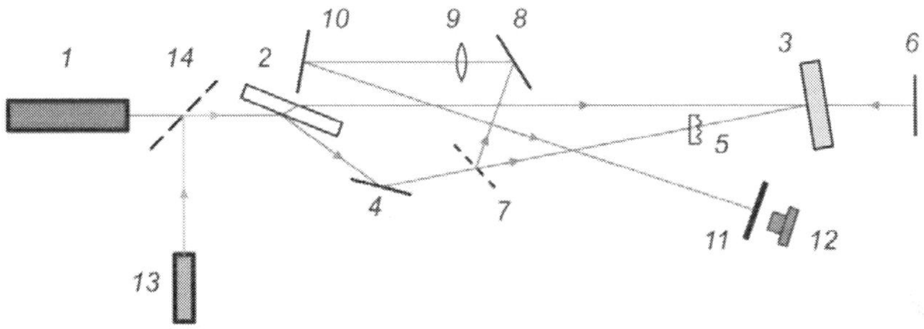

Figure 22. Optical scheme of the wavefront correction experiment. 1: pulsed ruby laser; 2: plane-parallel beam splitter plate; 3: CdF_2:In ($N_{In} = 4 \times 10^{18}$ cm^{-3}) sample; 4, 6, 8, 10, 14: plane mirrors; 5: model phase inhomogeneity; 7: beam splitter; 9: lens with 2 m focal length; 11: dull screen; 12: CCD-camera and line 13: auxiliary adjusting He-Ne laser.

The measuring scheme included a beam splitter, 7, and a lens, 9, with a focal length of 2 m, which focused the beam E_4 reflected from a wavefront-conjugative mirror in the plane of the screen, 11, after which a camera, 12, was located. Figure 22 also shows the position of a model phase inhomogeneity, 5, whose action was compensated for by a phase conjugation technique.

In the *Q*-switched mode of the source laser, the counter pump beam, E_1, and signal beam, E_3, had the energy of 40 mJ and their total power density in the crystal was 110 mJ/cm^2. The transmittance of CdF_2:In crystal at the ruby laser wavelength was 30%. Under these conditions, the reflectance of the wavefront-conjugating mirror in the absence of model phase distortion κ = 0.8%, while the diffraction efficiency of the reflection hologram recorded in the crystal η = 1.9%. The quality of the phase conjugated wave was found to be only insignificantly lower than that of the signal wave.

The time parameters of the wavefront-conjugating mirror were determined by analyzing the oscillograms of the laser and phase conjugated pulses. This analysis showed that the operating speed of the mirror under study was 10–20 ns.

In the case of wavefront-conjugating compensation for model phase distortions, the focal distribution of the phase conjugated wave was measured after the compensation for distortions. The similar distribution for the wave that passed twice through distortion without compensation was also recorded. In this case, the crystal, 3, was replaced by a planar auxiliary mirror reflecting the signal wave exactly backward. For comparison, the phase conjugated wave in the absence of distortion at the same parameters was also recorded. This dataset allows determination of the quality of distortion compensation and the gain in the divergence as a result of the compensation.

Figure 23 and Figure 24 illustrate the distortion compensation for spherical and astigmatic lenses, respectively. The focal length of the lenses is 0.8 m. It can be seen that distortion is so large that, in the absence of compensation, the focal intensity distribution is distorted beyond recognition (Figure 23a and Figure 24a). With phase conjugated correction, the focal distribution is close to that in the absence of distortion; in this case, effective compensation of distortion occurs with a large gain in divergence. It should be noted that every picture shown in Figure 23 and Figure 24 was registered during a single 20 ns pulse of the ruby laser.

Volume reflecting holograms recorded in CdF_2:Ga crystals were also used for correction of a model dynamic distortion of the wavefront [53].

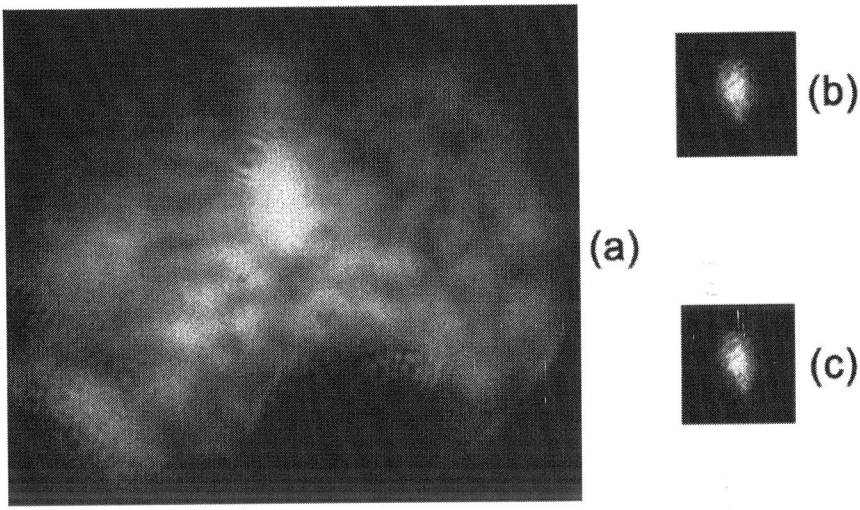

Figure 23. Results of compensation of a model phase distortion of the type of a spherical lens: (**a**) the focal distribution with a distortion of the type of a spherical lens (+1.25) and without phase conjugation; (**b**) the distribution with phase conjugation compensation of the lens, and (**c**) a phase conjugation wave without distortion. All images are given in the equal scale.

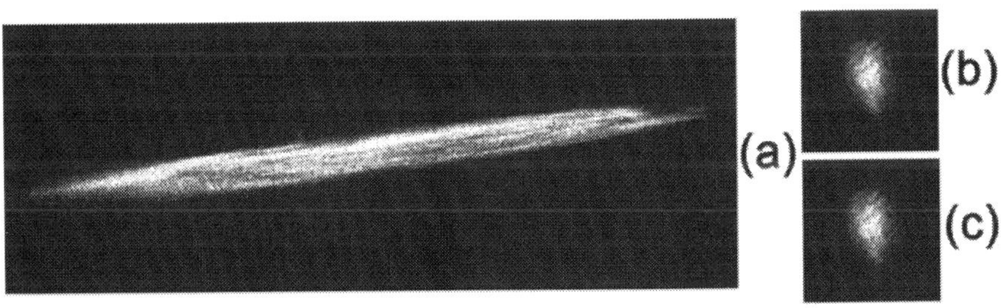

Figure 24. Results of compensation of a model phase distortion of the type of an astigmatic lens: (**a**) the focal distribution with a distortion of the type of an astigmatic lens (+1.25) and without phase conjugation; (**b**) the distribution with phase conjugation compensation of the lens and (**c**) a phase conjugation wave without distortion. All images are given in the equal scale.

4.2. Holographic Pattern Recognition

Another example of a CdF_2-based holographic element application is a spatial-frequency filtration for dynamic optical image correlation. The volume nature of these filters ensures higher resolution compared to plane holographic filters due to high angular selectivity. It allows recognition of compared images with small angular difference. To obtain sufficiently large diffraction efficiency of the holographic filter, one needs to apply good-quality fringe pattern. To get such pattern in a volume sample with finite absorption (even in the isobestic gap) it is reasonable to record transmission holograms; in this case, intensities of interfering beams are equal though they change simultaneously along the sample thickness. For an optically isotropic medium such as CdF_2 crystal, the use of parallel-polarized beams allows formation of high-contrast pattern at any beam convergence angle; in this case, polarization planes of the beams are perpendicular to the grating vector.

The dynamic matched spatial-frequency filter based on a CdF_2:In crystal was used in a van der Lugt correlator scheme; the filter record and readout proceeded simultaneously by the same laser pulse but with 90°-shifted polarization [51].

The joint transform correlator (JTC) using a CdF_2:Ga-based spatial-frequency filter [54] is shown in 25. A Figure 10× 10 × 10 mm^3 crystal with an antireflection coating on the working faces was used; the concentration of photochromic (optically active) centers was ~10^{17} cm^{-3}. The power density of the recording/readout argon laser operating at 488 nm in the recording channel was ~100 mW/cm^2, and the readout beam power was approximately 5% of the recording beam power. The hologram diffraction efficiency at room temperature was several percent, and the hologram decay time was several seconds.

In the recording channel of the JTC, the laser beam is expanded to the required diameter using a collimator, 6, and the plane wavefront of this beam illuminates an information pattern with the images of reference (r(x, y), 8) and analyzed (s(x, y), 9) objects, located at the front focus of lens 10. Lenses 10 and 12 perform Fourier transformation of the pattern and an interference pattern is recorded in the frequency plane of the JTC within a CdF_2:Ga crystal, 14; thus, a holographic spatial-frequency filter is formed in the crystal volume.

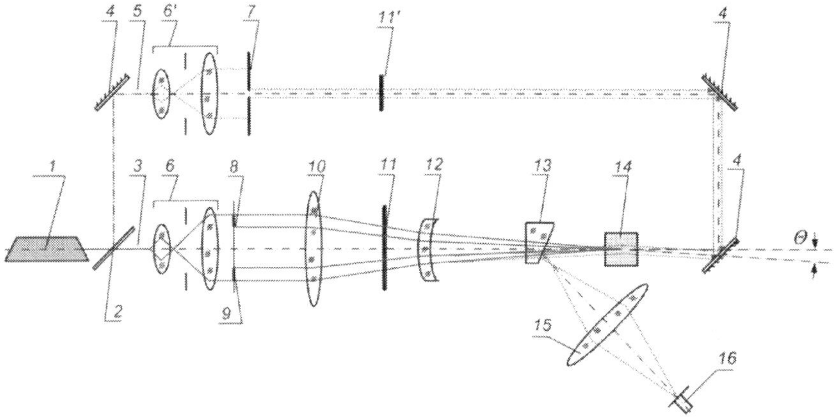

Figure 25. Optical scheme of the joint transform correlator based on a CdF_2:Ga ($N_{Ga} = 3 \times 10^{17}$ cm^{-3}) crystal: 1: Ar laser; 2: semitransparent mirror; 3: recording beam; 4: planar opaque mirror; 5: reading beam; 6 and 6': collimators; 7: diaphragm; 8 and 9: information pattern (images 8r(x, y) and 9s(x, y)); 10 and 15: Fourier transforming lens; 11 and 11': controlled gates; 12: negative lens; 13: optical wedge; *14* CdF_2:Ga crystal; and *16:* CCD-camera.

To recognize the reference image among analyzed ones, the plane wavefront, formed in the reading channel illuminates the crystal in the direction opposite to the recording beam from the $r(x, y)$ image field. The reading beam reconstructs conjugated wavefronts of images coincident with the reference one. An optical wedge, 13, directs the diffracted beam to the detection channel, in which lens 15 performs the second Fourier transformation and focuses the correlation signal at the correlator output plane, where it is detected by grayscale CCD-camera, 16. The dynamic mode of the JTC operation is provided by computer-controlled mechanical gates, 11 and 11'. These gates enable time separation of the recording and reading beams illuminating the crystal located in the frequency plane. This allows vanishing of scatter from the recording channel light in the detection channel. The gate operation time, determined by the hologram decay time, makes it possible to perform correlation at switching frequencies of the recording/reading channels in the range of 0.1–1 Hz.

The JTC optical scheme is based on a standard 4f correlator scheme; however, one should note its features: (1) introduction of an additional negative lens, 12, increases the effective focal length of the Fourier transform and dimensions of the recording beam caustic; it makes possible to expose a larger crystal volume during hologram recording, thus preventing overexposure of the crystal and decreasing recording beam distortions; (2) the dynamic nature of the holographic medium enables changing of information patterns with compared objects at the correlator input plane in real time; one does not need to perform forced erasure of a previously written hologram or change the recording medium.

Binary and grayscale amplitude information patterns (transparencies) were used in the JTC. Both classes of transparencies were 24 × 36 mm^2 in size (photographic frame format) and had a resolution of 25 pixels/mm. In Figure 26, the sample of selective object recognition, i.e., recognition of a specified object in a group of objects presented for recognition, is shown. In this case, the reference object r(x) is a binary image of the Chinese hieroglyph 学(2), located on the right-hand side of a transparency (Figure 26a) and the analyzed objects s(x) are three hieroglyphs placed on the left-hand side of the transparency (1, 2, 3).

According to the simulation data (Figure 26b), three peaks should be observed in the correlator output plane: one peak with the largest amplitude and the least half-width corresponds to the correlation signal of the reference object, 2, while the two other peaks arise due to the cross-correlation of the reference with the other two hieroglyphs (1 and 3). The experimentally obtained signal intensity distributions in the correlator output plane (Figure 26c) really contain three peaks. As one can see, the amplitude of the correlation peak is indeed larger and its half-width is, accordingly, smaller than that of the cross-correlation peaks; the positions of the experimental peaks in the output plane correspond to the positions of the hieroglyphs compared. The shapes of the experimentally observed peaks differ from the calculated ones; this difference may be related to nonuniformity of laser illumination of the transparency field as well as aberrations of the optical scheme elements. However, one can easily perform threshold processing of the output JTC signals and thus detect the reference object among others presented for recognition.

Another example of JTC performance is the recognition of an industrial object in the landscape environment or development elements in a satellite grayscale photograph, taken from the Google Map database (Figure 27).

The object under recognition is a round building with a rectangular extension. As one can see in Figure 27, the shape and the position of the experimentally observed correlation signal correspond sufficiently well to the calculated data.

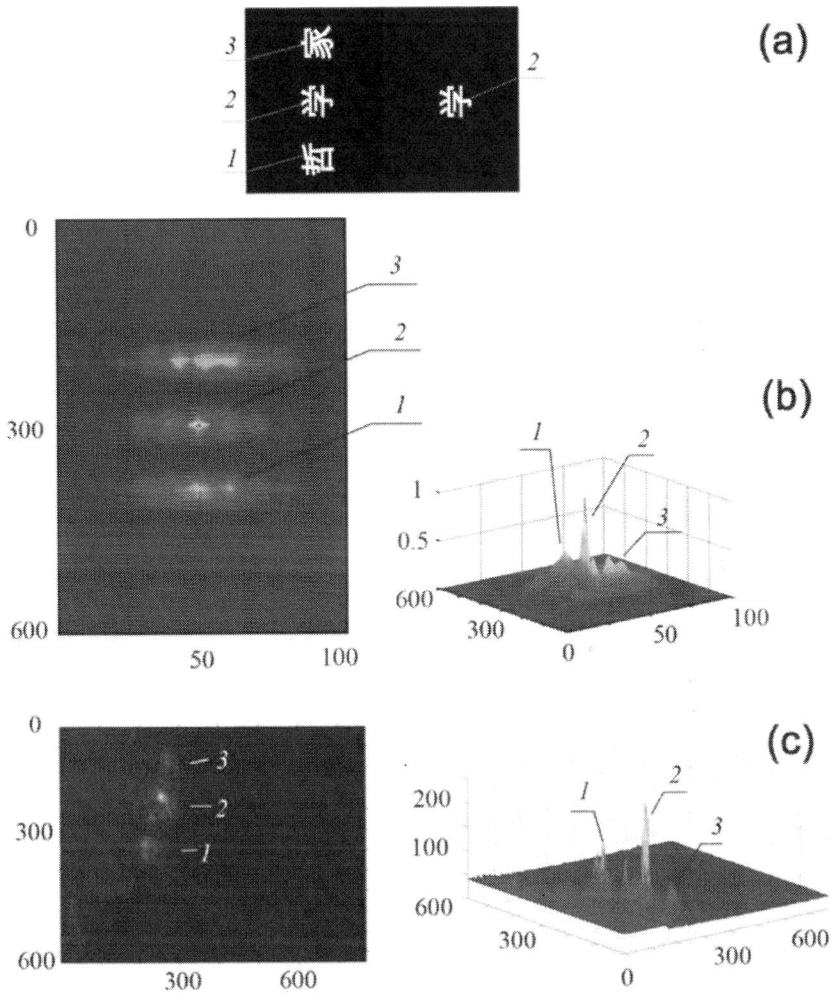

Figure 26. Recognition of hieroglyph 2 in a group of objects: (**a**) information pattern; (**b**) two-dimensional (left) and three-dimensional (right) images of calculated correlation field (signal coordinates are measured in pixels, signal intensities are measured in arbitrary units); (**c**) two-dimensional (left) and three-dimensional (right) experimental signal intensity distribution in the JTC output plane (signal coordinates are measured in pixels, signal intensities are measured in units of grayscale).

The experimental JTC based on CdF_2:Ga crystal demonstrates good ability for selective recognition when operating with objects of different complexity. It should be noted that the JTC operation rate can be increased by heating the CdF_2:Ga crystal. CdF_2 crystals possess higher resolution and operation speed as compared with widely used liquid crystals. To obtain high operating speed in combination with an acceptable hologram diffraction efficiency, it is not necessary to apply additional heating and/or external illumination or a high-voltage electric field to CdF_2 crystal as for photorefractive crystals.

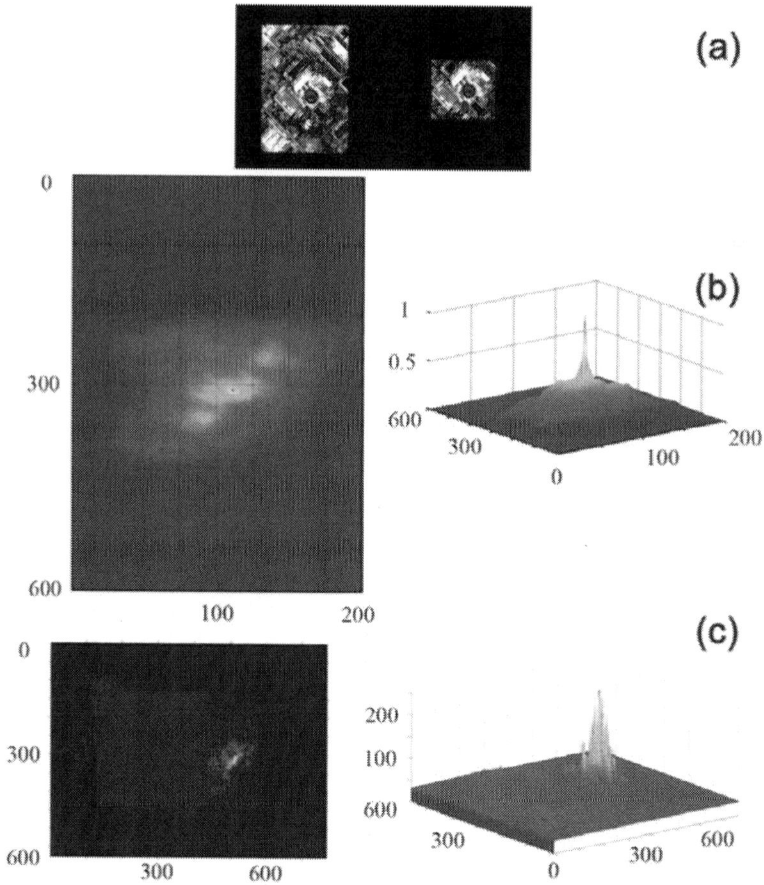

Figure 27. Correlation analysis of a satellite photograph of a commercial development region; the sequence order (**a–c**) and measuring units are as in Figure 26.

Bearing in mind the use of optical materials in correlators, one should note two modern trends of optical correlator buildup: the first involves the widespread use of liquid crystal light valves or spatial light modulators, which carry calculated spatial-frequency spectra of correlated patterns in the Fourier plane; the second is based on the use of a numerical correlation experiment instead of a physical one. The latter approach assumes that the correlation signal is calculated on the basis of known Fourier spectra of patterns and parameters of a virtual "optical scheme" of the correlator under study. In this case, one can easily implement additional spatial-frequency filters to the "scheme" and thus vary correlation signal characteristics [55]. However, such additional filters can increase the discrimination ability of traditional analogous correlators which use holographic media. For instance, one can make a hybrid wavelet-joint transform correlator in which additional wavelet-transform filters are employed, making the JTC less sensitive to scale and rotational non-invariances. Moreover, wavelet filtration enables to the increase of signal-to-noise ratio and correlation peak sharpness due to its edge enhancement nature which is useful for target recognition [56,57,58]. Thus, CdF_2 crystals with bistable impurity centers can be considered as appropriate holographic media for fast hybrid wavelet-joint transform correlators.

5. CONCLUSIONS

The photochromy of CdF_2 crystals with bistable impurity centers is based on the local modification of optical properties of the crystal that is stipulated by the reconstruction of the impurity center at the photoinduced change of its charge state. The potential barrier separating the ground (deep) and excited (shallow) states of the center determines the metastable nature of its excited state. The photoinduced deep-to-shallow center conversion underlies the hologram recording in these crystals. The hologram decay is tied to the thermally induced process of reverse (nonequilibrium shallow)-to-deep center conversion. These two processes have different natures and elapse in different time ranges.

The characteristic time of the photoinduced deep-to-shallow state conversion is determined by the lattice rearrangement around the free carrier and the impurity ion. The conversion process also includes the capture of the photoinduced free carrier by the trivalent impurity ion. The concentration of these ions is high and the capture occurs for a very short time interval as well as the above-mentioned rearrangement of the lattice. Hologram formation typically takes time of the order of nanoseconds at ambient temperature; this time weakly depends on the temperature.

The shallow-to-deep state conversion corresponds to the formation of one deep center from two shallow centers. This conversion includes the capture of thermally induced free carriers produced by the thermal ionization of a shallow center by another shallow center with its subsequent conversion into a deep center. The hologram decay that is determined by this process obeys bimolecular kinetics, degenerated into monomolecular kinetics at the latest stage of decay or at sufficiently high temperature. The deep center formation due to capture of the second electron by a shallow center occurs when the potential barrier is overcome. As a result, the time of hologram decay strongly depends on temperature. It varies in a wide limit from infinity (for low temperatures) down to times of the millisecond or nanosecond range, depending on the specific dopant ion.

Two bistable impurities, Ga and In, differ strongly in several respects (the barrier height, the deep state energy, the dependency of population of center states on the temperature). These distinctions determine the characteristics of phase holograms recorded in CdF_2:Ga and CdF_2:In crystals in the range of the overlapping tails of the UV-VIS and IR absorption bands (the isobestic gap), in particular, decay time of the hologram. Using the temperature as a managing parameter one can use holograms to follow up optical processes in the second-millisecond (CdF_2:Ga) or second-nanosecond (CdF_2:In) time ranges with diffraction efficiency from tenths down to units of percent. Such holograms can be recorded in several-millimeter-thick crystals.

The locality of the recording process ensures high resolution of the media that reaches at least 5000 mm^{-1}.

The cubic symmetry of the crystals allows the use of polarization of radiation, which is especially useful for high frequencies since it creates the opportunity to divide hologram recording and readout processes. An unlimited number of record/readout cycles are possible for these media.

Holographic optical elements based on materials with volume holographic gratings are of great interest for optical information storage and processing. They can serve as spatial-frequency filters for a variety of purposes, such as encrypted optical memory systems, neural network simulation, holographic interferometry, image distortion corrections using phase-conjugation or backward-wave four-wave mixing, and optical correlation and convolution operations for pattern recognition. Some of these applications demand both static and dynamic holographic elements, whereas others need elements operating only in real time.

We maintain that real-time holographic media based on CdF_2 crystals with bistable centers allow efficient operation within wide time scale and spectral ranges.

REFERENCES AND NOTES

1. Linke, R.A.; Thio, T.; Chadi, J.D.; Devlin, G.E. Diffraction from optically written persistent plasma gratings in doped compound semiconductors. Appl. Phys. Lett. **1994**, 65, 16–18.
2. Thio, T.; Bennett, J.W.; Chadi, J.D.; Linke, R.A.; Becla, P. DX centers in CdZnTe:Cl and their applications. J. Cryst. Growth **1996**, 159, 345–349.
3. McKenna, J.M.; Nolte, D.D.; Walukiewicz, W.; Becla, P. Persistent holographic absorption gratings in AlSb:Se. Appl. Phys. Lett. **1996**, 68, 735–737.
4. Ryskin, A.I.; Shcheulin, A.S.; Koziarska, B.; Langer, J.M.; Suchocki, A.; Buczinskaya, I.I.; Fedorov, P.P.; Sobolev, B.P. CdF_2:In: A novel material for optically written storage of information. Appl. Phys. Lett. **1995**, 67, 31–33.
5. Koziarska, B.; Langer, J.M.; Ryskin, A.I.; Shcheulin, A.S.; Suchocki, A. Use of bistable centers in CdF_2 in holographic recording. Mater. Sci. Forum **1995**, 1103–1108.
6. Shcheulin, A.S.; Angervaks, A.E.; Ryskin, A.I.; Zakirov, R.Y.; Serov, T.V. Additive coloration of cadmium fluoride crystals. J. Opt. Technol. **2006**, 73, 745–748.
7. Shcheulin, A.S.; Semenova, T.S.; Koryakina, L.F.; Petrova, M.A.; Kupchikov, A.K.; Ryskin, A.I. Additive coloration of calcium and cadmium fluorides. Opt. Spectrosc. **2007**, 103, 660–664.
8. Kingsley, J.D.; Prener, J.S. Free charge carrier effects in cadmium fluoride. Phys. Rev. Lett. **1962**, 8, 315–316.
9. Prener, J.S.; Kingsley, J.D. Mechanism of the conversion of CdF_2 from an isolator to a semiconductor. J. Chem. Phys. **1963**, 38, 667–671.
10. Khosla, R.P. Electrical properties of semiconducting CdF_2:Y. Phys. Rev. **1969**, 183, 695–703.
11. Langer, J.M.; Langer, T.; Pearson, G.L.; Krukowska-Fulde, B.; Piekara, U. Shallow donor states in semiconducting CdF_2. Phys. Status Solidi (b) **1974**, 66, 537–545.

12. Khosla, R.P.; Matz, D. Effective mass and scattering mechanisms in semiconducting CdF_2 at high temperatures. Solid State Commun. **1968**, 6, 859–864.
13. Kunze, I.; Ulrici, W. Electrical and optical studies of semiconducting CdF_2:In crystals. Phys. Status Solidi (b) **1973**, 55, 567–578.
14. Piekara, U.; Langer, J.M.; Krukowska-Fulde, B. Shallow versus deep In donors in CdF_2 crystals. Solid State Commun. **1977**, 23, 583–587.
15. Langer, J.M. Strong lattice relaxation at localized defects. J. Phys. Soc. Jpn. **1980**, 49, S207–S214.
16. Langer, J.M. DX-like centers in solids (metastability, bistability and negative-U). Rev. Solid State Sci. **1990**, 4, 297–317.
17. Toyozawa, Y. Symmetry breaking excitonic instabilities in deformable lattice. Phys. B+C **1983**, 117–118, 23–29. [Google Scholar] [CrossRef]
18. Here and further we use two equivalent notations: shallow/deep state of the center and shallow/deep center.
19. Shcheulin, A.S.; Ryskin, A.I.; Swiatek, K.; Langer, J.M. Deep-shallow transformation of bistable centers in semiconducting CdF_2 crystals. Phys. Lett. A **1996**, 222, 107–112.
20. Kazanskii, S.A.; Ryskin, A.I.; Romanov, V.V. Paramagnetic susceptibility of semiconducting CdF_2:In crystals: Direct evidence of the negative-U nature of the DX-like center. Appl. Phys. Lett. **1997**, 70, 1272–1274.
21. Kazanskii, S.A.; Ryskin, A.I.; Romanov, V.V. Paramagnetic susceptibility of additively colored CdF_2:In photochromic crystals. Phys. Solid State **1997**, 39, 1067–1071.
22. Park, C.H.; Chadi, D.J. First-principles study of structural bistability in Ga- and In-doped CdF_2. Phys. Rev. Lett. **1999**, 82, 113–116.
23. Nissilä, J.; Saarinen, K.; Hautojärvi, P.; Suchocki, A.; Langer, J. M. Universality of the bond-breaking mechanism in defect bistability: observation of open volume in the deep states of In and Ga in CdF_2. Phys. Rev. Lett. **1999**, 82, 3276–3279.
24. Ritus, A.I.; Pronin, A.V.; Volkov, A.A.; Lunkenheimer, P.; Loidl, A.; Shcheulin, A.S.; Ryskin, A.I. Determination of the parameters of semiconducting CdF_2:In with Schottky barriers from radio-frequency measurements. Phys. Rev. B **2002**, 65, 165209:1–165209:8.

25. Ionic radii of eight-coordinated Cd^{2+}, In^{3+}, and Ga^{3+} ions are 1.24 Å, 1.06 Å, and 0.78 Å, respectively.

26. The excitation in the IR absorption band does not influence the absorption spectrum of the crystals due to practically instantaneous capture of the free carrier by one of the Me^{3+} ions with formation of the shallow donor state.

27. Shcheulin, A.S.; Onopko, D.E.; Ryskin, A.I. Electronic level diagram and structure of metastable centers in CdF_2:Ga and CdF_2:In semiconducting crystals. Phys. Solid State **1997**, 39, 1906–1911.

28. Ryskin, A.I.; Shcheulin, A.S.; Onopko, D.E. DX Centers in ionic semiconductor CdF_2:Ga. Phys. Rev. Lett. **1998**, 80, 2949–2952.

29. Kazanskii, S.A.; Ryskin, A.I. Electron statistics in CdF_2 semiconductor crystals with DX centers. Phys. Solid State **2006**, 48, 1663–1671.

30. Fedorov, P.P.; Buchinskaya, I.I.; Ivanov, S.P.; Sobolev, B.P.; Shcheulin, A.S.; Ryskin, A.I. A new class of holographic materials based on semiconductor CdF_2 crystals with bistable centers. Part II. Growth of optically perfect crystals. Opt. Spectrosc. **2002**, 92, 125–132.

31. Kazanskii, S.A.; Klimin, S.A.; Shcheulin, A.S.; Narozhnyi, M.A.; Ryskin, A.I. Optical spectra of $(CdF_2)_{0.9}(InF_3)_{0.1}$ semiconductor solid solutions. Opt. Spectrosc. **2008**, 104, 861–865.

32. In Figure captions, N_{Ga} and N_{In} denote concentration of optically active bistable centers, whereas N_Y denotes concentration of reduced Y ions that form only shallow.

33. Kazanskii, S.A.; Warren, W.W.; Ryskin, A.I. Spin-lattice relaxation of ^{113}Cd and ^{19}F nuclear spins in the crystal lattice of CdF_2 semiconductor crystals with DX centers. Semiconductors **2009**, 43, 985–992.

34. Eisenberger, P.; Pershan, P.S.; Bosomworth, D.R. Polaron effects in semiconducting CdF_2: cyclotron resonance and far-infrared properties. Phys. Rev. **1969**, 188, 1197–1208.

35. Kazanskii, S.A.; Rumyantsev, D.S.; Ryskin, A.I. Ionized donor pairs and microwave and far-infrared absorption in semiconducting CdF_2. Phys. Rev. B **2002**, 65, 165214:1–165214:12.

36. Stasl'ko, D.I.; Tikhomirov, S.A.; Buganov, O.V.; Shcheulin, A.S.; Angervaks, A.E.; Ryskin, A.I. Dynamics of the transformation of

bistable centers and lattice in CdF_2 crystals under femtosecond pulsed photoexcitation. Opt. Spectrosc. **2011**, 110, 33–41.

37. The peak at the initial part of curve 3 in Figure 8 is due to free-carrier absorption.
38. Moser, F.; Matz, D.; Lyi, S. Infrared optical absorption in semiconducting CdF_2:Y crystals. Phys. Rev. **1969**, 182, 808–814.
39. Kazanskii, S.A.; Shcheulin, A.S.; Ryskin, A.I. Energy barrier between states of a bistable center in photochromic crystals CdF_2:Ga and CdF_2:In. Opt. Spectrosc. **2007**, 102, 419–425.
40. Shcheulin, A.S.; Angervaks, A.E.; Ryskin, A.I. Formation kinetics of volume holograms in reversible photochromic media. Opt. Spectrosc. **2011**, 110, 609–616.
41. Ryskin, A.I.; Shcheulin, A.S.; Miloglyadov, E.V.; Linke, R.A.; Redmond, I.; Buchinskaya, I.I.; Fedorov, P.P.; Sobolev, B.P. Mechanisms of writing and decay of holographic gratings in semiconducting CdF_2:Ga. J. Appl. Phys. **1998**, 83, 2215–2221.
42. Linke, R.A.; Redmond, I.; Thio, T.; Chadi, D.J. Holographic storage media based on optically active bistable defects. J. Appl. Phys. **1998**, 83, 661–673.
43. Kogelnik, H. Coupled wave theory for thick hologram gratings. AT&T Tech. J. **1969**, 48, 2909–2947.
44. Uesu, Y.; Yasukawa, K.; Saito, N.; Itoh, S.; Odoulov, S.; Shcherbin, K.; Ryskin, A.I. Dynamic grating recording in semiconductor CdF:Ga,Y. J. Opt. Soc. Am. Opt. Phys. B **2003**, 20, 1905–1911.
45. Linke, R.A.; Shcheulin, A.S.; Ryskin, A.I.; Buchinskaya, I.I.; Fedorov, P.P.; Sobolev, B.P. Properties of CdF_2:Ga as a medium for real-time holography. Appl. Phys. B **2001**, 72, 677–683.
46. Ryskin, A.I.; Shcheulin, A.S.; Kazanskii, S.A.; Angervaks, A.E. CdF_2 Crystals with bistable impurity centers as media of the real-time holography. J. Luminesc. **2007**, 125, 18–24.
47. Concentration of optically active bistable centers, i.e., centers that can form both deep and shallow state, was $\sim 10^{18}$ cm^{-3} for both crystals. One should note that holographic characteristics of CdF_2:Ga and CdF_2:Ga,Y crystals are similar to each other.
48. The cross-section of electron capture by Me^{3+} ion is so high that an electron formed after photoionization of the shallow center is captured by an ionized center or another Me^{3+} ion practically instantly.

49. Uesu, Y.; Yasukawa, K.; Saito, N.; Odoulov, S.; Shcherbin, K.; Ryskin, A.I.; Shcheulin, A.S. Backward-wave four-wave mixing and coherent oscillation in CdF_2:Ga, Y. Appl. Phys. B **2004**, 78, 601–605.
50. Nahata, A.; DiCaprio, C.J.; Yamada, H.; Ryskin, A.I.; Shcheulin, A.S.; Linke, R.A. Widely tunable distributed Bragg reflector laser using a dynamic holographic grating mirror. IEEE Photonic Technol. Lett. **2000**, 12, 1525–1527.
51. Shcheulin, A.S.; Verkhovskii, E.B.; Angervaks, A.E.; Ryskin, A.I. Recording of dynamic information holograms in a CdF_2:In crystal. Opt. Spectrosc. **2005**, 99, 803–805.
52. Angervaks, A.E.; Dimakov, S.A.; Kliment'ev, S.I.; Ryskin, A.I.; Shcheulin, A.S. Dynamic wavefront-conjugating mirror based on CdF_2 crystals with bistable in centers. Opt. Spectrosc. **2006**, 101, 131–135.
53. Angervaks, A.E.; Dimakov, S.A.; Kliment'ev, S.I.; Ryskin, A.I.; Shcheulin, A.S. Dynamic reflection holograms in CdF_2 crystals with bistable centers. Opt. Spectrosc. **2002**, 93, 307–314.
54. Fedorov, I.Yu.; Angervaks, A.E.; Sokolov, V.K.; Shcheulin, A.S. Dynamic holographic joint fourier transform correlator based on a CdF_2:Ga crystal. Opt. Spectrosc. **2008**, 105, 962–971.
55. Tripathi, R.; Singh, K. Pattern discrimination using a bank of wavelet filters in a joint transform correlator. Opt. Eng. **1998**, 37, 532–538.
56. Joseph, J.; Oura, T.; Minemoto, T. Optical implementation of a wavelet transform by the use of dynamic holographic recording in a photorefractive material. Appl. Opt. **1995**, 34, 3997–4003.
57. Zhang, H.; Cartwright, C.M.; Ding, M.S.; Gillespie, W.A. Optical implementation of a photorefractive joint transform correlator with wavelet filters. Opt. Commun. **2000**, 181, 223–230.
58. Widjaja, J. Wavelet filter for improving detection performance of compression-based joint transform correlator. Appl. Opt. **2010**, 49, 5768–5776.

9
CHAPTER

CHARACTERIZATION OF THE ELECTRONIC STRUCTURE OF SPINEL SUPERCONDUCTOR $LITI_2O_4$ USING SYNCHROTRON X-RAY SPECTROSCOPY

Chi-Liang Chen[1, 2] and Chung-Li Dong[1, 2]

[1] *Institute of Physics, Academia Sinica, Nankang, Taipei, Taiwan*

[2] *National Synchrotron Radiation Research Center (NSRRC), Hsinchu, Taiwan*

1. INTRODUCTION

Numerous studies on high-transition temperature (T_c) superconductivity were motivated primarily by the intention to explore the nature of cuprates. Experimental results, particularly those based on hard X-ray absorption spectra, show that Cu ions reveal a mixed valence and are accountable for high-T_c superconductivity. Recently discovered iron-based superconductors with perovskite blocking layers [i.e., LnFeAsO (Ln = lanthanide), $BaFe_2As_2$, KFeAs, FeSe, and FeAs] have received considerable attention similar to cuprate superconductivity, which was first explored in the 1980s. New superconducting compounds and non-cuprate superconductors with magnetic elements, particularly 3*d* transition metals, have been widely analyzed by various scholars. Spinel $LiTi_2O_4$ (LTO) is a titanium (Ti)-based superconductor that is considered an exotic BCS *s*-wave superconductor with a T_c of ~12 K [21]. This superconductor is regarded as such because it can be described using electron–phonon interaction within the framework of the BCS model [43]. Of the more than 200 typical compounds with an AB_2O_4 normal spinel structure, in which low-valent-state cation "A" occupies the tetrahedral (8a, T_d) interstitial sites and high-valent-state cation "B" occupies the octahedral (16d, O_h) sites, only a few exhibit superconductivity, including CuV_2S_4 (T_c = 4.5 K), $CuRhS_4$ (T_c = 4.8 K), and $CuRhSe_4$ (T_c = 3.5 K). LTO is the only known spinel oxide superconductor, and it has the highest T_c among superconductors with a spinel structure. The 3*d* transition metals occupying the O_h sites in spinel oxides may generally exhibit antiferromagnetism, ferromagnetism, and charge/orbital ordering. The nature of magnetic and electronic properties depends on the average valence of the cations. The closely related case of LiV_2O_4 (LVO), which is an isostructural compound, exhibits distinct physical properties. LVO is the closest neighbor to LTO, and it contains a mixed-valence system with an equal ratio of V^{3+} (S = 1) and V^{4+} (S = ½). LVO exhibits strong electronic correlation, resulting in a greatly enhanced effective mass. The resistivity of this compound displays a metallic character. Although LVO is not a superconductor, it exhibits a heavy fermionic behavior typically observed in a Ce-based (4*f* electron) system [2, 22].

Similar to cuprates [6, 33] and pnictides [9, 24], Ti-based superconductors (titanate) present the possibility of a mixed-valence state (with electronic configuration of $3d^{0.5}$, equal ratio of Ti^{3+} of spin $S = 1/2$ and Ti^{4+} with $S = 0$) in the ground state. Given that magnetic impurities can suppress the superconductivity of various materials, an investigation into whether the dilute doping of magnetic ions in LTO results in the complete suppression of superconductivity must be conducted. Such an undertaking may provide valuable insights into the mechanism of superconductivity. A previous research observed a rapidly suppressed superconducting T_c in magnetically doped system $LiTi_{1-x}M_xO_4$ (M = Cr and V) [19]. Among the explored dopants, vanadium (V) was determined to be more effective in decreasing T_c even at small proportions, and it was also observed in other 3*d* transition metals (e.g., Cr, Mn, Fe, etc.). The results of the study not only signified the magnetic influences and other mechanisms for T_c suppression, but also the charge density wave and structural distortion. For V-doped LTO, the T_c of LTO decreases with V at 2% proportion from ~13 K to ~5 K [18, 19, 24, 49]. These superconducting activities in such a series may unravel the mechanism of superconductivity. Material study on the atomic/electronic structure has essentially illuminated the understanding of its electron transportation properties. Studies have investigated the electronic structure of LTO [8, 39]. The decrease in superconductivity of V-doped LTO is discussed based on a pair-breaking mechanism [19] and a crystalline distortion from octahedral symmetry [34]. The change of Ti and V valence states is suggested to be thoroughly linked to their chemical and physical properties. Electron–electron interaction cannot be overlooked because magnetic ions occupy the structures [32]. In this event, the mechanism of superconductivity remains unknown.

The atomic and electronic structures of the novel spinel LTO superconductors have been investigated. X-ray spectroscopy, X-ray absorption near-edge structure (XANES) spectroscopy, and resonant inelastic soft X-ray scattering (RIXS) spectroscopy are powerful tools for obtaining information on the local orbital character of a specific element. These approaches are also used to probe occupied or unoccupied electronic states near the Fermi level (E_F) and the structural symmetry of mixed-oxide systems. However, these spectroscopic techniques are sensitive to atomic symmetry because edge-sharing charge and charge distribution induce electron–electron and electron-orbital interactions, which may be related to the magnetic nature of the transition metal systems.

XANES spectroscopy is a sensitive tool that can be employed to probe unoccupied electronic states above E_F and to analyze the structural symmetry of mixed-oxide systems [10, 12]. It can provide details on the electronic–orbital interaction from the transition metal 3*d*-O 2*p* hybridization states and the symmetry of the atomic structure. Therefore, the Ti L- and K-edges can be used to determine the valence states of Ti ions in $LiTi_{2-x}V_xO_4$ (LTVO). The V ions occupying and doped in the O_h site may distort the crystal structure of LTVO because of the bonding of O–Ti–O as well as the unoccupied states in the 3*d* orbitals [34].

RIXS spectroscopic technique is used to explore the electronic structure of materials and to associate the structure with the XANES spectrum. RIXS is a process that describes the de-excitation of the final state of X-ray absorption, providing information on ground state via the excited intermediate states. Consequently, by tuning various excitation energies, certain RIXS features can be enhanced, allowing the separation of different electronic configurations in the mixed ground states [11, 30, 42]. The above conditions indicate that RIXS is a complementary tool compared with XANES. Unlike XANES, RIXS can be used to investigate a forbidden electronic transition (e.g., *d-d* or *f-f* excitation) because it involves two dipole-allowed transitions, given that its final state has symmetry similar to that of the initial state of the former. The RIXS of TM L-edge can reflect the 3*d* partial density of states and is regarded as a useful approach for analyzing the electron correlation among strongly correlated materials [31, 36, 45] as well as the charge transfer between TM 3*d* and O 2*p* orbitals.

In this research, XANES and RIXS spectra were applied to understand the influence of V doping on the atomic/electronic properties of the LTVO system and to discuss the basis of the quick suppression of superconductivity. The occupied or unoccupied states near E_F of a solid-state solution may be a useful starting point in obtaining fundamental information on the complicated behavior of materials, including ferrites, high-T_c superconductors, and strongly correlated and ladder systems. These properties demonstrate that magnetic behavior and atypical physical and chemical characteristics can be synthesized and analyzed. The obtained X-ray spectra reveal that the rapidly suppressed superconductivity is associated with the variation of Ti-O hybridization mainly at e_g bands rather than with the magnetic nature of the substituted V ion.

2. EXPERIMENTS

All the LTVO samples were synthesized through conventional two-step solid-state reactions with highly pure oxides of TiO_2 (99%), Li_2CO_3 (99.99%), and V_2O_3 (99%) [19]. The samples were dried at 150 °C for at least 2 h. Additional Li_2CO_3 (5 mol%) was added into the samples to compensate for lithium evaporation at high temperature. These processes were performed in a dynamic vacuum environment. In the first step, a mixture of Li_2CO_3 and TiO_2 was pulverized and calcined at 750 °C for 8 h in ambient air. The mixture was pulverized and calcined again at 800 °C for 12 h to form the intermediate compound $Li_2Ti_2O_5$. In the second step, the stoichiometric powders of $Li_2Ti_2O_5$, Ti_2O_3, and V_2O_3 were mixed and ground and then cold-pressed into pellets. The pellets were wrapped with gold foil and sintered in an alumina crucible at 880 °C for 24 h in a dynamic vacuum environment at less than 10^{-5} torr. The final products were stored in an Ar-filled glove box or vacuum desiccator to prevent aging. Powder X-ray diffraction (XRD) patterns [19] with X-ray (Cu, K_α=1.5418 Å) radiation in a diffractometer (Philips PW3040/60) confirmed the phase purity of the product and showed the variation of the lattice parameter as a function of V dopant. Different proportions of the dopant up to 2% were selected to understand the effect of dilute doping. Oxygen stoichiometry was expected to remain constant for all small concentrations given that the preparations were all the same.

XANES spectra were obtained from the National Synchrotron Radiation Research Center (Taiwan) and were operated at 1.5 GeV with a current of 360 mA. The K-edge of Ti and V K-edge of LTVO were measured with a wiggler beamline BL17C1 equipped with Si (111) crystal monochromators. The absorption spectra were recorded in the fluorescence-yield mode with a Lytle detector [27]. These hard X-ray absorption spectra, which could provide information on the unoccupied states with transition metal *p* state, were normalized to a unit step height in the absorption coefficient from below to above the edges. The oxide powders, namely, TiO_2, VO_2, Ti_2O_3, and V_2O_3, and standard metal foils were used for energy calibration and for comparing the electronic valence states. The XANES spectra at the Ti L- and O K-edges were determined at beamline BL20A in the total-electron-yield mode using an ultrahigh-vacuum chamber with pressure of $\sim 5\times10^{-9}$ torr. All spectra were normalized with the standard procedure.

The experiments for X-ray emission spectroscopy (XES) were conducted at beamline 7.0.1 at the Advanced Light Source, Lawrence Berkeley National Laboratory. The beamline is armed with a spherical grating monochromator and an undulator (period 5 cm in 99 poles) [3]. The RIXS spectra were recorded with a high-resolution grating spectrometer at grazing incidence with a 2D detector [17]. The resolution for Ti L emission spectra was about 0.4 eV. The monochromator resolution was set up similar to that of the emission measurements.

X-ray absorption at the Ti L-edge determined that the electron in the Ti 2*p* core level was excited to the unoccupied Ti 3*d* and 4*s* orbitals. XES recorded the signal from the decay process related to the partial densities of Ti 4*s* and 3*d* states. RIXS spectra were acquired by tuning different excitation energies according to the x-ray absorption spectral profile to measure XES. Sample current mode was employed to record XANES with an energy resolution of ~0.15 eV. Meanwhile, a high-resolution grazing-incidence grating spectrometer with a 2D multi-channel plate detector was used to record XES at a resolution of 0.6 eV [36].

3. RESULTS AND DISCUSSION

3.1. Structure of LTVO

The crystal structure of cubic spinel LTO, which belongs to the Fd3m space group, is illustrated in Fig. 1(a). The structure demonstrates that the lattice parameter of LTO is 8.404 Å (a = b = c) and contains eight AB_2O_4 units per unit cell. In general, a spinel structure has a total of 56 ions, 32 anions, and 24 cations per unit cell. The Li, Ti, and oxide ions are located at tetrahedral (T_d) A-sites, octahedral (O_h) B-sites, and 32e sites, respectively. The XRD patterns of $LiTi_{2-x}V_xO_4$ (x = 0, 0.05, 0.01, 0.015, and 0.02) are shown in Fig. 2(a). The result of the chemical analysis (determined with ICP-AES) is consistent with the nominal composition [19, 18, 24, 49]. These patterns reflect that the crystal structure is of pure spinel structure without an impure phase. The lattice parameter of LTVO samples displayed in Fig. 2(b) linearly decreases with the increasing concentration of the dopant. Doping V, which has a smaller ionic radius, induces a change in the lattice parameter when in a Ti site. This lattice shrinkage can be attributed to the fact that the ionic radius of V^{3+} (0.64 Å) is smaller than that of Ti^{3+} (0.67 Å) in the O_h site [29]. The rate of shrinkage of the lattice parameter is ~3.6 Å/at% based on the occupancy of O_h sites. This rate is also related to the bond length between the cations (Ti and V) and oxygen. Electronic structure strongly depends on the hybridization states of the transition metal 3*d*-O 2*p* orbitals nearby E_F when electronic exchange normally occurs. Thus, the changed electronic structure around the transition metal ions greatly affects the oxidation state and physical properties of the latter. Considering these physical properties in AB_2O_4 with mixed-valence states require wide-ranging knowledge of their electronic structure, particularly the *p*-*d* hybridization and spin-orbital symmetry. Several experiments and theories show that AB_2O_4 exhibits strong electron–hole correlations because of the TM ions located in various sites [8, 39].

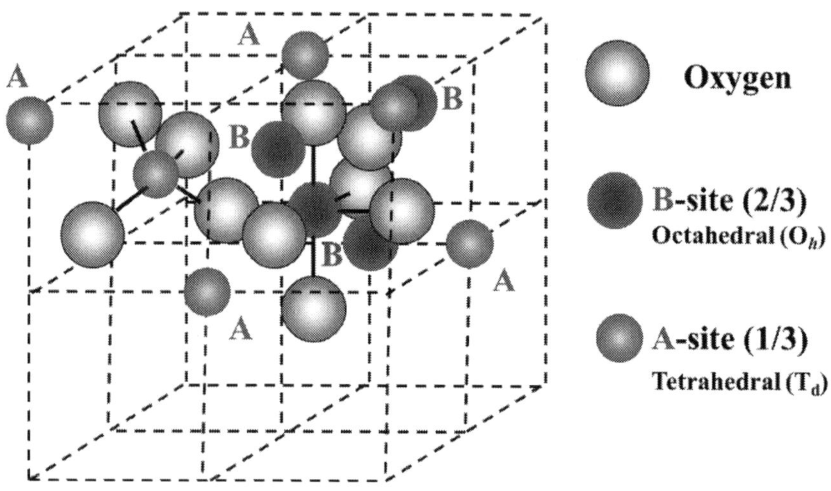

Figure 1. Crystal structure of LTO [Li (green), Ti (blue), and O (grey)].

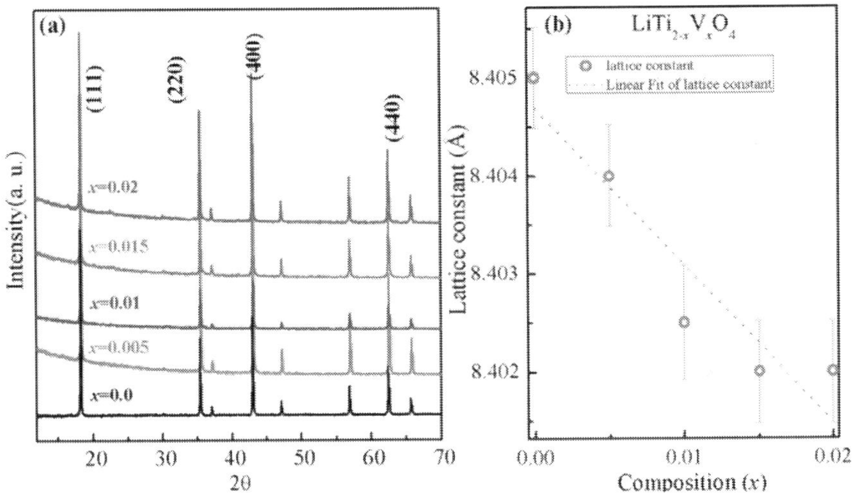

Figure 2. (a) Powder XRD patterns of LTVO crystals with x = 0, 0.005, 0.01, 0.015, and 0.02. The patterns are fitted to the Fd3m space group and are indexed. (b) Chemical composition of V dependence of the lattice constant of LTVO.

3.2. Resistive properties

The transport properties of the samples are specified in Fig. 3(a) with V doping of less than 0.025. The findings indicate that the superconducting T_c is suppressed with an increasing proportion of the dopant. Fig. 3(c) reveals that the resistivity of the samples gradually increases. The un-doped LTO has a superconducting T_c of ~12 K with a transition width (ΔT_c) of ~0.4 K. Fig. 3(b) indicates that when the Ti ion is replaced with V ions, T_c exhibits an abrupt but linear suppression with respect to the level of doping. Given the magnetic pair-breaking effects [47], the doped magnetic ions suppress superconductivity. Considerable research has been devoted to exploring cuprate [48]; Tarascon et al., 1987) and pnictide [41] superconducting systems based on resistivity measurements. For the LTO series, the effect of doping with both nonmagnetic and magnetic impurities at both T_d and O_h sites on $(Li_{1-x}A_x)(Ti_{2-y}B_y)O_4$ [A = Mg and Mn; B = Al and Cr] compounds has been previously reported [26]. The substitution of nonmagnetic ions (e.g., Mg^{2+} at the T_d site and Al^{3+} at the O_h site) slightly suppresses superconductivity similar to the case when Li^+ ions occupy the T_d site. However, the effect of superconductivity suppression is greater when the magnetic Mn^{2+} ions occupy the T_d site. The suppression of superconductivity is attributed to pair-breaking when the Cooper pairs are scattered with magnetic impurities. The effect of substituting Ti^{3+} with Cr^{3+} at the O_h site is even more significant.

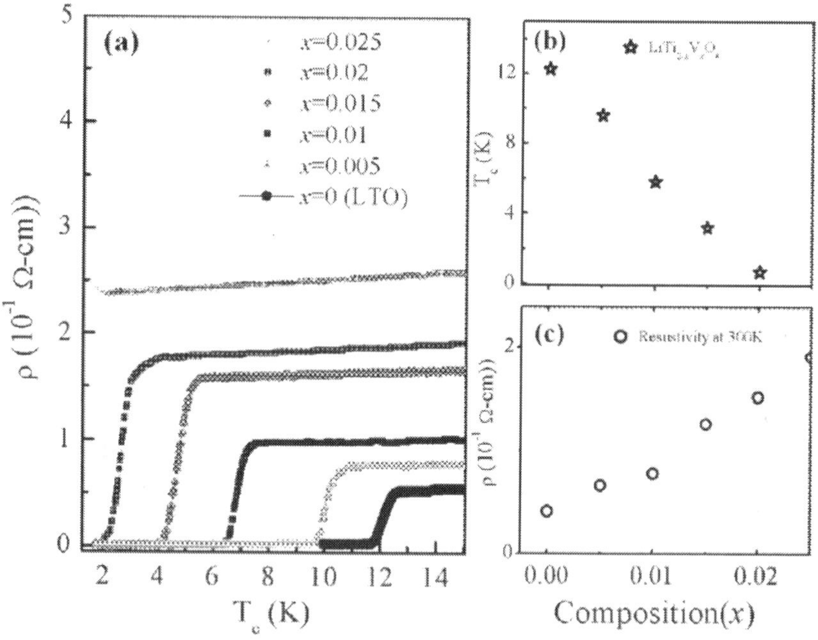

Figure 3. (a) Transport properties of LTO and LTVO samples at low temperature without magnetic field. (b) Chemical composition of V-dependence of superconducting T_c. (c) Resistivity at room temperature.

The resistivity near 300 K increases as x increases [Fig. 3(c)], indicating the extreme sensitivity of the transport properties to the electronic effect of 3d transition metals. The itinerant carrier density can be attributed to the cation (Ti^{3+}) in LTO. Similar to the reports of a previous study on Al-doped LTO [16], the results of the current study clarify the effect of carrier density on superconductivity, such that the resistivity increases with the level of doping in the normal state. Except for the semiconducting behavior at the normal state, a similar T_c has been determined in $LiTi_{1.7}$, $Al_{0.3}O_4$, and LTO samples [19]. This observation implies that T_c is slightly affected by substituting a considerable proportion of the trivalent non-magnetic ions.

The ratio Ti^{4+}/Ti^{3+} in O_h symmetry can then be modulated with impurity doping to vary the superconducting properties of LTO. Electron–electron interaction may be a key factor in causing changes in the transport properties of the samples. By distorting the local symmetry of the spinel structure, V substitution will not only alter the valence of Ti, but the hole/electron effect as well. This phenomenon is discussed in the next section.

3.3. Electronic structure results based on x-ray spectra

With the aim to investigate the effects of V dilute doping on LTO, this research measures XANES and RIXS at Ti L-edges. As shown in the inset in Fig. 4(a), the Ti L-edge XANES spectra exhibit several well-resolved features because of the excitations of a $2p$ core electron into the Ti $3d$ empty states, that is, a transition from the ground state with configuration $2p^6 3d^n$ to an excited electronic configuration $2p^5 3d^{n+1}$ with numerous multiplet excitations. When the spin-orbital coupling is in the transition metal $2p$ state, the spectra reveal two prominent features located at the energy ranges of 455 eV to 461 eV and 461.2 eV to 468 eV, corresponding to the absorptions L_3 ($2p_{3/2} \rightarrow 3d$) and L_2 ($2p_{1/2} \rightarrow 3d$) edge respectively. These prominent features are due to a strong Coulombic interaction between the poorly screened Ti $3d$ electrons and the Ti $2p$ core hole [12]. L_2 edge features are normally broadened compared with the L_3 edge because of the lesser lifetime of the $2p_{1/2}$ core hole (i.e., a radiationless electron transition from energy level $2p_{3/2}$ to the $2p_{1/2}$), accompanied by the promotion of a valence electron into the unoccupied states (conduction band). In the O_h crystal field, the $3d$ band splits into t_{2g} (d_{xy}, d_{xz}, and d_{yz}) and e_g ($d_{x^2-y^2}$ and $3d_{3z^2-r^2}$) subbands [$\Delta = (e_g) - (t_{2g}) = 10$ Dq]. Given the O_h crystal-field splitting, the Ti L_3-edge feature possesses t_{2g} and e_g bands. The RIXS Ti $3d$ spectra of LTO and doped LTO (V 2 %) are exhibited with energy-loss scales in Fig. 4(a). Four distinct spectral features are observed in RIXS spectra. These features are as follows: elastic peak at the energy loss at zero, d–d excitations at about 4 eV, a broad-band feature at around 7 eV, and a large energy dispersed-feature above 10 eV. Letters a to k denote the different exciting energies based on the XANES spectra, as displayed in the inset of Fig. 4(a). For example, XES spectrum b is collected by tuning the incident photon energy at 457 eV in XANES. The inelastic scattering features ranging from 5 eV to 10 eV originate from the complicated charge-transfer excitation from O $2p$ to Ti $3dt_{2g}$ and e_g subbands [1, 31]. The low-energy inelastic scattering features absent in TiO_2 are observed at energy of less than 5 eV [36]. The spectra within this energy range significantly differ from that of TiO_2. The appearance of Ti^{3+} implies that an electron occupies the empty $3dt_{2g}$

orbitals, thus inducing the energy-loss features. Strong *d*-electron correlation is revealed because the relative intensities of peaks A_4 and B_4 (*d-d* excitations) are observed to be markedly different at varying excitation energies. An enhanced energy-loss peak is observed at 1 eV in spectrum *a*, and the intensity of this peak intensively drops when the excitation energy is tuned to a higher energy (peak *b* in XANES). With the increase of excitation energy at *c*, another intense peak at energy loss of about 2.9 eV is observed. An excitation energy tuned below the L_3t_{2g} peak in the spectrum (letter *a*) enhances inelastic scattering A_4, which is uncommon. This enhanced inelasticity is not observed at either t_{2g} (*b*) or e_g (*d*) resonant energy. This phenomenon can be explained by considering the presence of Ti^{3+}, as revealed by the constant-initial-state absorption spectrum [3]. The spectral shapes in XANES spectra are similar for both LTO and LTVO, wherein spectral differences are slightly small. Given that the spectral change of XANES is small with V doping, the spectral deconvolution and specific electronic state can be measured separately from RIXS [11]. The result at a specific energy (*c*) in Figs. 4(b) and 4(c) indicates that an enhanced trivalent Ti contribution is observed at a dip quadrivalence within the spectrum (between t_{2g} and e_g). The $Ti^{3+}t_{2g}$- and e_g-resonance energies correspond to the excitation energies 456 and 457.5 eV respectively. [3].

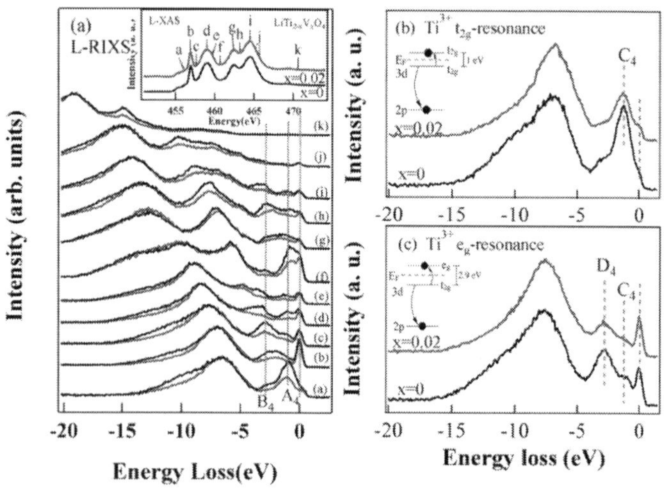

Figure 4. (a) Ti L-RIXS spectra of LTVO (x = 0 and 0.02) recorded with several excitation energies labeled with letters *a* to *k* in the Ti XANES L-edge spectrum. Ti L_3 RIXS spectra at (b) Ti^{3+} t_{2g}-resonance and at (c) Ti^{3+} e_g-resonance. The insets of (b) and (c) display the energy diagram of the *d-d* excitation.

When excitation energy is tuned to the t_{2g} resonance, the low-energy-excited feature C_4 at ~1 eV demonstrated in Figs. 4(b) and 4(c) is resonantly enhanced. This feature corresponds to the electron-hole pairs within the t_{2g} band. When the spectrum is acquired with energy at about e_g resonance energy [Fig. 4(c)], the intensity of peak D_4 increases. Such an increase can be attributed to the transition between the occupied t_{2g} and the unoccupied e_g states. The presence of this RIXS feature indicates a strong electron–electron association. The *d-d* excitation peak at an energy loss of about 2.9 eV corresponds to the crystal-field splitting (i.e., 10 Dq) and refers to the ground state without the core hole. The comparison of the spectra of each set shows that the intensity of the loss feature decreases with the increasing concentration of dopant (x = 0.02). Hence, fewer electrons are distributed in the t_{2g} band. In a previous study on TiO_2 [31], no obvious feature was exhibited in the region located below the elastic signal (peak). The change in intensity is due to the variation of the incompletely filled t_{2g} band arising from the V doped effect. This result forcefully indicates that the Ti valence is enhanced when V is doped. The active Ti ions in LTO have a formal oxidation state trivalence and quadrivalence, which display nearly 0.5 electrons in the 3*d* orbital, and possess a lesser electronic density. Doping with V reduces the number of Ti 3*d* electrons, suggesting an increase in the formal oxidation number of the Ti ion. The Ti ions show a mixed-valence state and contain some t_{2g} electrons in LTO. With no t_{2g} electron, *d-d* excitation is absent because no electron subsists in t_{2g} to be excited to the t_{2g}/e_g unoccupied state. A little variation in valence of Ti is then reflected in the RIXS spectra. The diminution or absence of features in the energy range from 0 eV to 5 eV indicates a decrease in *d–d* transitions. More electrons occupy the t_{2g} orbital when the oxidation number of Ti changes from trivalence to quadrivalence. The probability of exciting the t_{2g} electron into the unoccupied states (t_{2g} or e_g) is therefore increased. Only RIXS can observe the *d-d* excitations (t_{2g}-t_{2g} and t_{2g}-e_g) in the spectra. The significance of observing this *d-d* excitation is that the t_{2g} occupation number changes with V doping, whereas the number of t_{2g} electrons decreases with V doping. Therefore, the significant RIXS spectrum can observe this kind of valence change.

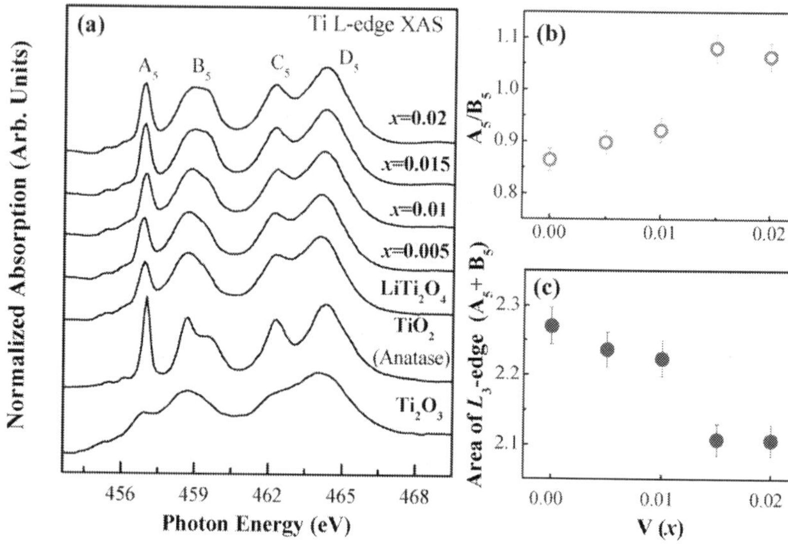

Figure 5. (a) XANES Ti $L_{3,2}$-edge spectra for LTVO (x = 0.005 to 0.02), LTO (x = 0), Ti_2O_3, and anatase TiO_2. (b) Integrated area ratio of A_5/B_5. This ratio increase implies the presence of $3d^0$ states. (c) Total area of L_3-edge (A_5+B_5) decreases with the increase of V concentration

Complementary information is also acquired for this research by thoroughly analyzing the XANES Ti $L_{3,2}$-edge spectra. The spectra of $LiTi_{2-x}V_xO_4$ (or LTVO) (x = 0 to 0.02) are presented in Fig. 5(a). The white line shapes of the spectra of LTO (undoped) and slight V doping (x = 0.005) in LTO are similar. The A_5 and C_5 features (B_5 and D_5) are previously assigned as the t_{2g} (e_g) states of the 10 Dq crystal-field-split in 3d orbitals. This O_h crystal-field splitting is ~1.8 eV at LTO, but increases from 1.8 eV to 2.0 eV when the V concentration increases from x = 0.005 to x = 0.02. This phenomenon reveals the distortion of O_h symmetry when Ti is replaced with slight V doping. This finding is consistent with the powder XRD results and the theoretical calculations [12, 13, 15]. Moreover, the results specify that B_5 peak broadens with its gradual V doping. As previously reported [12, 13, 15], and according to the Jahn-Teller distortion with ΔE_{J-T} energy splitting in the e_g band, the e_g orbitals point directly toward the Ti 2p orbitals of the octahedrally coordinated O atoms. In general, e_g band is sensitive to the local environment, producing a changed bonding distance and a O–Ti–O angle in the presence of V doping. These e_g-related peaks are also broader than the t_{2g} peak because of the larger hybridization between e_g orbitals and O ligand states, and because of the associated effects of solid-state broadening [12]. The e_g feature appears as a narrow and symmetric profile in LTO, representing symmetric octahedrally coordinated Ti–O bonds. This feature becomes broad and asymmetric upon doping, suggesting a distortion. In particular, distortion may arise from the uneven Ti-O bonds in the O_h symmetry. The e_g peak at the side with greater energy originates from the short Ti-O bonds because of the increase in hybridization (relative to the long Ti-O bonds); hence, the intensity ratio of the high-energy to low-energy e_g peaks is increased with V doping. This result implies that long Ti–O bonds become shorter [17]. An enhancement in the integrated area under A_5 peak implies the increase in unoccupied states of Ti 3*d* states, thereby indicating that the Ti valence increases with slight V doping. From the Ti–O bond, the electronic configuration of Ti exhibits a combination of $3d^0$ ($t_{2g}{}^0e_g{}^0$) and $3d^1$ ($t_{2g}{}^1e^0$) in the ground state. Fig. 5(b) shows the integrated area of ratio A_5/B_5, which clearly increases with V doping. Consequently, the more intense A_5 (t_{2g}) feature implies a higher oxidation number and exhibits the presence of Ti^{4+} ($3d^0$) [28, 40]. Moreover, the area of the L_3-edge that corresponds to (A_5+B_5) progressively decreases when V concentration increases [Fig. 5(c)] because V replaces Ti. This trend is in good agreement with the XRD results [18]. A small but actual increase in Ti valence with slight V doping is observed in XANES Ti $L_{3,2}$-edge spectra. Such an increase is significantly obvious in the Ti L-RIXS spectra.

The pre-edge features of Ti K-edge XANES spectra are displayed in Fig. 6(a). These spectral features at K-edge are due to the transitions from the Ti 1*s* core level to the 4*p*-derived final states based on the dipole selection rule. The pre-edge features are a combination of strongly hybridized Ti 4*sp* and 3*d* and O 2*p* orbitals. Quadrupole-allowed transitions generally occur at the pre-edge region in the transition metal oxides, which correspond to the contribution from 3*d* orbitals through 4*sp*-3*d* hybridization [14, 25, 37]. The inset of Fig. 6(a) reveals that a Gaussian function is subtracted from the original Ti K-edge spectrum for a detailed comparison of the pre-edge spectra. Fig. 4(a) shows the Ti K-edge spectra of LTO and LTVO (0.005 to 0.02). As marked by the black arrow in Fig. 6(a), the spectra of LTVO with a small concentration of *x* from 0.005 to 0.01 are similar to that of LTO with the same photon energy at the main peak A_6. The Ti valence performs to maintain +3.5 owing to the smaller concentration of V doping. Nevertheless, the intensity of the pre-peak spectrum increases relative to that of pure LTO as the doping level is increased to x = 0.015 and x= 0.02. Meanwhile, as directed by the red arrow in Fig. 6(a), chemical shift is also observed as V concentration increases. Fig. 6(b) displays the pre-edge region between 4969 and 4977 eV, as well as the three main features (i.e., C_6, D_6, and E_6) in XANES Ti K-edge spectra of TiO_2 (anatase, Ti^{4+}) and LTVO (x = 0, 0.02) samples. The origin of the splitting of pre-peaks is caused by local excitations (1*s* to 3*d*t_{2g} and e_g) [5, 14, 38, 46]. Conversely, the greatest contribution to this splitting is suggested to be the corner- and edge-sharing Ti octahedra that yield non-local, intersite hybrid excitations. The next-nearest neighbor 3*d* states (t_{2g} and e_g) are related Ti 4*p* states that absorb atom via the anion O 2*p* states [7, 23, 44]. The local structure of Ti for LTO and LTVO also possesses O_h symmetry; hence, the pre-peak region (A_6 and B_6) can be described in terms of a similar scenario [23, 44, 7]. The intensity of these pre-peaks varies with V concentration. In the V-doped LTO, the substitution of V for Ti slightly decreases the lattice parameter [18]. A reduced bond length (Ti–O–Ti) subsequently increases the overlap of the first-nearest-neighbors and the absorbing atom (e.g., Ti 4*p*-Ti 3*d* orbitals) mediated by O ion. Therefore, these pre-peaks are closely related to the increase in the number of the first-nearest-neighbor unoccupying the Ti 3*d* states. These intersite hybrid peaks intensities increase with the increase of V concentration. The inset of Fig. 6(b) shows that the variation in the area under the pre-peak suggests that the 3*d* unoccupied states are altered via the interaction of the Ti 4*p*-O 2*p*-Ti 3*d* states. These results strengthen the conclusion that Ti is in mixed-valence states between LTO and LTVO [14, 24]. The analysis of the spectra of LTO and LTVO (x = 0.02) reveals that pre-peaks A_6 and B_6 are stronger in LTVO than in LTO, attended by a chemical shift of the main peak to higher energy. Therefore, Ti valence is increased when V is doped. The average valence of Ti in the case of doped LTVO (x = 0.015 and 0.02) is approximately +3.6, which is obtained with a simple calculation and a fit of the ratio of areas under the spectral lines.

The O K-edge XANES spectra [Fig. 6(c)] provide useful information about the unoccupied density of states in TMO because of the covalent mixing between O and TM ions. The two pre-edge features are at about 530 eV to 536 eV because of the strong hybridizations in Ti 3*d*–O 2*p* states, and correspond to the t_{2g} and e_g states of Ti 3*d* orbitals respectively. The intensity of the t_{2g} and e_g peaks at the O K-edge changes is similar to an increased peak ratio t_{2g}/e_g in Ti L-edge XANES. In sum, the above results specified by the XANES and RIXS spectra strongly confirm that the valence of Ti is increased when the doping level of x increases. Thus, the Ti-O hybridization and electron–electron correlation are modified when a slight doping of V ions completely suppresses the superconductivity of LTO.

Previous studies have indicated that LTO is a spinel superconductor, whereas spinel LVO reveals a heavy fermionic behavior with a Curie-Weiss spin susceptibility and a large electronic specific heat. Conduction arises on a Ti sublattice via the t_{2g} orbitals with a narrow bandwidth (2 eV to 3 eV), displaying possible strong electronic correlations. The origin of the physics and electronic properties of the spinel structure are yet to be resolved. Our current research demonstrates the role of 3*d* electrons in the properties of LTO and LTVO. The full solid solution of $LiTi_{2-x}V_xO_4$ ($0 \leq x \leq 2$) has been investigated [24]. The replacement of V ions for Ti site modifies the bandgap and *d* electron/hole exchange in Ti energy levels in order to maintain electrical neutrality. Several mechanisms for understanding the quickly suppressed superconductivity with a slight doping of LTVO have been proposed. The slight doping of LTVO may incur two major electronic effects. The first one is a simple pair-breaking effect with electron spin $S = 1$ on superconductivity [20, 24]. The second one refers to the effect of carrier doping on electron correlation via band filling. The 3*d* electrons of V are assumed to hybridize the Ti conduction electrons, making them itinerant. A simple magnetic pair-breaking of electron spins induced by additional impurity is commonly assumed to be responsible for suppressing superconductivity based on the localized moments of 3*d* electrons of V [24]. The localized magnetic moment per V atom is estimated to be 1.7 μB in the V^{4+} state with $S = 1/2$ [24]. The XANES measurements at V K-edges plotted in Fig. 7 show that V is more likely to be in state 3+ (Bordage et al., 2110), which contains a d^2 electronic configuration and is supposed to offer a weak magnetic moment probably because of the antiparallel orientation of the electron spins. The decreased V magnetic moment is also obvious in $LiZnV_2O_4$ [35]. A second possible reason for a weak magnetic moment is the decreased charge carriers at Ti t_{2g} bands. In this case, V electrons are expected to be localized at the Ti conduction electrons. If the Ti ions in LTO possess oxidation numbers 3+ and 4+, then ~0.5 (t_{2g}) electrons are expected to be in the conduction band. This condition results in the observation of a *d-d* excitation within the energy range of 0 eV to 5 eV owing to the presence of electrons in the t_{2g} occupied state that are excited to either the t_{2g} or e_g unoccupied state. The number of electrons occupying the t_{2g} orbital is expected to decrease if the oxidation number of Ti increases from trivalence to quadrivalence because of the doping of V ions. The spectra in RIXS reveal a significant variation in the *d-d* excitation feature, which suggestively decreases when LTO becomes LTVO, even with slight doping of V ions. This situation shows the decreased Ti 3*d* occupied states with the increase of *x* concentration. The Ti t_{2g} occupation number is then altered upon V doping. This valence variation is small in XANES Ti K- or L-edges spectra, but weighty in RIXS spectra. The results, particularly the reflection of the decreased intensities of t_{2g}-t_{2g}, and t_{2g}-e_g transitions from the RIXS spectra, support the role of these electronic states in rapidly suppressing T_c in slight V doping. Accordingly, the findings of this research support the conclusion that the decreased density of the states of Ti 3*d* electrons at E_F is responsible for rapidly suppressing superconductivity.

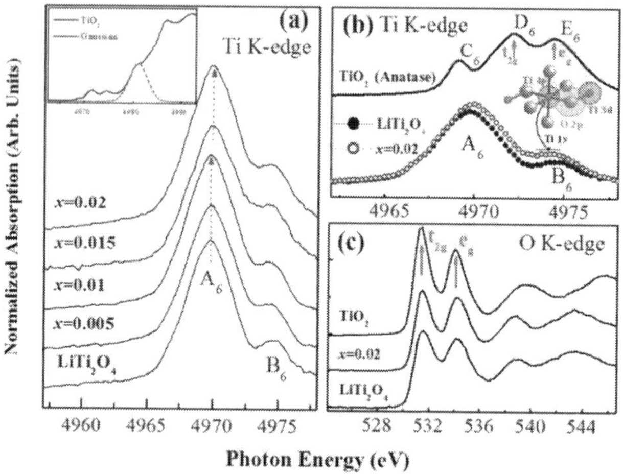

Figure 6. (a) XANES Ti K-edge spectra of LTVO (x = 0 to 0.2) and Anatase-TiO_2 in the pre-edge region. The inset shows that a Gaussian function was subtracted from the original TiO_2 spectrum. (b) Detailed comparison of the pre-edge region for LTO and LTVO (x = 0.02). (c) XANES O K-edge spectra shown with two main peaks in the pre-edge region corresponding to the hybridization of Ti 3d (t_{2g} and e_g)-O 2p states.

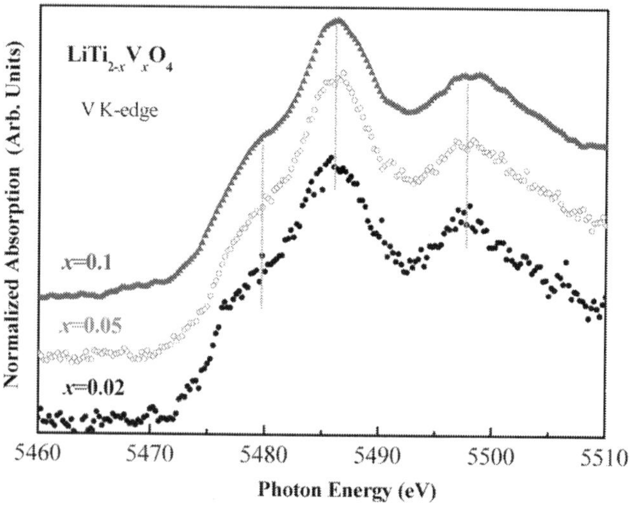

Figure 7. XANES V K-edge of LTVO (x = 0.02) compared with the larger doping of LTVO (x = 0.05 and 0.1). No significant variation is observed in the spectra

4. CONCLUSION

The results on the effect of V doping on the atomic and electronic structures of LTVO, XAS, and RIXS spectra reveal that the valence of V ions remains constant, but the hybridization of Ti-O considerably varies. In particular, the results show the mixed valency nature of Ti ions and a significant variation in the hybridization of Ti 3d-O 2p states. The sharp decrease in the superconductivity of V-doped samples is ascribed to the electron–orbital interaction arising from the hybridization of TM and O orbitals. Low-energy excitation because of d-d excitation indicates that electron correlation and Ti oxidation number are enhanced, which support the XAS observation. Meanwhile, the observed properties of superconductivity are attributed to an altered density of the states of Ti 3d electrons and Ti-O hybridization. The Ti electronic configuration and lattice distortion induce the rapidly suppressed superconductivity and are attributed to hybridization rather than to the magnetic nature of the substituted ion. This research also demonstrates that RIXS is a powerful tool for investigating the electronic states and electron correlations of Ti compounds, in which the XANES spectral features are subdued.

REFERENCES

1. Agui, A.; Uozumi, T.; Mizumaki, M. & Käämbra, T. (2009). Intermetallic charge transfer in FeTiO3 probed by resonant inelastic soft x-ray scattering. *Physical Review B*, 79: 092402.
2. Anisimov, V. I.; Korotin, M. A.; Zölfl, M.; Pruschke, T.; Hur, K. Le & Rice, T. M. (1999). Electronic Structure of the Heavy Fermion Metal LiV2O4. *Physical Review Letters*, 83: 364.
3. Augustsson, A.; Henningsson, A.; Butorin, S. M.; Siegbahn, H.; Nordgren, J. & Guo, J.-H. (2003). Lithium ion insertion in nanoporous anatase TiO2 studied with RIXS. *The Journal of Chemical Physics*, 119: 3983.
4. Bordage, A.; Balan, E.; de Villiers, J. P. R.; Cromarty, R.; Juhin, A.; Carvallo, C.; Calas, G.; Sunder Raju, P. V. & Glatzel, P. (2011). V oxidation state in Fe–Ti oxides by high-energy resolution fluorescence-detected x-ray absorption spectroscopy. *Physics and Chemistry of Minerals*, 38: 449-458.

5. Brydson, R.; Sauer, H.; Engel, W.; Thomass, J. M.; Zeitler, E.; Kosugi, N. & Kuroda, H. (1989). Electron energy loss and x-ray absorption spectroscopy of rutile and anatase: a test of structural sensitivity. *Journal of Physics: Condensed Matter*, 1: 797.
6. Bussmann-Holder; Müller, K. A. & Keller, H. (2007). High *Tc* Superconductors and Related Transition Metal Oxides ISBN 3540710221.
7. Cabaret, D; Bordage, A; Juhin, A; Arfaoui, M. & Gaudry, E. (2010). First-principles calculations of x-ray absorption spectra at the K-edge of 3*d* transition metals: an electronic structure analysis of the pre-edge. *Phys. Physical Chemistry Chemical Physics*, 12: 5619-5633.
8. Cava, R. J.; Murphy, D. W.; Zahurak, S.; Santoro, A. & Roth, R. S. (1984). The crystal structures of the lithium-inserted metal oxides Li0.5TiO2 anatase, LiTi2O4 spinel, and Li2Ti2O4. *Journal of Solid State Chemistry*, 53: 64-75.
9. Chen, C. L. & Dong, C. L. (20012). SUPERCONDUCTOR-Materials, Properties and Applications, 2012 ISBN: 979-953-307-798-6 Chapter 2 'X-ray spectroscopy studies of the iron chalcogenides', 21-44.
10. Chen, C. L.; Yeh, K. W., Huang, D. J., Hsu, F. C.; Lee, Y. C., Huang, S. W., Guo, G. Y.; Lin, H. J.; Rao, S. M. & Wu, M. K. (2008). Orbital polarization of the unoccupied states in multiferroic LiCu2O2. *Physical Review B*, 78: 214105.
11. Dallera, C.; Grioni, M.; Shukla, A.; Vankó, G.; Sarrao, J. L.; Rueff, J. P. & Cox, D. L. (2002). New Spectroscopy Solves an Old Puzzle: The Kondo Scale in Heavy Fermions. *Physical Review Letters*, 88: 196403.
12. De Groot, F. M. F.; Fuggle, J. C.; Thole, B. T. & Sawatzky, G. A. (1990). 2*p* x-ray absorption of 3d transition-metal compounds: A atomic multiplet descriptiom including the crystal field. *Physical Review B*, 42: 5459. (1990). L2,3 x-ray-absorption edges of *d*0 compounds: K+, Ca2+, Sc3+, and Ti4+ in Oh (octahedral) symmetry. *Physical Review B*, 41: 928.
13. De Groot, F. M. F.; Faber, J. C.; Michile, J. J. M.; Czyzyk, M. T.; Abbate, M. & Fuggle, J. C. (1993). Oxygen 1*s* x-ray absorption of tetravalent titanium oxides: A comparison with single-particle calculations. *Physical Review B*, 48: 2074

14. Durmeyer, O.; Kappler, J. P.; Beaurepaire, E.; Heintz, J. M. & Drillon, M. (1990). TiK XANES in superconducting LiTi2O4 and related compounds. *Journal of Physics: Condensed Matter*, 2: 6127

15. Finkelstein, L. D.; Zabolotzky, E. I.; Korotin, M. A.; Shamin, S. N.; Butorin, S. M.; Kurmaev, E. Z. & Nordgren, J. (2002). Vacant states of TiO2 with rutile structure and their reflection in different-type x-ray absorption spectra. *X-ray Spectrometry*, 31: 414-418.

16. Fazileh, F.; Gooding, R. J.; Atkinson, W. A. & Johnston, D. C. (2006). Role of Strong Electronic Correlations in the Metal-To-Insulator Transitionin Disordered LiAlyTi2-yO4. *Physical Review Letters*, 96: 046410.

17. Harada, Y.; Kinugasa, T.; Eguchi, R.; Matsubara, M.; Kotani, A.; Watanabe, M.; Yagishita, A. & Shin, S. (2000). Polarization dependence of soft-x-ray Raman scattering at the L edge of TiO2. *Physical Review B*, 61: 12854.

18. Hsu, F. C. Liao, Y. C.; Yan, D. C.; Gu, S. Y.; Wu, M. K.; Tang, H. Y. & Perng, T. P. (2007). Observation of metal–insulator transition in vanadium-doped superconducting Li(Ti2-xVx)O4. *Physica C*, 460-462: 546-548.

19. Hsu, F. C. (2009), Investigation of the electrical, magnetic and thermal properties of spinel LiTi2-xMxO4 (M=V and Cr). Department of Materials Science and Engineering, *National Tsing Hua University, Taiwan.* Ph. D. Thesis.

20. Itoh, Y,; Moritsu, N. & Yoshimura, K. (2008). Emergence of Antiferromagnetic Correlation in LiTi2-xVxO4 via 7Li NMR. *Journal of the Physical Society of Japan*, 77: 123713.

21. Johnston, D. C. (1976). Superconducting and normal state properties of Li1+xTi2−xO4 spinel compounds. I. Preparation, crystallography, superconducting properties, electrical resistivity, dielectric behavior, and magnetic susceptibility. *Journal of Low Temperature Physics*, 25: 145-175

22. Johnston, D. C.; Swenson,C. A. & Kondo, S. (1999). Specific heat (1.2–108 K) and thermal expansion (4.4–297 K) measurements of the 3*d* heavy-fermion compound LiV2O4. *Physical Review B*, 59: 2627.

23. Joly, Y; Cabaret, D.; Renevier, H. & Natoli, C. R. (1999). Electron Population Analysis by Full-Potential X-Ray Absorption Simulations. *Physical Review Letters*, 82: 2398.

24. Kichambare, P. Kijima, N.; Honma, H., Ebisu, S. & Nagata, S. (1996). Suppression of superconductivity in Li(Ti1−xVx)2O4. *Journal of Physics and Chemistry of Solids*, 57: 1615-1620.
25. Kucheyev, S. O.; van Buuren, T.; Baumann, T. F.; Satcher, J. H.; Willey, Jr., T. M.; Meulenberg, R. W., Felter, T. E.; Poco, J. F.; Gammon, S. A. & Terminello L. J. (2004). Electronic structure of titania aerogels from soft x-ray absorption spectroscopy. *Physical Review B*, 69: 245102.
26. Lambert, P. M.; Edwards, P. P. & Harrison, M. R. (1990). Magnetism, superconductivity, and the metal-nonmetal transition in the spinel LiMxTi2−xO4; M = Al3+, Cr3+. *Journal of Solid State Chemistry*, 89: 345-360.
27. Lytle, F. W., Greegor, R.B., Sandstrom, D.R., Marques, E.C., Wong, J., Spiro, C.L., Huffman, G.P. & Huggins, F. E. (1984). Measurement of soft x-ray absorption spectra with a fluorescent ion chamber detector. Nuclear Instruments and Methods in Physics Research Section A: Accelerators, Spectrometers, Detectors and Associated Equipment, 226:542-548.
28. Le Fèvre, P.; Danger, J.; Magnan, H.; Chandesris, D.; Jupille, J.; Bourgeois, S.; Arrio, M. A.; Gotter, R.; Verdini, A. & Morgante, A. (2004). Stoichiometry-related Auger lineshapes in titanium oxides: Influence of valence-band profile and of Coster-Kronig processes. *Physical Review B*, 69: 155421.
29. Lide, D. R. (2007-2008) CRC handbook of Chemistry and Physics, 88th edition, Taylor & Francis Group.
30. Maddox, B. R.; Lazicki, A.; Yoo, C. S.; Iota V.; Chen, M.; McMahan, A. K.; Hu, M. Y.; Chow, P.; Scalettar, R. T. & Pickett, W. E. (2006). 4*f* delocalization in Gd: Inelastic x-ray scattering at ultrahigh pressure. *Physical Review Letters*, 96: 215701.
31. Matsubara, M.; Uozumi, T.; Kotani, A.; Harada, Y. & Shin, S. (2002). Polarization dependence of resonant x-ray emission spectra in 3*d*n transition metal compounds with n = 0, 1, 2, 3. *Journal of the Physical Society of Japan*, 71: 347-356.
32. Massidda, S.; Yu, J. & Freeman, A. J. (1988). Electronic structure and properties of superconducting LiTi2O4. *Physical Review B*, 38: 11352.

33. Merz, M.; Nücker, M.; Schweiss, P.; Schuppler, S.; Chen, C. T.; Chakarian, V.; Freeland, J.; Idzerda, Y. U,; Kläser M.; Müller-Vogt, M. & Wolf, Th. (1998). Site-Specific X-Ray Absorption Spectroscopy of Y1−xCaxBa2Cu3O7−y: Overdoping and role of apical oxygen for high temperature superconductivity. *Physical Review Letters*, 80: 5192.

34. Millis, A. J. (1998). Lattice effects in magnetoresistive manganese perovskites. *Nature*, 392: 147.

35. Muhtar; Takagi, F.; Kawakami, K. & Tsuda, N. (1988). Magnetic Susceptibility of Metal-Insulator System LixZn1-xV2O4. *Journal of the Physical Society of Japan*, 57: 3119-3127.

36. Nordgren, J.; Bray, G.; Cramm, S.; Nyholm, R.; Rubensson, J. E. & Wassdahl, N. (1989). Soft x-ray emission spectroscopy using monochromatized synchrotron radiation (invited). *Review of Scientific Instruments*, 60: 1690.

37. Poumellec, B.; Marucco, J. F. & Touzelin, B. (1987). X-ray-absorption near-edge structure of titanium and vanadium in (Ti,V)O2 rutile solid solutions. *Physical Review B*, 35: 2284.

38. Parlebas, J. C.; Khan, M. A.; Uozumi, T.; Okada, K. & Kotani, A. (1995). Theory of many-body effects in valence, core-level and isochromat spectroscopies along the 3*d* transition metal series of oxides. *Journal of Electron Spectroscopy and Related Phenomena*, 71: 117.

39. Ra, W.; Nakayama, M.; Uchimoto, Y. & Wakihara, M. (2005). Experimental and computational study of the electronic structural changes in LiTi2O4 Spinel Compounds upon Electrochemical Li insertion reactions. *The Journal of Physical Chemisity B*, 109(3): 1130-1134.

40. Richter, J. H.; Henningsson, A.; Sanyal, B.; Karlsson, P. G.; Andersson, M. P.; Uvdal, P.; Siegbahn, H.; Eriksson, O. & Sandell, A. (2005). Phase separation and charge localization in UHV-lithiated anatase TiO2 nanoparticles. *Physical Review B*, 71: 235419.

41. Rotter, M,; Tegel, M. and Johrendt, D. (2008). Superconductivity at 38 K in the Iron Arsenide (Ba1−xKx)Fe2As2. *Physical Review Letters*, 101: 107006.

42. Rueff, J. P.; Hague, C. F.; Mariot, J. M.; Journel, L.; Delaunay, R.; Kappler, J. P.; Schmerber, G.; Derory, A.; Jaouen, N. & Kril,l G.

(2004). f-state occupancy at the γ-α phase transition of Ce-Th and Ce-Sc alloys. *Physical Review Letters*, 93: 067402.

43. Sun, C. P.; Lin, J. Y.; Mollah, S.; Ho, P. L.; Yang, H. D.; Hsu, F. C.; Liao, Y. C. & Wu, M. K. (2004). Magnetic field dependence of low-temperature specific heat of the spinel oxide superconductor LiTi2O4. *Physical Review B*, 70: 054519.
44. Uozumi, T.; Okada, K.; Kotani,; Durmeyer, A. O.; Kappler, J. P.; Beaurepaire, E. & Parlebas, J. C. (1992). Experimental and Theoretical Investigation of the Pre-Peaks at the Ti K-Edge Absorption Spectra in TiO2. *EPL (Europhysics Letters)*, 18: 85.
45. Warwick, T.; Heimann, P.; Mossessian, D.; McKinney, W. & Padmore, H. (1995). Performance of a high resolution, high flux density SGM undulator beamline at the ALS (invited). *Review of Scientific Instruments*, 66: 2037.
46. Wu, Z. Y.; Ouvrard, G.; Gressier, P. & Natoli, C. R. (1997). Ti and O K edges for titanium oxides by multiple scattering calculations: Comparison to XAS and EELS spectra. *Physical Review B*, 55: 10382.
47. Xiao, G.; Cieplak, M. Z.; Xiao, J. Q. & Chien, C. L. (1990). Magnetic pair-breaking effects: Moment formation and critical doping level in superconducting La1.85Sr0.15Cu1-xAxO4 systems (A=Fe, Co, Ni, Zn, Ga, Al). *Physical Review B*, 42: 8752(R)
48. Xiao, G.; Streitz, F. H.; Gavrin, A.; Du, Y. W. & Chien C. L.(1987). Effect of transition-metal elements on the superconductivity of Y-Ba-Cu-O. *Physical Review B*, 35: 8782(R)
49. Xu, F.; Liao, Y. C.; Wang, M. J.; Wu, C. T.; Chiu, K. F. & Wu, M. K. (2003). The Preparation Effect of Li1+x Ti2O4 and Its Aging Effect. *Journal of Low Temperature Physics*, 131: 569-574.

10
CHAPTER

SPINTRONICS DRIVEN BY SUPERCONDUCTING PROXIMITY EFFECT

Guoxing Miao[1]

[1] *Institute for Quantum Computing and Electrical and Computer Engineering, University of Waterloo, Waterloo ON, Canada*

1. INTRODUCTION

In this chapter, we will discuss a few selected topics on the applications of superconducting proximity effect, and the related inverse proximity effect, in the field of spintronics. Superconducting proximity effect occurs when Cooper pairs from a superconductor propagate into the adjacent metallic systems and induce superconducting correlations in the otherwise non-superconducting materials. Due to the limited coherence length of Cooper pairs, this effect is confined to the very interface between the two materials and can be used as a method to trigger supercurrent flow, or to create particle-hole symmetry, in a wide range of devices. The inverse proximity effect can be viewed as the counteraction of the above-mentioned effect. Superconductivity in the original superconductor material inevitably weakens when it drives superconductivity into its neighbours; in addition, back flow of unpaired electrons and sometimes spin polarized electrons, will create more pair breaking within the superconductor and weaken it further. These effects can, however, be used as effective ways to control superconductivity through spin manipulations.

This chapter is organized in the following way. After a brief introduction on superconducting proximity effect and inverse proximity effects, we will continue our discussions from a few device points of view. 1. Proximity induced superconductivity in low dimensional electron systems, such as in the surface states of a 3D topological insulator and in the Rashba-split bands of a heavy metal/semiconductor nanowire. These form the most promising platforms that can host the elusive Majorana fermions for quantum computing applications. 2. Inverse proximity from multiple ferromagnetic neighbours can controllably turn superconductivity On and Off. The change in superconductivity states also leads to a large change in the device resistance, known as the superconducting spin valve effect.

i. Superconducting proximity effect — the example of Majorana fermion creation
ii. Inverse proximity on superconductors — the example of superconducting spin valves

2. SUPERCONDUCTING PROXIMITY EFFECT — THE EXAMPLE OF MAJORANA FERMION CREATION

The term superconducting proximity effect [1, 2], by default, refers to the leak-out of a superconductor's Cooper pair wave function into an otherwise non-superconducting material, and the generation of induced superconductivity in that material. This is a spontaneous process that occurs whenever the superconductor makes clean enough contact with a metallic system. Fig.1 illustrates the distribution of superconductor pairing potential Δ across the interface between a superconductor and a normal metal. Far away from the interface, Δ simply sits at the respective bulk values, being Δ_0 on the superconductor side and vanishing on the normal metal side. The proximity effect shows up clearly near the interface: Δ is weakened on the superconductor side but gradually emerges in the normal metal towards the interface. Naturally, the two systems have to be in atomic contact for such wave function overlap to happen effectively. The interface transparency is critical for the strength of the proximity, and is often characterized with an impedance parameter γ_B such that the interfacial transparency $T = \frac{1}{1+\gamma_B}$. The transparency captures the interfacial quality as well as the wave function matching between the materials. In Fig.1, a clear discontinuity is seen across the interface which indicates a less-than-perfect interfacial transparency. A cleaner interface means stronger proximity, which adds some quite stringent requirements to the materials' development.

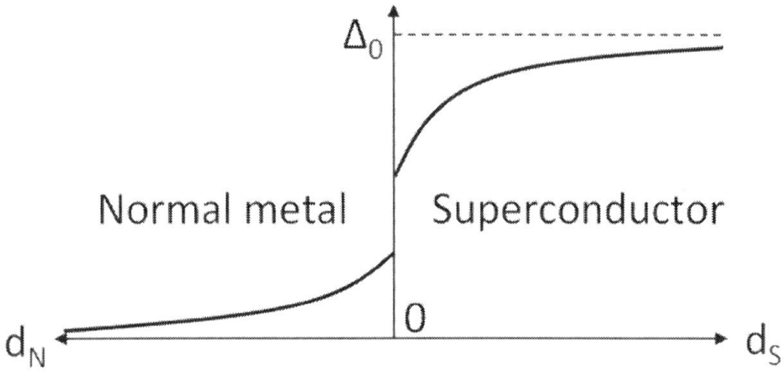

Figure 1. Evolution of the superconductor pairing potential Δ near an S/N interface. Δ_0 is the bulk pairing potential of the superconductor.

Such superconducting proximity effect is of particular interest to the spintronics community because it not only allows for spontaneously driving a non-superconducting material into superconductivity, but also enables actively pumping supercurrents over a prolonged distance across a non-superconducting material. Their fundamental principles are the same: extension of the superconductor wave function into the non-superconducting material. Naturally, the proximity effect can propagate into a material as long as there are available electronic states near the Fermi level: i.e., a metallic material. There have been some quite important reports in the field demonstrating superconductor coupling through unconventional media: for example, the manifestation in 2D semiconductor electron gases [3, 4, 5] and topological insulator surface states [6, 7, 8]; the two-valley nature with relativistic Dirac electrons of graphene [9]; the gate tunability when coupled to 1D carbon nanotubes [10, 11], as well as with 0D C_{60} molecules [12] and InAs quantum dots [13]; the flow of supercurrents through double-stranded DNA molecules [14]; and the conversion into spin triplet supercurrents in half-metals [15, 16].

Whereas the superconducting proximity effect can essentially drive any electron system into superconducting, we focus our attention on a very specific topic: generation of the long-sought-after Majorana fermions on superconductor platforms and application of these particles to topological quantum information processing [17]. A Majorana fermion, by definition, is a spin-1/2 particle identical to its own antiparticle, as opposed to a conventional Dirac fermion. Being its own antiparticle suggests that the particle has to be charge neutral as well. A superconductor's wavefunction has perfect electron-hole symmetry, making it an ideal platform for creating objects with similar properties. Other than being electrically charged, the quasiparticle excitations in a superconductor come really close to the Majorana fermions. Naturally, many of the proposals for creating Majorana fermions in solid-state systems involve superconductivity as one of the core ingredients. There are a number of ingenious device concepts for realizing Majorana fermions in a superconductor platform. The most practical route of synthesizing a localized Majorana fermion is to build on the platform of p-wave superconductors [18, 19] such as strontium ruthenate (Sr_2RuO_4): a spinless p_x+ip_y superconductor [20]. A p-wave superconductor's wavefunction is two-fold degenerate and has chirality on the angular momentum; it is therefore often termed the chiral p-wave state. It differs from the more conventional s-wave BCS superconductors and d-wave high-temperature superconductors in the sense that the electrons prefer to pair into spin-triplet configurations. Since ferromagnetism is observed in many closely related strontium ruthenate compounds (such as $SrRuO_3$), it is natural to consider the possibility of triplet pairing on these superconducting species. There has been compelling evidence confirming that Sr_2RuO_4 is indeed a p-wave superconductor [21]. This spontaneous time-reversal symmetry breaking ensures that the system is stable against magnetic impurity scattering, but not so much against normal elastic impurity scatterings, such as from non-magnetic defects, grain boundaries, or surfaces. As a result, it is very hard to obtain materials with a high enough quality for device purposes. With Sr_2RuO_4 as an example, while its T_C can reach as much as 1.5 K in the best-quality single crystals, it is practically non-superconducting in any of the thin films deposited so far [21], which is a significant obstacle for its potential application in the topological quantum computer architectures.

The superconductor proximity effect comes in as an experimentally feasible route to "simulate" a p-wave superconductor that can subsequently host Majorana modes on the resulting 2D platforms [22, 23]. Now that the particle-hole symmetry has already been guaranteed on the given platforms by the superconductivity induced from the superconducting proximity effect, what about the spinless nature of the p-wave superconductors? The signature we need to look for is the so-called helical Dirac spin state, as illustrated in Fig.2, which readily shows up on the surface of a 3D topological insulator [24]. A 3D topological insulator is a quantum spin Hall insulator with an inside that is insulating and a surface that is conducting [25]. Due to the nontrivial topology of the material, its interface with a topologically trivial insulator, such as a normal insulator or vacuum, always has the conducting surface states present, which are strictly protected by the topological orders. We will not go into too much detail about topological insulator itself as it is not the focus of this article, except to point out that: a) the electron dispersion relation shows a helical spin texture — the electrons' spin orientation and momentum direction are tightly locked due to strong spin-orbit coupling; b) the dispersion is linear for the energy range we are interested in — known as the relativistic Dirac dispersion; and c) there is no energy gap and no spin degeneracy — a single band structure, different from a zero-gap semimetal or the spin and valley degenerate graphene. As the ingenious proposal from Fu and Kane [22] demonstrates, by coupling the helical surface states of a topological insulator with a conventional s-wave BCS superconductor through proximity effect, one can essentially simulate the properties of a p-wave superconductor and perform any subsequent operations with the composite states. Due to the two-dimensional nature of the topological surface states, the resultant "p-wave superconductor" is also two-dimensional. When a vortex nucleates on such a surface, it defines a non-superconducting region which essentially creates an "edge" on the surface localized to a tiny circle. A closed contour around this circle has a Berry phase shift of π due to the rotation of spin orientation around this circle. A Majorana zero mode is trapped within this vortex core, and can therefore be moved coherently when one physically moves this vortex around, as if it is a standalone physical particle. Once these localized, non-Abelian Majorana bound states are established, one can then perform the necessary braiding operation on them which covers all the quantum gates required for quantum information processing [17].

Though a topological insulator makes a convenient platform for creating a p-wave superconductor and the subsequent Majorana fermions, the above proposal remains an experimental challenge mainly due to the difficulties associated with perfecting the topological materials themselves, which requires significantly more effort along the road [24, 26, 27, 28]. Experimentalists have progressed remarkably further on a much more mature material platform: the surface of a semiconductor with strong Rashba spin-orbit coupling. Evidence of Majorana bound states has already been reported in this direction [29, 30, 31]. The basic idea here is to also look for an electronic state that resembles the helical spin states as

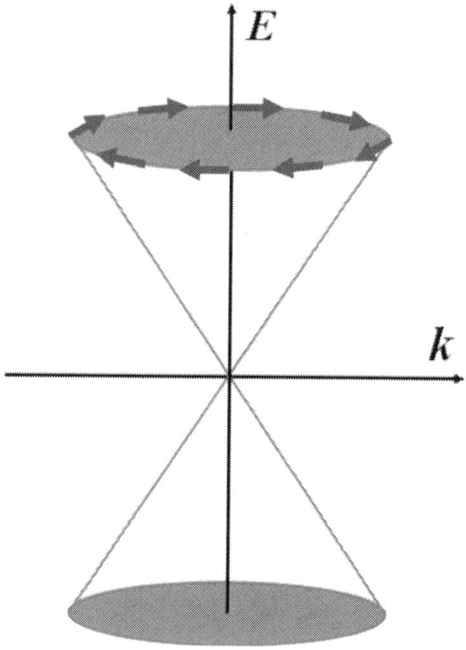

Figure 2. Electron energy dispersion with a helical spin texture similar to the surface states of a topological insulator. Blue arrows indicate the directions of the spins, and red cones indicate the linear dispersion near the Dirac point.

described above, and the Rashba spin-orbit coupling, resulting from structural change at interfaces; this idea can become a convenient solution to the problem. Fig.3 illustrates the effect of Rashba spin-orbit coupling on a "free" electron band. Due to the relativistic effect, electrons moving in the interfacial electric fields also experience effective magnetic fields on their rest frame. Forward-moving electrons and backward-moving electrons see opposite in-plane effective magnetic fields. In the forward-moving direction, the spin-"inward" electrons are lifted higher in energy than the spin-"outward" electrons, and vice versa in the opposite-moving direction. Here the inward and outward spin orientations are with respect to the paper of drawing, and which spin configuration has lower energy is also determined by the Rashba field direction of that interface. Note that the spins are still degenerate at the k=0 point where the spin-orbit coupling vanishes, and it creates a crossing between the two bands. The dispersion relation is nearly linear in k close to this crossing point. Fig.3 only depicts the situation of one-dimensional electron motions (x-direction). Imagine that we are actually dealing with a two-dimensional surface, and the k space extends in both x and y directions. This will revolve the spin-split bands around the centre axis of k=0, and the crossing point evolves into a Dirac point under the electrons' two-dimensional motion. Because the Rashba field is normal to the xy-plane electron motion, the resulting effective magnetic fields — and therefore the principle spin quantum axes —always lie within the xy-plane and transverse to the direction of motion. This leads to a helical spin texture as depicted in Fig.3. Note that two helical spin-textured Fermi surfaces exist in this structure and a perpendicular Zeeman field along z-direction is necessary to open a gap at the Dirac point and remove the spin degeneracy from this system, eventually rendering it spinless. With this construction of a helical spin state through Rashba spin-orbit coupling, we can again proceed with superconducting proximity effect to simulate the properties of a p-wave superconductor.

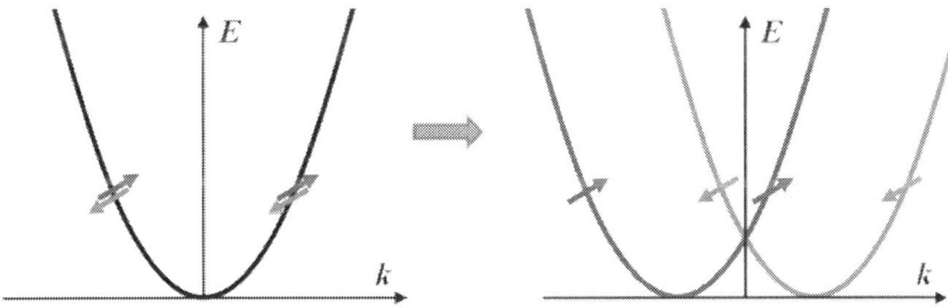

Figure 3. Illustration of a Rashba type spin-orbit splitting. Left: a typical parabolic "free" electron band with two-fold spin degeneracy. Right: In the presence of Rashba spin-orbit coupling, moving electrons feel an effective magnetic field on their rest frame, and the two spin channels are therefore split in energy. The effect is linear with respect to the wave vector *k*, and opposite for the two types of spins. Therefore, the Rashba spin-orbit coupling adds a linear component to the quadratic function, creating a uniform shift of the parabolas from the origin but in opposite directions for the opposite spin channels.

The proposal from Das Sarma et al. [23] is based on a 2D semiconductor electron system governed by the following equation of motion,

$$H_0 = \frac{\hbar^2 p^2}{2m^*} - \mu + V_z\sigma_z + \alpha(\vec{\sigma} \times \vec{p}) \cdot \hat{z} \tag{1}$$

Here, m^*, μ, V_z, α, are the electrons' effective mass, the chemical potential, the out-of-plane Zeeman splitting, and the Rashba spin-orbit coupling strength, respectively. Without the Zeeman term, the band structure would look just like that in Fig.3: revolving around the centre axis, as described above. We can easily see that the Zeeman term has a negligible influence on the total wavefunction when $\vec{p}$

is large, but becomes dominant close to $p\rightarrow$

=0 and opens up an energy gap right at the Dirac point, with the gap size $2V_z$ determined by the electrons' g-factor and the applied Zeeman field strength. This perpendicular Zeeman field is critical in this proposal for removing the spin degeneracy from the structure, leaving only one band present within the Zeeman gap. One does need to carefully tune the Fermi level of the system to within this gap to take advantage of this feature. The next ingredient is the superconducting proximity effect, which will induce superconductivity into the semiconductor's Rashba-split surface states. In order for the proximity effect to be effective, the semiconductor needs to be sufficiently conducting, in addition to possessing strong Rashba spin-orbit coupling. These requirements readily identify some of the heavy-element, low-gap semiconductors as the most promising choices, such as InSb or InAs. They tend to also have very large conduction electron g-factors, dozens of times larger than 2, making them even more ideal in applied magnetic fields. The superconductors, on the other hand, tend to have lower critical magnetic fields if the field direction is pointing out of the film plane, and in many cases the superconductivity will be completely destroyed by a field of a few Tesla. In order to induce appreciable Zeeman splitting, hopefully a few meV or higher, without suppressing the superconductivity, magnetic insulators are often used in these proposals to induce the necessary perpendicular Zeeman field. These materials are known to produce large exchange Zeeman fields onto adjacent free electrons, and are most effective on low-dimensional 2D or 1D electron systems [32]. The induced effective field can reach hundreds of Tesla with this approach. With exchange fields being short ranged, they have negligible influence on the superconductivity because the superconductor and the magnetic insulator are not in direct contact. Later, Alicea et al. [33] offered a clever twist on the above proposals by combining the Dresselhaus spin-orbit coupling, intrinsic to III-V semiconductors because of their crystallographic inversion asymmetry, and the Rashba spin-orbit coupling, due to the structural inversion asymmetry at interfaces, in a semiconductor quantum well structure with selected crystalline orientations. Now that the principle spin quantum axis no longer lies within the electrons' motion plane, a magnetic field with reasonable strength applied parallel to the plane surface, coupled with the relatively large g-factors in these semiconductors, can also open up an appreciable energy gap at the centre point, removing the spin degeneracy found there. Once the spinless platform is constructed, the next important practice is to tune the chemical potential of the system right to the middle of this gap through chemical doping or electric field gating. One can then place a conventional s-wave superconductor on top, which induces superconductivity into the system through proximity effect. The induced superconducting correlation opens up another energy gap on the conduction electrons and the system becomes fully gapped. With all these ingredients in place, the system can be described with exactly the same Bogoliubov-de Gennes (BdG) equations as the topological p-wave superconductors, and Majorana bound states are expected to exist at vortex cores on this special platform.

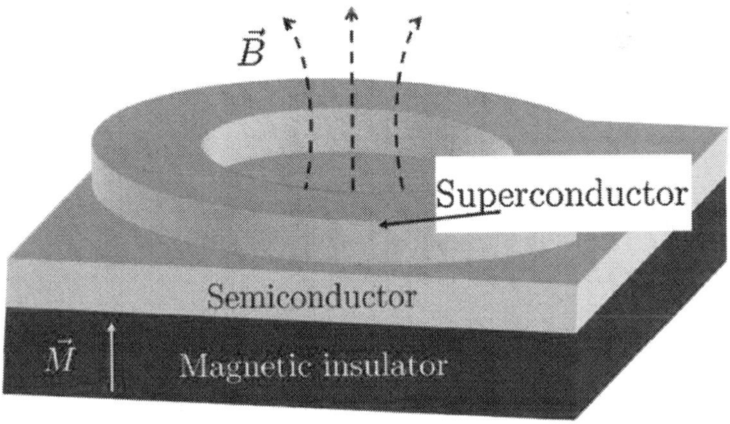

Figure 4. Majorana state bound to a vortex core on a heterostructure of a Rashba semiconductor platform, with both superconducting and magnetic proximity effects [23].

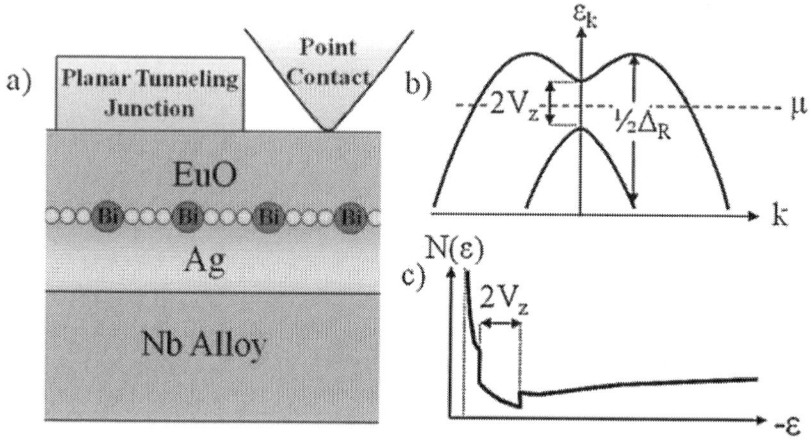

Figure 5. The proposed sample layout for the heavy-metal-based Majorana fermion creation [34]. (a) Ag(111) surface with 1/3 monolayer of Bi coverage in the (3√×3√) R30° reconstruction. Superconducting proximity from an s-wave superconductor (Nb alloy), and magnetic proximity from a magnetic insulator (EuO), together create a topological superconductor channel on the system. (b) Electrons confined on the metal surface experience strong Rashba spin-orbit coupling and form two parabolas crossing at k=0. The gap of $2V_Z$ comes from the exchange fields from EuO, and the chemical potential μ of the system needs to be tuned to the mid-gap. (c) Potential verification of the surface states through tunnelling spectrum, where one should see a dip corresponding to the opening of the exchange gap and a van Hove singularity corresponding to the very top of the Rashba bands.

Another route of synthesizing Majorana fermions is to perform superconducting proximity on top of a heavy metal surface with strong Rashba spin-orbit coupling [34, 35]. The Rashba splitting tends to be much larger than in the semiconductor systems. Under the strong Rashba spin-orbit coupling, the metal's surface states also show two spin-split parabolas, with a crossing at k=0 similar to those shown in Fig.3. A perpendicular external magnetic field, or a perpendicular exchange field from magnetic insulators, will open up a Zeeman gap at the cross point, and an s-wave superconductor will induce superconductivity into the system through superconducting proximity effect. While the superconducting gap stays with the chemical potential of the system, one still needs to match the chemical potential to the middle of the Zeeman gap. Due to the abundance of carriers in the metallic systems, however, it is extremely difficult to move the chemical potential through electric field gating. Therefore, one has to carefully select a suitable electron system to begin with, with the crossing point of the surface states not too far from the metal's Fermi level. Fig.5 shows one such set-up. Tunnelling spectrum to the interface can reveal the existence of the desired, gapped Rashba surface states. Once the superconducting proximity effect kicks in, appearance of Majorana zero-energy modes can mediate resonant Andreev reflection between two leads, with perfect conductance of $2e^2/h$ independent on the coupling strength. As a comparison, conventional resonant Andreev reflection only shows the maximum conductance when the two coupling amplitudes are identical.

The construction of the p-wave-superconductor-like electron systems makes the search for Majorana fermions much more tangible for experimentalists. It is now possible to revisit Kitaev's initial concept of 1D wires hosting Majorana end modes [36], which has been subsequently extended onto semiconductor nanowire platforms coupled with strong superconducting proximity effect [37, 38, 39]. The two ends of a 1D wire form a protected pair of Majorana fermions. This is analogous to a vortex on a 2D topological superconductor, which is also a boundary of the 2D platform but a physically moveable entity. To create a 1D p-wave superconductor wire, one can start with a 1D semiconductor wire and drive it into superconductivity with superconducting proximity effect. The regions surrounding the nanowire are just topologically trivial superconductors/insulators, leaving the enclosed topological superconductor nanowire intact. One can also start with the constructed 2D p-wave superconductor platforms as described earlier, and use selected electric and magnetic manipulation to suppress or flip the polarity of the topological superconductivity in order to define the desired 1D nanowire region with a distinct topology. Specifically, magnetic insulators can exert very strong interfacial exchange fields onto the desired 2D platforms, and can therefore be used as a local control to open and close the Zeeman gap of the system without much adverse influence on the superconductor system. Electric field manipulation from gate electrodes allows for local and active tuning of chemical potentials in the system. The topologically nontrivial superconductor state is enabled when the chemical potential is tuned into the mid gap, and will be destroyed when it is too far off. Therefore, we see that we can conveniently perform the fusion and braiding operations on the Majorana end modes with controlled gating on the 1D wire networks [39]. Similar manipulations can also be achieved on 2D platforms by controllably moving the vortices. Quantum information is stored non-locally on these non-Abelian systems and therefore is largely immune to any local perturbations, as long as the topology of the above systems remains protected by the superconducting and exchange gaps. These properties ensure that topological quantum computing on these solid-state platforms is inherently error-tolerant. We will not discuss in this article the details of braiding operations for topological quantum information processing, except to point out that quantum braids cover all the necessary quantum gates required for quantum computing. One can refer to the original article by Kitaev et al. [17] for an accurate account for topological quantum computation with non-Abelian anyons.

There have been a number of reports about successfully observing the signatures of Majorana fermions along the route of the proposed proximity-driven platforms [29,30,31, 40, 41, 42]. Semiconductors offer the most opportunities so far because of their maturity in technology. The first convincing report [29] was based on InSb nanowires with strong spin-orbit coupling and a very large g-factor ($g \approx 50$), and the device layout is shown in Fig.6. The nanowire portion under the superconductor is the region of interest, where the superconducting proximity effect drives this nanowire region into superconducting. Note that the superconductor only covers half of the wire in order that the electric fields coming from the buried gate electrodes are not completely screened. This practice is very important to ensure that the chemical potential of this portion of wire can be tuned into the middle of the Zeeman-split gap with controlled gating. The semiconductor nanowire with strong Rashba spin-orbit coupling can be treated as a quasi-1D electron system, and its dispersion relation is essentially the same as that illustrated in Fig.3's right-hand panel, with two Rashba-spin-split parabolas in zero magnetic fields. In the nanowire geometry, one can find that when the applied magnetic field is along the wire direction, a Zeeman energy gap opens up at the Dirac point of the dispersion curve. The above actions ensure that Majorana bound states will emerge at the ends of the wire when the topological superconductor phase disappears and the normal gapped semiconductor/superconductor phase emerges. A quantum mechanical tunnelling measurement on this Majorana mode confirmed that it is bound strictly at zero energy throughout the field range in which it is supported. It emerges when the field applied along the wire opens up the Zeeman gap, and disappears when the field becomes too strong and destroys the topological superconductor phase. The behaviour is distinctly different from other phenomena such as Andreev reflection and Kondo effect, and is therefore widely accepted as the first observation of Majorana bound states in solid-state systems. The bound state vanishes under elevated temperatures, because thermal excitations start to overcome the gap protections. Strictly speaking, a semiconductor nanowire is not exactly a 1D electron system, and there are multiple transverse

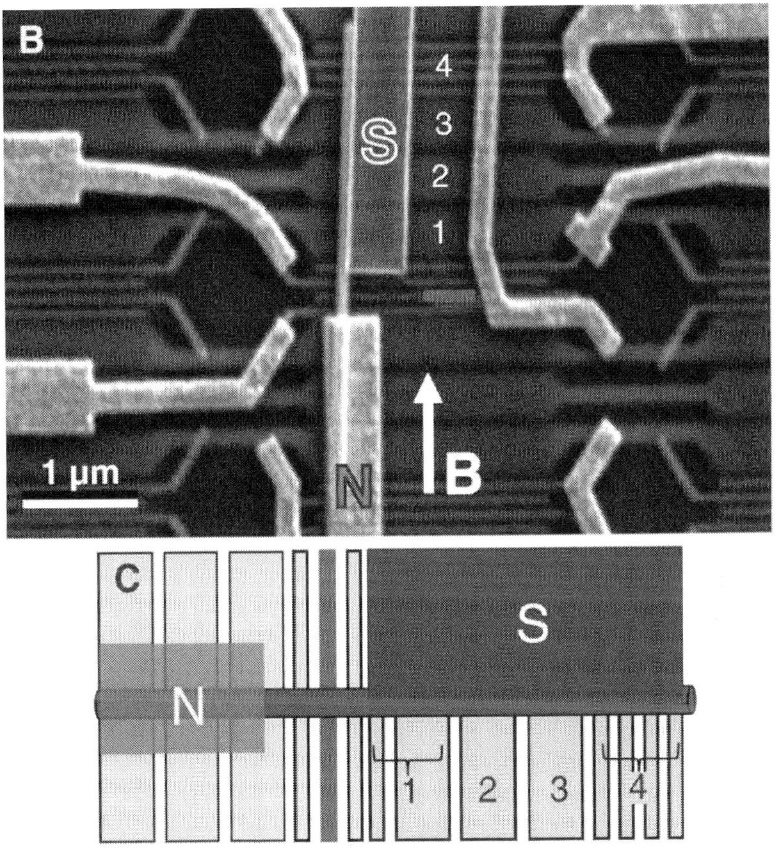

Figure 6. Device layout for the experimental evidence of Majorana fermion in a solid-state system [29]. The electrodes seen on the background, numbered 1, 2, 3, 4, etc., are the gate electrodes, separated from the nanowire and top electrodes by Si_3N_4 dielectrics. The nanowire is in contact with a normal metal (N) on one end, and a superconductor (S) on the other. The gate electrode labelled in dark green is the one used to create a tunnel barrier between the two sides of the nanowire. This effectively forms a S-NW-N tunnel junction for spectroscopy measurement. The nanowire region covered under the superconductor would have proximity-induced superconductivity in it, and a Majorana bound state is therefore trapped at the very edge of the superconductor region when the topology changes. It shows up as a zero-bias peak on the tunnel spectroscopy when a suitable magnetic field is applied along the wire direction.

subbands due to the finite width. The above 1D wire theory was expanded to the more general multichannel situations [34, 43], and the protected Majorana modes exist in the quasi-1D wires as long as there are only an odd number of bands crossing the Fermi level and the wire is not substantially wider than the superconducting coherence length. These conditions are both satisfied in this experiment. The magnetic field was applied along the wire direction in this experiment, and a magnetic field perpendicular to the device plane (out-of-the-page) is no longer a suitable choice. The circular geometry of the nanowires leads to a circular distribution of the Rashba electric fields, and for some parts of the wire surface the applied magnetic field would be aligned parallel to the wire surface and perpendicular to the wire direction. Magnetic field in such a configuration is only able to shift the Dirac point a little bit in energy and momentum, but is not able to open up the desired exchange gap. Therefore, some parts of the wire would remain gapless if a perpendicular-to-the-device-plane magnetic field is chosen.

3. INVERSE PROXIMITY ON SUPERCONDUCTORS — THE EXAMPLE OF SUPERCONDUCTING SPIN VALVES

As we have described above, the superconducting proximity effect describes the superconductor's wavefunction propagating into non-superconducting materials and inducing superconducting paring in there. Because of the leak-out of paired electrons, and the influx of unpaired electrons, the proximity effect would naturally generate a counter effect that weakens the superconductor itself. This is known as the inverse proximity effect. We will only focus our attention on a specific aspect of it — inverse proximity from ferromagnetic materials. The presence of spin information in the ferromagnets allows for spintronic manipulation on the superconductor system. For a detailed review on superconducting proximity with ferromagnets, please see the thorough review articles by Buzdin and Bergeret et al. [44, 45]. Here we proceed to demonstrate the inverse proximity effect from ferromagnets by illustrating with superconducting spin valve devices that can essentially turn the superconductivity On and Off through spin manipulations.

The propagation of superconductor wavefunction in a ferromagnetic metal is qualitatively different from that in a normal metal. This originates from the mutual incompatibility between a BCS superconductor, in which opposite spins pair up into Cooper pairs, and a ferromagnet, in which spins prefer to align parallel to each other to minimize the exchange energy. The presence of magnetic species in or near a superconductor material could be a drag if one wants to maximize the performance of the superconductor, because magnetic interactions break time reversal symmetry and become a strong source of Cooper pair breaking. We can, however, achieve desired controllability on the superconductor devices through manipulating the proximity-coupled magnetic systems. Fig.7 illustrates the propagation of superconducting wavefunction into a ferromagnetic metal. We note that instead of a simple exponential decay as in the normal metal/superconductor situation (Fig.1), the superconducting order parameter picks up a fast oscillatory behaviour. This can be qualitatively understood by considering the two electrons of a Cooper pair with opposite spins sitting inside an exchange field. One of the electrons, with its spin aligned along the exchange field direction, has its kinetic energy lowered by δE, while the opposite electron sees an increase in its kinetic energy by δE in the exchange field. Therefore, they start to propagate with different wavevector k, and the Cooper pair as a whole no longer has zero net momentum as it used to when it was inside the superconductor. For an electron near the Fermi level of a parabolic band, the kinetic energy is quadratic to the wavevector k and an increase (or decrease) in the kinetic energy by δE leads to a change in the wavevector by $\delta E/\hbar v_F$, where v_F is the Fermi velocity. The Cooper pair now gains a centre-of-mass momentum of $2\delta E/v_F$, and therefore the order parameter of the Cooper pairs will oscillate in space with a period of $hv_F/2\delta E$. This roughly describes the system in the clean limit. For the dirty limit of a diffusive system, the Cooper pair motion can be described well with the Usadel equation [46]. Under the assumption that the exchange energy is much larger than the pairing energy, which is true for most ferromagnets relative to superconductors, the pair wave function can be solved and has a form of $\Delta \exp\left(-\frac{x}{\xi_f}\right)\cos\left(\frac{x}{\xi_f}\right)$ [44], i.e., an oscillatory function with exponential damping. In this special situation, the decay length and the oscillation period coincide (but they do not have to in more general situations). Here $\xi_f = \sqrt{D_f / \delta E}$, $D_f = \frac{1}{3} v_F l$ is the diffusion coefficient and l is the electrons' mean-free-path in the ferromagnet. The oscillatory behaviour of the pair wave function adds an oscillatory dependence to the structure's overall critical temperature [47] when the ferromagnet layer thickness is comparable or longer than the pair oscillating period ξ_F. This has been verified in various superconductor-ferromagnet bilayer structures [48, 49, 50, 51]. One can readily see that when the ferromagnet layer thickness is equal to $\pi\xi_F$, the wavefunction has a phase shift of π across the ferromagnet layer, and the superconductivity is therefore weakened. As the ferromagnet increases in thickness, the phase oscillates back and forth and the system shows an oscillatory, non-monotonic behaviour in the critical temperature. The parameter ξ_F can be approximately read out from the oscillation period of T_C. The oscillatory behaviour of the pair wavefunction shows up not only in the phases, but also in the electron density of states inside the ferromagnets. At certain distance away from the interface, the density of states can even become larger than the normal value in the absence of superconducting proximity [52].

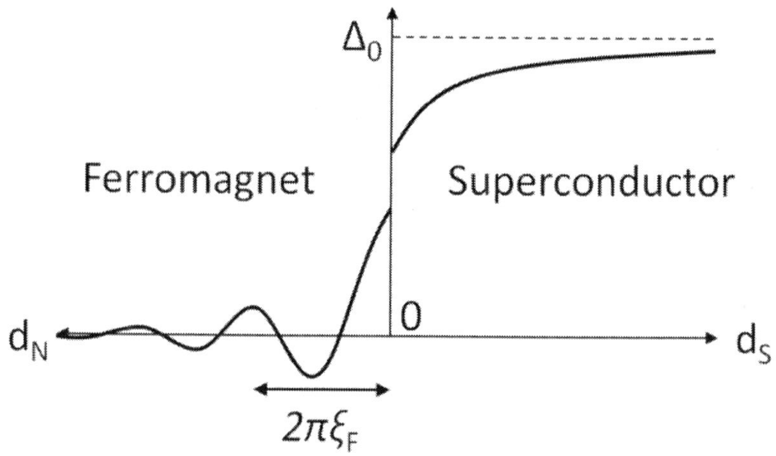

Figure 7. Evolution of the superconductor pairing potential near a S/F interface. The pairing potential inside the ferromagnet shows a strong oscillatory and damping behaviour due to the large internal exchange field. ξ_F characterizes the superconducting correlation oscillation length as well as the correlation decay length within a strong ferromagnet, where the exchange energy δE far exceeds the pairing potential Δ.

Though it is not immediately clear, the above discussions already imply that superconductivity within the superconductor itself will also be significantly weakened in the presence of a strong ferromagnet. A critical difference of the proximity with a strong ferromagnet, compared to that with a normal metal, is the extremely short decay length of the pairing potential. By connecting the superconducting wavefunction across the interface, one can already see that the steep slope of the pairing potential on the ferromagnet side influences that on the superconductor side and pulls the whole pairing potential profile downward. The exact boundary condition matching is subject to the interfacial transparency and the respective diffusion coefficients. A stronger ferromagnet creates stronger exchange interaction and shorter ξ_F: therefore, stronger suppression to the superconductivity. Now we can consider constructing a sandwich of ferromagnet/superconductor/ferromagnet: i.e., a superconducting spin valve structure [53, 54]. Although the two ferromagnet layers will both strongly suppress superconductivity in the sandwiched middle layer, we can try to cancel out their influence through spin manipulation. Recall that when one of the ferromagnet layers is placed on top of the superconductor, strong suppression of superconductivity is present. If we place the second ferromagnet layer with the same spin polarity, the inverse proximity effects coming from both ferromagnets strengthen each other and suppress superconductivity even more. This spin-parallel configuration ensures that the wavefunction over the whole structure is symmetric with respect to the film centre (mid-plane of the superconductor). We can conceptually fold the wavefunction in the middle and it becomes identical to that of a bilayer situation. Because the equivalent thickness of the superconductor layer is only half of its actual thickness, the layer suffers even stronger suppression from the inverse proximity effect. If we instead place the second ferromagnet with an opposite spin polarity relative to the first one, i.e., configuring the system into a spin-antiparallel state, the proximity effects from the two ferromagnets differ precisely by a phase of π and will largely cancel each other out. Superconductivity is therefore restored in this configuration.

We now use an epitaxial pristine structure of Fe/V/Fe as an example to illustrate the above mentioned superconducting spin valve effect [51]. The structure was deposited with molecular beam epitaxy (MBE) onto MgO-buffered Si(100) substrates. X-ray diffraction verified the epitaxy of the films showing clear four-fold symmetry in the off-axis diffraction patterns. The quality of the films and interfaces can be verified with the critical temperature T_C and upper critical field H_{C2} measurements on V single layer films and Fe/V bilayer films. Fig.8 (a) shows that T_C varies linearly with respect to the inverse film thickness. Such dependence indicates clearly that there exists a superconducting "dead" layer on the film surface due to the lowering of electron density and weakening of phonon coupling there [55]. This surface layer is superconductingly inactive, and amounts to about 1 atomic layer (0.18 nm) on the pure V films (interfaced with MgO on both sides), confirming the superior quality of these films. On the other hand, the series of bilayer samples of V with proximity to 6 nm Fe show a significantly steeper slope, corresponding to a much thicker inactive layer of 13.5 nm. For V of the same thickness, presence of Fe pulls T_C much lower than MgO does in pure V films. This is a clear manifestation of the above-described inverse superconducting proximity effect: the presence of a ferromagnet strongly suppresses superconductivity in the adjacent superconductor. Fig.8 (b) shows the upper critical fields of the set of films with 30 nm V. The addition of proximity from Fe again significantly suppresses superconductivity. The linear dependence of H_{C2} square with respect to temperature is an indication of two-dimensional superconductivity behaviour, which follows $H_{C2}(T)=H_{C2}(0)(1-\frac{T}{T_C})^{1/2}$ [56]. From the determined slope of the plot, we can identify the superconducting coherence length ξ_S, being about 8.2 nm for the 30 nm V films. It is interesting to note that the slope does not change when the proximity effect from Fe kicks in, indicating that no change happens to the Cooper pair correlations. This is quite expected because the proximity effect and inverse proximity effect do not modify the pairing mechanisms within the superconductor, and the addition of a few nm metals has little influence on the electrons' mean-free-paths either. Fig.8 (c) shows the most important aspects of the inverse superconducting proximity effect: the quick oscillation and damping of T_C with thickness. As we have described earlier, T_C oscillates with the ferromagnetic layer thickness because of the Cooper pairs' phase oscillation inside an exchange field. We can roughly estimate the superconducting correlation length inside the ferromagnet ξ_F to be about 1 nm. We can also clearly see that the superconductivity suppression from inverse proximity effect is quite dramatic: with a few monolayers of Fe overlayer, the critical temperature already drops from the original value of over 5K to only slightly above 3K. The oscillation quickly damps out after several ξ_F, when the Cooper pair wavefunction is almost fully suppressed. After the first few nm, further increase of the ferromagnetic layer thickness does not continue to influence the overall wavefunction. On the other hand, we notice that T_C does not really drop to zero before recovering, as we have depicted before. This actually is also a consequence of the fast damping behaviour which significantly limits the transparency of the Cooper pairs into the ferromagnet. Beyond the first few nm, the remaining portion of the ferromagnet has no significant further contribution to the T_C suppression. In addition, because the measurement of T_C is over the whole structure, the region of the superconductor far away from the interface essentially retains the original superconductivity strength, and renders the whole structure superconducting under conventional DC measurements. Detailed calculations on T_C of such ferromagnet/superconductor bilayer structures confirm that the oscillation indeed quickly reaches a saturation value after the ferromagnet thickness increases beyond the first few oscillations [57].

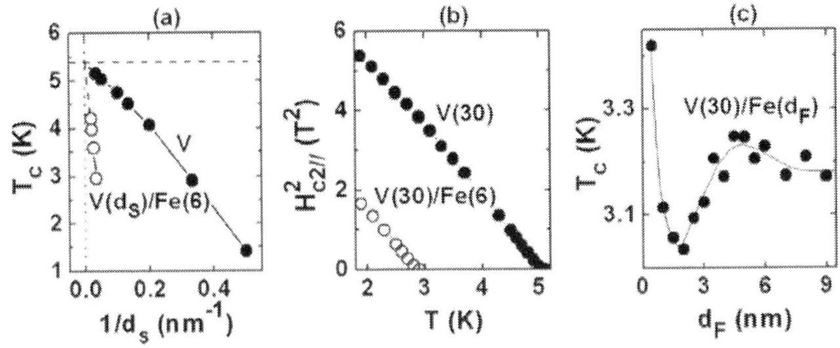

Figure 8. Critical temperature T_C and upper critical field H_{C2} for epitaxial V and Fe/V films [51]. (a) T_C as a function of inverse V layer thickness; (b) H_{C2} square as a function of temperature; (c) T_C damping and oscillation with respect to the Fe layer thickness, the solid line is provided as a guide for the eye. ξ_F can be estimated to be about 1 nm.

Next, we consider the magnetoresistance of these superconducting spin valves. The structure is fully epitaxial, with high interfacial quality. In order to magnetically separate the two Fe layers, a CoO antiferromagnetic layer was introduced above the top Fe layer, which exerts exchange bias [58] on the top Fe layer and essentially pins the magnetization of this layer to a chosen direction. When the applied magnetic field only sweeps in a range not exceeding the exchange bias pinning strength, one can only observe magnetic switching of the "free" layer, which is the bottom Fe layer in this structure. Fig.9 (a) shows the magnetoresistance results across temperatures close to T_C. When the magnetic field varies, clear resistance high and low states are observed, indicating that the structure toggles between the non-superconducting and superconducting states. In the spin-parallel configuration, the inverse proximity effects from both ferromagnet layers strengthen each other and strongly suppress superconductivity of the sandwiched V layer, leading to the high-resistance, non-superconducting state; in the spin-antiparallel configuration, the inverse proximity effects largely cancel each other out and the system returns to the low resistance, superconducting state. For certain temperatures the ratio between the normal resistance and the superconducting resistance is essentially infinite, making the magnetoresistance ratio essentially infinite. The effect is only pronounced in a very narrow temperature window close to T_C, because this is when the superconductivity is already quite weakened by thermal energy, and slight perturbations from the environment (such as from the inverse proximity effects) can readily drive the system in and out of its superconducting state.

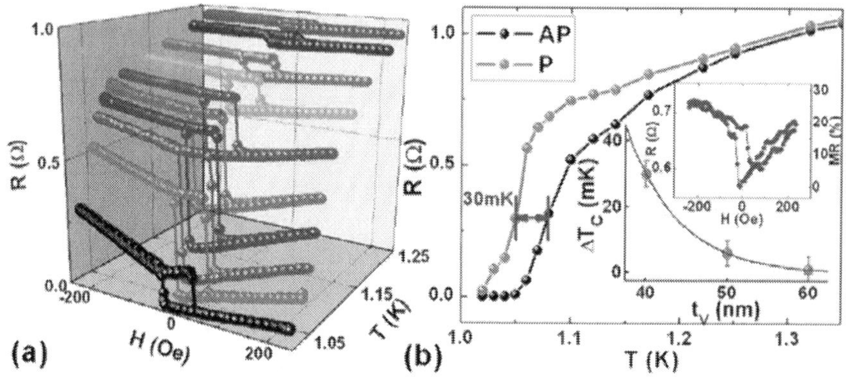

Figure 9. Measured superconducting spin valve effect in the epitaxial Fe/V/Fe structure [51]. (a) Device resistance as a function of applied magnetic field H and temperature T. A large magnetoresistance is associated with the switching of the free Fe layer in the temperature range close to T_C. (b) Resistance as a function of temperature for the structure in its spin-parallel (P) and spin-antiparallel (AP) configurations. There is a clear offset of 30mK between their corresponding T_C's. Inset: T_C offset as a function of the V layer thickness, and an example of a magnetoresistance curve for the sample with 50 nm V.

Fig.9 (b) shows T_C determined for the two spin configurations. Clearly, spin-parallel state has weaker superconductivity and lower T_C, while spin-antiparallel state has stronger superconductivity and higher T_C. In the temperature range between these two T_C, a large resistance difference exists between the two states leading to the large magnetoresistance described above. The superconducting spin valve effect relies on the fact that the two ferromagnet layers are effectively coupled to each other through the superconductor. Therefore, the effect decreases as the superconductor layer thickness increases, and essentially drops to zero when the superconductor layer is much thicker than the superconducting coherence length ξ_S of about 8 nm in this system. On the other hand, if the superconductor layer is too thin, its superconductivity would be fully suppressed by the Fe layers and the superconducting spin valve effect is also no longer achievable. The T_C shift as seen in Fig.9 (b) is clear indication of the operation temperature range of this effect, which is, however, limited to only tens of mK across many different systems [59, 60, 61, 62,51].

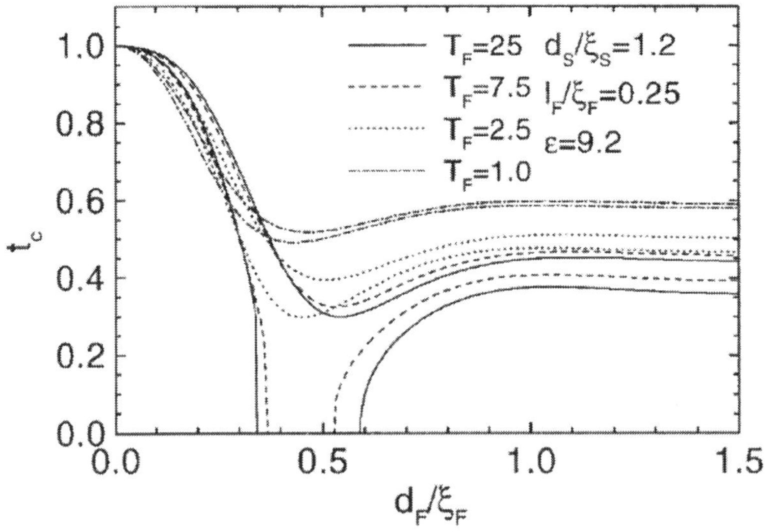

Figure 10. Expected T_C shift in superconducting spin valves when the interface transparency is varied [53]. Here d_S and d_F are the superconductor and ferromagnet layer thicknesses, l_F is the electron mean-free-path in the ferromagnet, ξ_S and ξ_F are the superconducting correlation lengths in the superconductor and ferromagnet, and ε is a parameter depending on the properties of both the superconductor and the ferromagnet, getting smaller with stronger ferromagnets. Two curves are shown for each choice of interfacial transparency, with the upper curve corresponding to the AP state and the lower one to the P state.

Improving interfacial transparency is the key for further improving the performance of these devices. Fig.10 shows the calculated variation of the P and AP T_C's with the change of the interfacial transparency [53]. The curves with the highest transparency (solid lines, T_F=25) clearly show a region with very large T_C shift. For example, around $d_F/\xi_F=0.4$, T_C of the AP state is around half the bulk value while that of the P state is fully suppressed. For samples of this structure, infinite magnetoresistance would show up for any temperatures below half the bulk T_C, making it a practically usable device for turning the superconductivity On and Off. As a comparison, for the curves corresponding to the lowest interfacial transparency, T_F = 1, the T_C difference between P and AP states is very small throughout all the d_F

choices. This is easy to understand because poor interface transparency breaks up the correlation between the superconductor and the ferromagnets: therefore, the correlation between the two ferromagnets also vanishes. The two ferromagnets do not feel much influence from each other and the P and AP configurations make little difference. Although a weak superconducting spin valve effect is still present, the resultant magnetoresistance is very small and there is no truly On and Off tuning of superconductivity, making it less useful in practice.

We next examine a special type of superconducting spin valve, where the ferromagnet layers are not metallic but insulating [63]. The above described inverse superconducting proximity mechanism does not apply because the Cooper pair wavefunction cannot penetrate into the insulators. The inverse proximity happens due to the interfacial exchange interactions from the localized moments of the magnetic insulator, which exert a large effective Zeeman field onto the superconductor and can also suppress the superconductivity [32]. The interaction between the magnetic insulator and the carriers in the superconductor is through indirect exchange interaction, where the free electrons communicate with multiple localized magnetic moments of the magnetic insulator, and become spin-polarized in the process. This behaves as an effective Zeeman field on the electron system and favours one type of spins over the other, therefore counteracts the tendency of Cooper pairing and suppresses superconductivity. Now that the magnetism-superconductivity interaction is established, it is again possible to use spin manipulations to control superconductivity in a sandwich structure [64]. The magnetic insulator material used in this study was EuS, which has been shown to generate exchange fields as large as a few Tesla on thin superconducting Al films [65, 66]. Under this effective Zeeman field on the conduction electrons, Cooper paring in the superconductor becomes less stable and superconductivity is thus weakened accordingly. A second magnetic insulator layer, when configured with its magnetic orientation opposite to the first one, will serve to cancel the exchange fields from the first one, provided that the electrons' mean-free-paths are long enough relative to the superconductor layer thickness such that the Zeeman fields from one side can propagate across the whole superconductor layer and influence the other side. As a result, we again recover a spin valve performance: when the two magnetic insulators are aligned parallel to each other, their exchange fields stack up and destroy the superconductivity in the system; when they are aligned antiparallel to each other, their effects cancel and the system goes back to the superconducting state. Experimentally, toggling the superconductivity On and Off leads to a very large magnetoresistance response, as shown in Fig.11. Here the system is in the spin-parallel state (finite resistance, non-superconducting state) when the external field is large and both EuS layers are saturated in the same direction; and in the spin-antiparallel state (zero resistance, superconducting state) when the external field is tuned between the magnetic switching fields of the two EuS layers, such that one of the layers flips while the other does not. Similar to the superconducting spin valves with magnetic metals, this effect only shows up in a very narrow temperature window close to T_C, when the superconductivity is already very weak to begin with. Though this effect appears very similar to the previously described superconducting spin valve effect, there is one critical difference: in this structure, there are negligible superconducting proximity and inverse proximity effects because of the insulating nature of the ferromagnets. The observed magnetic tuning of superconductivity is in fact a result of the propagation of magnetic exchange interaction into the superconductor — the magnetic proximity effect, which is especially pronounced with magnetic insulators [32]. As a result, the system is described with the Cooper pairs feeling the average exchange fields *inside the superconductor* rather than with Cooper pairs feeling the exchange fields *inside the ferromagnets.*

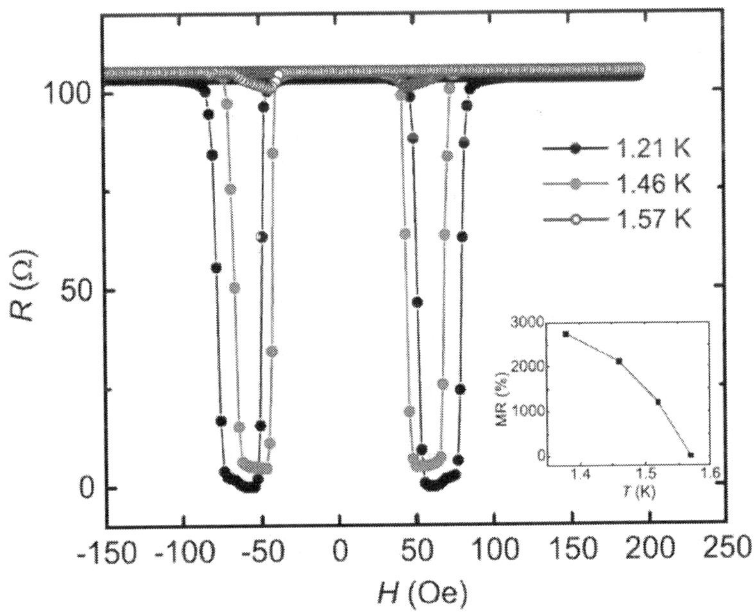

Figure 11. Superconducting spin valve effect in a magnetic insulator/superconductor/magnetic insulator sandwich structure (EuS/Al/EuS) [64]. Inset shows the variation of magnetoresistance ratio over the temperatures close to T_C.

4. SUMMARY

In summary, we have used a few very specific examples (the creation of Majorana fermions on 2D surface states and the superconducting spin valves with ferromagnets) to illustrate the application of superconducting proximity and inverse proximity effects in spintronics. These examples can be viewed as passive devices of superconducting proximity/inverse proximity. One can also use active pumping of Cooper pairs to drive supercurrents into non-superconducting materials, which is yet another manifestation of the superconducting proximity effect and can induce superconducting Josephson coupling on many spintronics platforms, such as graphene and topological insulators. The Josephson effect will be covered in other chapters of this book and we will not discuss it in this chapter. Overall, we see that the superconducting proximity and inverse proximity effects are convenient approaches to couple superconductivity with many other types of spin systems, and allow us to create hybrid devices that can benefit from these very distinct spin states: superconductivity, magnetism, and topological quantum spin Hall state. Such manipulation of superconductivity offers important new routes for information storage and processing, and rapid advances are expected to happen in these directions taking information processing to the quantum level. We, as well as many other groups in the world, are actively working toward such goals.

5. ACKNOWLEDGEMENTS

The author wish to thank the Natural Sciences and Engineering Research Council of Canada (NSERC) for support on this work.

REFERENCES

1. Meissner H, Superconductivity of Contacts with Interposed Barriers, Phys. Rev. 1960; 117: 672.
2. de Gennes P G, Boundary Effects in Superconductors, Rev. Mod. Phys. 1964; 36: 225.
3. Takayanagi H, Kawakami T, Superconducting Proximity Effect in the Native Inversion Layer on InAs, Phys. Rev. Lett. 1985; 54: 2449.
4. Nitta J, Akazaki T, Takayanagi H, Arai K, Transport Properties in an InAs-inserted-channel In0.52Al0.48As / In0.53Ga0.47As

Heterostructure Coupled Superconducting Junction, Phys. Rev. B 1992; 46: 14286(R).
5. Nguyen C, Kroemer H, Hu E L, Anomalous Andreev Conductance in InAs-AlSb Quantum Well Structures with Nb Electrodes, Phys. Rev. Lett. 1992; 69: 2847.
6. Zhang D, Wang J, DaSilva A M, Lee J S, Gutierrez H R, Chan M H W, Jain J, Samarth N, Superconducting Proximity Effect and Possible Evidence for Pearl Vortices in a Candidate Topological Insulator, Phys. Rev. B 2011; 84: 165120.
7. Williams J R, Bestwick A J, Gallagher P, Hong S S, Cui Y, Bleich A S, Analytis J G, Fisher I R, Goldhaber-Gordon D, Unconventional Josephson Effect in Hybrid Superconductor-Topological Insulator Devices, Phys. Rev. Lett. 2012; 109: 056803.
8. Qu F, Yang F, Shen J, Ding Y, Chen J, Ji Z, Liu G, Fan J, Jing X, Yang C, Lua L, Strong Superconducting Proximity Effect in Pb-Bi2Te3 Hybrid Structures, Sci Rep. 2012; 2: 339.
9. Heersche H B, Jarillo-Herrero P, Oostinga J B, Vandersypen L M K, Morpurgo A F, Bipolar Supercurrent in Graphene, Nature 2007; 446: 56–59.
10. Morpurgo A F, Kong J, Marcus C M, Dai H, Gate-Controlled Superconducting Proximity Effect in Carbon Nanotubes, Science 1999; 286 (5438): 263–265.
11. Jarillo-Herrero P, van Dam J A, Kouwenhoven L P, Quantum Supercurrent Transistors in Carbon Nanotubes, Nature 2006; 439: 953–956.
12. Winkelmann C B, Roch N, Wernsdorfer W, Bouchiat V, Balestro F, Superconductivity in a Single-C60 Transistor, Nature Physics 2009; 5: 876–879.
13. van Dam J A, Nazarov Y V, Bakkers E P A M, De Franceschi S, Kouwenhoven L P, Supercurrent Reversal in Quantum Dots, Nature 2006; 442: 667–670.
14. Kasumov A Y, Kociak M, Guéron S, Reulet B, Volkov V T, Klinov D V, Bouchiat H, Proximity-Induced Superconductivity in DNA, Science 2001; 291 (5502): 280–282.
15. Peña V, Sefrioui Z, Arias D, Leon C, Santamaria J, Varela M, Pennycook S J, Martinez J L, Coupling of Superconductors Through a Half-Metallic Ferromagnet: Evidence for a Long-Range Proximity Effect, Phys. Rev. B 2004; 69: 224502.

16. Keizer R S, Goennenwein S T B, Klapwijk T M, Miao G X, Xiao G, Gupta A, A Spin Triplet Supercurrent Through the Half-Metallic Ferromagnet CrO2, Nature 2006; 439(7078): 825.
17. Kitaev A Yu, Fault-Tolerant Quantum Computation by Anyons, Annals. Phys. 2003; 303(1): 2–30.
18. Read N, Green D, Paired States of Fermions in Two Dimensions with Breaking of Parity and Time-Reversal Symmetries and the Fractional Quantum Hall Effect, Phys. Rev. B 2000; 61: 10267.
19. Ivanov D A, Non-Abelian Statistics of Half-Quantum Vortices in p-Wave Superconductors, Phys. Rev. Lett. 2001; 86: 268.
20. Das Sarma S, Nayak C, Tewari S, Proposal to Stabilize and Detect Half-Quantum Vortices in Strontium Ruthenate Thin Films: Non-Abelian Braiding Statistics of Vortices in a px+ipy Superconductor, Phys. Rev. B 2006; 73: 220502(R).
21. Mackenzie A P, Maeno Y, The Superconductivity of Sr2RuO4 and the Physics of Spin-Triplet Pairing, Rev. Mod. Phys. 2003; 75: 657.
22. Fu L, Kane C L, Superconducting Proximity Effect and Majorana Fermions at the Surface of a Topological Insulator, Phys. Rev. Lett. 2008; 10 (9): 096407.
23. Sau J D, Lutchyn R M, Tewari S, Das Sarma S, Generic New Platform for Topological Quantum Computation Using Semiconductor Heterostructures, Phys. Rev. Lett. 2010; 104(4): 040502.
24. Hasan M Z, Kane C L, Colloquium: Topological Insulators, Rev. Mod. Phys. 2010; 82: 3045.
25. Fu L, Kane C L, Mele E J, Topological Insulators in Three Dimensions, Phys. Rev. Lett. 2007; 98: 106803.
26. He L, Kou X, Wang K L, Review of 3D Topological Insulator Thin-Film Growth by Molecular Beam Epitaxy and Potential Applications, phys. stat. sol. - Rap. Res. Lett. 2013; 7: 50–63.
27. Sasaki S, Kriener M, Segawa K, Yada K, Tanaka Y, Sato M, Ando Y, Topological Superconductivity in CuxBi2Se3
28. Wang M X, Liu C, Xu J P, Yang F, Miao L, Yao M Y, Gao C L, Shen C, Ma X, Chen X, Xu Z A, Liu Y, Zhang S H, Qian D, Jia J F, Xue Q K, The Coexistence of Superconductivity and Topological Order in the Bi2Se3 Thin Films, Science 6 April 2012; 336: 52–55.
29. Mourik V, Zuo K, Frolov S M, Plissard S R, Bakkers E P A M, Kouwenhoven L P, Signatures of Majorana Fermions in Hybrid

Superconductor-Semiconductor Nanowire Devices, Science 2012; 336 (6084): 1003–1007.

30. Deng M T, Yu C L, Huang G Y, Larsson M, Caroff P, Xu H Q, Anomalous Zero-Bias Conductance Peak in a Nb–InSb Nanowire–Nb Hybrid Device, Nano Lett. 2012; 12, 6414–6419.
31. Das A, Ronen Y, Most Y, Oreg Y, Heiblum M, Shtrikman H, Zero-bias Peaks and Splitting in an Al-InAs Nanowire Topological Superconductor as a Signature of Majorana Fermions, Nature Physics 2012; 8: 887–895.
32. Miao G X, Moodera J S, Spin Manipulation with Magnetic Semiconductor Barriers, Phys. Chem. Chem. Phys. 2014; DOI: 10.1039/c4cp04599h, in print.
33. Alicea J, Majorana Fermions in a Tunable Semiconductor Device, Phys. Rev. B 2010; 81: 125318.
34. Potter A C, Lee P A, Multichannel Generalization of Kitaev's Majorana End States and a Practical Route to Realize Them in Thin Films, Phys. Rev. Lett. 2010; 105: 227003.
35. Law K T, Lee P A, Ng T K, Majorana Fermion Induced Resonant Andreev Reflection, Phys. Rev. Lett. 2009; 103: 237001.
36. Kitaev A Y, Unpaired Majorana Fermions in Quantum Wires, Physics-Uspekhi 2001; 44: 131.
37. Lutchyn R M, Sau J D, Das Sarma S Majorana Fermions and a Topological Phase Transition in Semiconductor-Superconductor Heterostructures. Phys. Rev. Lett. 2010; 105 (7): 077001.
38. Oreg Y, Refael G, von Oppen F, Helical Liquids and Majorana Bound States in Quantum Wires, Phys. Rev. Lett. 2010; 105: 177002.
39. Alicea J, Oreg Y, Refael G, von Oppen F, Fisher M P A, Non-Abelian Statistics and Topological Quantum Information Processing in 1D Wire Networks, Nature Physics 2011; 7: 412–417.
40. Nadj-Perge S, Drozdov I K, Li J, Chen H, Jeon S, Seo J, MacDonald A H, Bernevig B A, Yazdani A, Observation of Majorana Fermions in Ferromagnetic Atomic Chains on a Superconductor, Science 2014; 346(6209): 602–607.
41. Stanescu T D, Tewari S, Majorana Fermions in SemiconductorNanowires: Fundamentals, Modeling, and Experiment, J. Phys.: Condens. Matter 2013; 25: 233201.
42. Beenakker C W J, Search for Majorana Fermions in Superconductors. Annu. Rev. Condens. Matter Phys. 2013; 4: 113.

43. Lutchyn R M, Stanescu T D, Das Sarma S, Search for Majorana Fermions in Multiband Semiconducting Nanowires, Phys. Rev. Lett. 2011; 106: 127001.
44. Buzdin A I, Proximity Effects in Superconductor-Ferromagnet Heterostructures, Rev. Mod. Phys. 2005; 77: 935–976.
45. Bergeret F S, Volkov A F, Efetov K B, Odd Triplet Superconductivity and Related Phenomena in Superconductor-Ferromagnet Structures, Rev. Mod. Phys. 2005; 77: 1321–1373.
46. Usadel K D, Generalized Diffusion Equation for Superconducting Alloys, Phys. Rev. Lett. 1970; 25: 507.
47. Buzdin A I, Kuprianov M Y, Transition Temperature of a Superconductor-Ferromagnet Superlattice, Pis'ma Zh. Eksp. Teor. Fiz. 1990; 52, 1089–1091.
48. Jiang J S, Davidović D, Reich D H, Chien C L, Oscillatory Superconducting Transition Temperature in Nb/Gd Multilayers, Phys. Rev. Lett. 1995; 74: 314.
49. Mercaldo L V, Attanasio C, Coccorese C, Maritato L, Prischepa S L, Salvato M, Superconducting-Critical-Temperature Oscillations in Nb/CuMn Multilayers, Phys. Rev. B 1996; 53: 14040.
50. Sidorenko A S, Zdravkov V I, Prepelitsa A A, Helbig C, Luo Y, Gsell S, Schreck M, Klimm S, Horn S, Tagirov L R, Tidecks R, Oscillations of the Critical Temperature in Superconducting Nb/Ni bilayers, Annalen der Physik 2003; 12: 37–50.
51. Miao G X, Ramos A V, Moodera J S, Infinite Magnetoresistance from the Spin Dependent Proximity Effect in Symmetry Driven bcc-Fe/V/Fe Heteroepitaxial Superconducting Spin Valves,Phys. Rev. Lett. 2008; 101: 137001.
52. Kontos T, Aprili M, Lesueur J, Grison X, Inhomogeneous Superconductivity Induced in a Ferromagnet by Proximity Effect, Phys. Rev. Lett. 2001; 86: 304.
53. Tagirov L R, Low-Field Superconducting Spin Switch Based on a Superconductor/Ferromagnet Multilayer, Phys. Rev. Lett. 1999; 83: 2058.
54. Buzdin A I, Vedyayev A V, Ryzhanova N V, Spin-Orientation-Dependent Superconductivity in F/S/F Structures, Europhys. Lett. 1999; 48: 686.
55. Wolf S A, Kennedy J J, Nisenoff M, Properties of Superconducting rf Sputtered Ultrathin Films of Nb,J. Vac. Sci. Technol. 1976; 13: 145.

56. Tinkham M, Introduction to Superconductivity, McGraw-Hill, New York, 1978.
57. Demler E A, Arnold G B, Beasley M R, Superconducting Proximity Effects in Magnetic Metals, Phys. Rev. B 1997; 55: 15174.
58. Berkowitz A E, Takano K, Exchange Anisotropy — A Review, J. Magn. Magn. Mater. 1999; 200: 552–570.
59. Gu J Y, You C Y, Jiang J S, Pearson J, Bazaliy Ya B, Bader S D, Magnetization-Orientation Dependence of the Superconducting Transition Temperature in the Ferromagnet-Superconductor-Ferromagnet System: CuNi/Nb/CuNi, Phys. Rev. Lett. 2002; 89: 267001.
60. Potenza A, Marrows C H, Superconductor-Ferromagnet CuNi/Nb/CuNi Trilayers as Superconducting Spin-Valve Core Structures, Phys. Rev. B 2005; 71: 180503(R).
61. Moraru I C, Pratt W P, Birge N O, Magnetization-Dependent Tc Shift in Ferromagnet/Superconductor/Ferromagnet Trilayers with a Strong Ferromagnet, Phys. Rev. Lett. 2006; 96: 037004.
62. Nowak G, Zabel H, Westerholt K, Garifullin I, Marcellini M, Liebig A, Hjörvarsson B, Superconducting Spin Valves Based on Epitaxial Fe/V Superlattices, Phys. Rev. B 2008; 78: 134520.
63. de Gennes P G, Coupling Between Ferromagnets Through a Superconducting Layer, Phys. Lett. 1966; 23: 10–11.
64. Li B, Roschewsky N, Assaf B A, Eich M, Epstein-Martin M, Heiman D, Munzenberg M, Moodera J S, Superconducting Spin Switch with Infinite Magnetoresistance Induced by an Internal Exchange Field, Phys. Rev. Lett. 2013; 110: 097001.
65. Moodera J S, Hao X, Gibson G A, Meservey R, Electron-Spin Polarization in Tunnel Junctions in Zero Applied Field with Ferromagnetic EuS Barriers, Phys. Rev. Lett. 1988; 61: 637.
66. Hao X, Moodera J S, Meservey R, Spin-Filter Effect of Ferromagnetic Europium Sulfide Tunnel Barriers, Phys. Rev. B 1990; 42: 8235

INDEX